Springer Theses

Recognizing Outstanding Ph.D. Research

For further volumes:
http://www.springer.com/series/8790

Aims and Scope

The series "Springer Theses" brings together a selection of the very best Ph.D. theses from around the world and across the physical sciences. Nominated and endorsed by two recognized specialists, each published volume has been selected for its scientific excellence and the high impact of its contents for the pertinent field of research. For greater accessibility to non-specialists, the published versions include an extended introduction, as well as a foreword by the student's supervisor explaining the special relevance of the work for the field. As a whole, the series will provide a valuable resource both for newcomers to the research fields described, and for other scientists seeking detailed background information on special questions. Finally, it provides an accredited documentation of the valuable contributions made by today's younger generation of scientists.

Theses are accepted into the series by invited nomination only and must fulfill all of the following criteria

- They must be written in good English.
- The topic should fall within the confines of Chemistry, Physics, Earth Sciences, Engineering and related interdisciplinary fields such as Materials, Nanoscience, Chemical Engineering, Complex Systems and Biophysics.
- The work reported in the thesis must represent a significant scientific advance.
- If the thesis includes previously published material, permission to reproduce this must be gained from the respective copyright holder.
- They must have been examined and passed during the 12 months prior to nomination.
- Each thesis should include a foreword by the supervisor outlining the significance of its content.
- The theses should have a clearly defined structure including an introduction accessible to scientists not expert in that particular field.

Katharina Fricke

Analysis and Modelling of Water Supply and Demand Under Climate Change, Land Use Transformation and Socio-Economic Development

The Water Resource Challenge and Adaptation Measures for Urumqi Region, Northwest China

Doctoral Thesis accepted by
Heidelberg University, Germany

Author
Dr. Katharina Fricke
Institute of Geography
Heidelberg University
Heidelberg
Germany

Supervisor
Prof. Dr. Olaf Bubenzer
Institute of Geography
University of Cologne
Cologne
Germany

ISSN 2190-5053 ISSN 2190-5061 (electronic)
ISBN 978-3-319-01609-2 ISBN 978-3-319-01610-8 (eBook)
DOI 10.1007/978-3-319-01610-8
Springer Cham Heidelberg New York Dordrecht London

Library of Congress Control Number: 2013948742

Printed on acid-free paper

Springer is part of Springer Science+Business Media (www.springer.com)

Parts of this thesis have been published in the following journal articles:

Fricke, K. & Bubenzer O. (2011). Available Water Resources and Water Use Efficiency in Urumqi, PR China. In: DAAD (eds.): Future Megacities in Balance, Young Researchers' Symposium in Essen October 9–10, 2010. DAAD-Series Dok&Mat, 66: 134–140

Fricke, K., Sterr, Th., Bubenzer, O., & Eitel B. (2009). The oasis as a mega city: Urumqi's fast urbanization in a semiarid environment. *Die Erde, 140*(4), 449–463.

Fricke, K. (2009). Integriertes Wassermanagement—Strategien für das Industriegebiet Midong in Urumqi, NW-China. *UmweltWirtschaftsForum, 17*(3), 291–298.

Fricke, K., Sterr, Th., Bubenzer, O., & Eitel, B. (2009). Das Beziehungsgeflecht "Megacity/Hinterland" am Beispiel der Wasserproblematik der chinesischen Megacity Urumqi. *Technikfolgenabschätzung—Theorie und Praxis, 18*(1), 62–70.

Supervisor's Foreword

The analysis and modelling of water supply and demand under climate change of very fast growing cities that experienced significant land use and socio-economic changes over the last few decades are challenging. This holds true especially for Urumqi, capital of the Xinjiang Uyghur Autonomous Region which has developed into the biggest urban growth pole among the provincial Capitals of Western China. Research on the sensible 'good' water in this part of the world must be constant and long-term, and requires not only scientific but also diplomatic and language skills.

Katharina Fricke was involved in the project "RECAST Urumqi: Meeting the Resource Efficiency Challenge in a Climate Sensitive Dryland Megacity Environment—Urumqi as a Model City for Central Asia," funded by the German Federal Ministry of Education and Research (BMBF) as part of the program "Future Megacities" and conducted at the Institute of Geography, Heidelberg University, since 2007. She had already completed her Diploma in Geography in 2008 with a thesis about the development of the urban district Midong in Urumqi, and had learned Chinese and how to negotiate with both scientists and administrative stakeholders. Within the subproject "Facilitation of Sustainable Megacity Development Through Water Resource Efficiency in a Semiarid Climate", she completed the present study. One of the most challenging questions in our project was how to obtain and produce consistent, objective and reliable data about the water resources of Urumqi. After reviewing all available official data, we decided to use, in addition, remote sensing techniques. With this concept, Katharina was able to measure missing hydrological and meteorological data for her complex model and to achieve a spatial and area-wide picture.

Urumqi constitutes, in many respects, an excellent example of city development in Central Asia. It changed from a relatively small oasis city with less than 100,000 inhabitants in 1950 to the actual size with a population of approximately 4 million people. Although agriculture is still by far the major 'water consumer', households and the industrial sector are requiring more and more water. In addition, Urumqi's fast development is the result of its strategic location in Northwestern China and its wealth in natural resources (e.g., oil and coal). Located in a narrow semiarid belt between the glaciated Tian Shan Mountains and the arid Gurbantüngütt Desert within the Junggar Basin, rain and meltwater from the surrounding mountains are indispensable to human activities.

Although, or even because, Katharina had to overcome great obstacles, she produced an outstanding study. In summary, she comes to very convincing and transferable results. Her models about Urumqi's recent and future water balance and socio-economic development now allow, for the first time, the deduction of sound scenarios as a base for recommendations and action plans concerning the recent water supply and future water demand under climate changes and land use transformation. I congratulate Katharina to this excellent work and I am proud to announce that she received the "Young Researchers Award" from the German Federal Ministry of Education and Research (BMBF) in May 2013.

Cologne, July 2013 Prof. Dr. Olaf Bubenzer

Acknowledgments

When visiting China for the first time to study in the bustling coastal city of Hong Kong I never imagined that I would spend quite some time in the far Northwest, in Urumqi, a city so far away from the sea during the following years. But when in 2007 I got the chance to write my diploma thesis during the preplanning phase of a research project in the capital of Xinjiang Province, my interest was awaken.

The multidisciplinary project "RECAST Urumqi—Meeting the Resource Efficiency Challenge in a Climate Sensitive Dryland Megacity Environment: Urumqi as a Model City for Central Asia" aims at developing strategies and tools for the sustainable development of the fast growing metropolises of arid Central Asia. The research project and my thesis was funded by the Federal Ministry of Education and Research (BMBF), Germany in a larger research program on 'Future Megacities' (funding code 01LG 0502A). The enormous socio-economic and urban development in such a special geographical location between the semi-desert and mountains poses enormous challenges to the sustainability of the core cycles water, materials and energy. The present thesis deals with the hydrological system of Urumqi Region, its future development due to climate change and land use transformation as well as adaptation strategies. From 2007 to 2011 a number of field trips were made within the sub-project "Facilitation of Sustainable Megacity Development Through Water Resource Efficiency". These and the consecutive results would have not been possible without the support of numerous persons.

I would like to especially thank my supervisor Prof. Dr. Olaf Bubenzer, for the extensive scientific supervision of my thesis, fruitful discussions and scientific and non-scientific support in Heidelberg and in Urumqi. Likewise, I would like to thank Prof. Dr. Bernhard Eitel, the intellectual initiator of the project and this research work. Despite his appointment as Rector of Heidelberg University in 2007, he continuously supported the project and was available as the second supervisor.

The organisation of the research work and the achieved results would not have been possible without the Sino-German project coordinator Dr. Thomas Sterr, Sha Xia, and Mr. Jiaerhen Ahati. Their steady and untiring efforts in organizing contacts, meetings and workshops in China and Germany, helped to solve problems and opened many doors. I would like to thank Prof. Dr. Kurt Roth, leader of the task group 'Water' for the support during workshops and project meetings as well as together with Dr. Patrick Klenk for the interdisciplinary insights from the environmental physics. Furthermore, I would also like to acknowledge the colleagues and cooperation within the research project and the opportunities to see more than just the geographical topics. My special thanks go to Antonia Koch, Yuan Yao und Ying Li for their valuable assistance with the translation work necessary for this project. Valuable information about the water resources in Urumqi Region were provided by Yunyun Wu from the Water Bureau Urumqi, Dr. Ping Chen from the Academy for Environmental Protection Sciences Xinjiang and Prof. Dr. Huifang Jiang from the College for Civil and Hydraulic Engineering, Xinjiang Agricultural University.

I would also like to thank Franziska Brohmeyer, Johannes Fuchs, Nina Barth and Cyrill Reinl, who wrote their theses on important aspects of the subproject 'Water' and extended the database. My thanks go also to Fabian Löw for the technical support when analysing the remote sensing data sets of the region. My colleagues Tobias Törnros, Dr. Patrick Klenk, Dr. Andreas Bolten, Dr. Christoph Siart, Barbara Brilmayer Bakti and Markus Forbriger I would like to thank for the scientific and technical suggestions with regard to methods and corrections as well as Ines Funke, Andreas Dorbach, and Dr. Christa Mahn-Fricke for the language control of the thesis.

I owe a special debt to my family and Andreas Dorbach who constantly supported me during the more intense phases of the dissertation.

Heidelberg, August 2012 Katharina Fricke

Contents

Abbreviations

A	Area [m^2]
ABIMO	AbflussBildungsMOdell (runoff formation model)
AOI	Area of interest
AON	Area of noninterest
α	Albedo [-]
α_{mun}/ α_{ind}/ α_{irr}	Progress factor (municipal, industrial, agricultural) [%]
a_s	Regression constant for the calculation of ET_p [-]
ASTER	Advanced spaceborne thermal emission and reflection
AVHRR	Advanced very high resolution radiometer
α_{w}	Albedo of water bodies [-]
AWWA	American Water Works Association
β_{mun}/ β_{ind}/ β_{irr}	Best practice factor (municipal, industrial, agricultural) [%]
b_s	Regression constant for the calculation of ET [-]
C	Heat capacity [J K^{-1}]
CAMS	Climate assessment and monitoring system
CRU	Climate research unit
c_s	Specific heat capacity [J K^{-1} kg^{-1}]
c_v	Volumetric soil heat capacity [J K^{-1} m^{-3}]
CWP	Crop water productivity
CWRF	Per capita crop water requirement for food
Δ	Slope vapour pressure curve [kPa °C^{-1}]
ΔP	Difference between P_{st} and distributed precipitation P [mm]
Δt	Time interval
$\Delta TRMM$	Difference between TRMM$_{\mathrm{st}}$ and the distributed TRMM values
Δz	Effective soil depth [m]
d	Zero plane displacement height of the surface; calculation of ET_p [m]
DDF	Degree-day-factor [mm °C^{-1} day^{-1}]
DEM	Digital elevation model
DGM 50	Digital elevation model
DLR	German Aerospace Center
DN	Digital number
DSM	Digital surface models

DVWK	German Association for Hydroengineering and Cultivation
ε	Weighing factor [-]
E	Evaporation
e_0	Vapour pressure [kPa]
e_a	Actual vapour pressure [kPa]
EROS	Earth Resources Observation and Science
e_s	Saturation vapour pressure [kPa]
$e_s - e_a$	Saturation vapour pressure deficit [kPa]
ET	Evapotranspiration
ET_a	Actual evapotranspiration [mm month^{-1}]
ET_p	Potential evapotranspiration [mm month^{-1}]
ET_{pm}	Potential evapotranspiration according to Penman-Monteith equation [mm day^{-1}]
FAO	Food and agricultural organisation
γ	Psychrometric constant [kPa °C^{-1}]
γ_{mun}/ γ_{ind}/ γ_{agr}	Growth rate (municipal, industrial, agricultural) [% yr^{-1}]
γ_{pop}/ γ_{GDP}	Growth rate (population or GDP) [% yr^{-1}]
G	Soil heat flux density [MJ m^{-2} day^{-1}]
GCM	Global climate model
GDEM	Global DEM (digital elevation model)
GDP	Gross domestic product
GES DISC DAAC	Goddard Earth Sciences Data and Information Center Distributed Active Archive Center
GHCN	Global Historical Climatology Network
G_i	Average daily soil heat flux density of month i [MJ m^{-2} day^{-1}]
GLCC	Global land cover characterisation
GLUE	Generalised likelihood uncertainty estimation
GMAO MERRA	Global Modelling and Assimilation Office Modern Era-Retrospective Analysis for Research and Applications (NASA)
GPCC	Global Precipitation Climatology Centre
GPS	Global positioning system
GSA	Global sensitivity analysis
h	Crop or canopy height; calculation of evapotranspiration [m]
HRU	Hydrologic response unit
I	Sensitivity index [-]
I	Rate of heat flow [W or J s^{-1}]
IPCC	International Panel on Climate Change
IW	Interval width
IWRM	Integrated water resource management
k	Karman's constant [-]
K	Thermal conductivity [MJ m^{-1} K^{-1}]
λ	Latent heat of vaporisation [2.45 MJ kg^{-1}]
LAI	Leaf area index [-]

LAI_{active}	Active leaf area index of the area [-]
Landsat ETM+	Landsat Enhanced Thematic Mapper Plus
Landsat TM	Landsat Thematic Mapper
LARSIM	Large Area Runoff Simulation Model
LCA	Life cycle analysis
LCI	Life cycle inventory
LCIA	Life cycle impact assessment
LP DAAC	Land Processes Distributed Active Archive Center
LSM	Land surface model
LST	Land surface temperature [K]
LST/E	Land surface temperature and emissivity
LST_{st}	Land surface temperature at climate station [K]
LULC	Land use and land cover
$\mu_{\mathrm{mun}}/ \mu_{\mathrm{ind}}/ \mu_{\mathrm{irr}}$	End use efficiency enhancement (municipal, industrial, agricultural) [% yr^{-1}]
m	Mass [kg]
MC	Monte Carlo
MENA	Middle East and North Africa
METI	Ministry of Economy, Trade and Industry, Japan
MMD	Multi model dataset
MME	Multi model ensembles
MODIS	Moderate-resolution Imaging Spectroradiometer
$\eta_{\mathrm{mun}}/ \eta_{\mathrm{ind}}/ \eta_{\mathrm{agr}}$	Efficiency of the water distribution system (municipal, industrial, agricultural) [%]
n	Actual duration of sunshine [hr day^{-1}]
N	Maximum possible duration of sunshine [hr day^{-1}]
NASA	National Aeronautics and Space Administration
NASIM	Precipitation-runoff model
NDVI	Normalised difference vegetation index
NESDIS	U.S. National Environmental Satellite, Data, and Information Service
NGA	National Geospatial Intelligence Agency
NOAA NCDC	National Oceanic and Atmospheric Administration (U.S.) National Climatic Data Centre
NRW	Non-revenue water
NSE	Nash-Sutcliffe coefficient of model efficiency [-]
$\omega_{\mathrm{mun}}/ \omega_{\mathrm{ind}}/ \omega_{\mathrm{agr}}$	Water demand (municipal, industrial, agricultural) [m^3]
OAT	One-at-a-time
OECD	Organisation of Economic Co-operation and Development
P	Precipitation [mm]
p	Atmospheric pressure [kPa]
ParaSol	Parameter Solution
PCA	Principal component analysis
PET	Potential evapotranspiration
P_n	Net precipitation [mm]

P_r	Precipitation fallen as rain (liquid) [mm]
p_{runoff}	Runoff coefficient, infiltration coefficient [%]
P_s	Precipitation fallen as snow (solid) [mm]
$\mathrm{P_{st}}$	Precipitation at the climate station [mm]
PUB	Prediction of ungauged basins
Q	Quantity of heat [J]
Q_{gw}	Ground water flow, ground water recharge [mm]
$\mathrm{Q_m}$	Runoff from snow melt, melt water flow [mm]
$\mathrm{Q_m}$	Modelled runoff [mm, m^3]
Q_n	Total water flow [mm]
$\mathrm{Q_o}$	Observed runoff [mm, m^3]
$\overline{Q}_0$	Mean of observed values [mm, m^3]
Q_{sf}	Surface water flow and runoff [mm]
ρ	Mass density [kg m^{-3}]
R^2	Model coefficient of determination
R_0	Incoming extraterrestrial solar radiation [MJ m^{-2} day^{-1}]
r_a	Aerodynamic resistance [s m^{-1}]
r_c	Correlation coefficient between modelled and observed flow for cumulative monthly values [-]
r_l	Average bulk stomatal resistance of an individual leaf [s m^{-1}]
r_m	Correlation coefficient between modelled and observed flow for monthly values [-]
RMB	Renminbi, currency of PR China
RMS	Root-mean-square
RMSE	Root-mean-square-error, mean difference between observed and simulated values
R_n	Net radiation at the land surface [MJ m^{-2}day^{-1}]
R_{nir}	Nearinfrared reflectances
R_{nl}	Net outgoing longwave radiation [MJ m^{-2}day^{-1}]
R_{ns}	Net solar or shortwave radiation [MJ m^{-2}day^{-1}]
R_{nw}	Net radiation over water bodies [MJ m^{-2}day^{-1}]
R_{red}	Red reflectances
r_s	Bulk surface resistance, stomatal resistance [s m^{-1}]
R_s	Solar radiation [MJ m^{-2}day^{-1}]
R_{so}	Clear-sky radiation [MJ m^{-2}day^{-1}]
SAR	Synthetic aperture radar
σ	Stefan-Boltzmann constant [4.903 $\times$ 10^{-9} MJ K^{-4} m^{-2} day^{-1}]
σ/std	Standard deviation
$\mathrm{SC_{2050,\ 50\ \%}}$	Climate scenario for 2050, 50 % projection of precipitation
$\mathrm{SC_{2050,\ 75\ \%}}$	Climate scenario for 2050, 75 % projection of precipitation
$\mathrm{SC_{2050,\ max}}$	Climate scenario for 2050, maximum projection of precipitation

SCE-UA	Shuffled complex evolution algorithm
SC_{LT}	Long-term climate scenario
SCS	Soil conservation service runoff curve number
SC_{ST}	Long-term climate scenario
SEBS/SEBAL	Surface energy balance algorithm for land
SPOT	Satellite Pour l'Observation de la Terre, Earth-observing satellite family
SRM	Snow melt runoff model
SRTM	Shuttle Radar Topography Mission
STREAM	Hydrological model
SUNGLASSES	Sources of uncertainty global assessment unsing split samples (uncertainty analysis)
SVAT	Soil-Vegetation-Atmosphere Transfer Models
SWAT	Soil and Water Assessment Tool
SWCI	Soil water content index
SWE	Snow water equivalent [mm]
t	Time [s]
T	Air temperature [°C or K]
T/s-value	Trend value
TC	Tasseled cap
T_{dew}	Dewpoint temperature [°C]
t_E	Ending time
T_i	Average temperature of the month i [°C]
T_{i+1}	Average temperature of the month $i+1$ (before) [°C]
T_{i-1}	Average temperature of the month $i-1$ (after) [°C]
TIN	Triangulated irregular network
T_K	Temperature in Kelvin [K]
T_m	Snow melting threshold temperature [°C]
T_{max}	Maximum temperature [°C]
T_{mean}	Mean air temperature [°C]
T_{mi}	Average temperature for month i [°C]
T_{min}	Minimum temperature [°C]
T_r	Upper threshold temperatures for snowfall [°C]
TRMM TMI	Tropical Rainfall Measuring Mission Microwave Imager
$TRMM_{st}$	TRMM value of the climate station [K]
T_s	Lower threshold temperatures for snowfall [°C]
t_S	Starting time
Ts/VI	Surface temperature/vegetation index
$T_{st,\ K}$	Termperture measured at the climate station [K]
u	Wind speed [m s^{-1}]
u_2	Wind speed measured at height 2 m [m s^{-1}]
UFW	Unaccounted for water
USGS	U.S. Geological Survey
UTM	Universal Transverse Mercator coordination system
V	Volume [m^3]

VE_{abs}	Absolute volume error
VE_{rel}	Relative volume error [%]
VE_{sys}	Systematic volume error
VWC	Virtual water content of the crop
WGMS	World Glacier Monitoring System
z_h	Height of humidity measurements [m]
z_m	Height of wind measurements [m]
z_{oh}	Roughness length governing the transfer of heat and vapour [m]
z_{om}	Roughness length governing momentum transfer [m]
z_{st}	Station elevation above sea level [m]

Chapter 1
Introduction

One of the most important problems on our planet earth is the contrast between natural resources and human demand and consumption of these resources. In this study, the natural resource in question is water. Water is an environmental factor, essential for most processes in the environment and also indispensible to human development and well-being (a.o. [35]). Almost all Millennium Goals issued by the United Nations are directly or indirectly connected to water as a resource. Water is a prerequisite and the key for poverty reduction, nutrition, health, improvement of living and housing standards, and education (cf. [31, 52]).

But with a continuous growth of the global population, water demand will grow disproportionally: the world population grows about 1.5 % per year while water demand increases about 3 % [34]. Human development requires additional water for an increasing population and for the production of food and industrial goods. Besides this dramatic increasing demand of water one has to consider all other human activities affecting the water supply. The reclamation and urbanisation of formerly natural land surfaces of watersheds changes all components of the water cycle and hydrological processes. Modern hydrologic constructions like dams and the sealing of the surfaces are particularly contributing to this development; as a result, the available water resources per capita will decrease, the burden on the existing water resources will grow, and storage and availability will be modified [39].

Besides the direct human interventions, the water resources are also affected by climate change. Due to the non-linear effects of 'global warming' on the hydrologic factors, the consequences are difficult to predict, but will most probably lead to a changed distribution and availability of water resources (cf. [32, 40]). Human activities are impacting on the Earth's ecosystems, human dominated cities in particular [20]. Examples for the significant effects on the hydrologic system are given by Wagener et al. [54] in a comparative analysis of human activities to geologic-scale forces, in which they cite human-induced climate change, shrinking aquifer storage due to excessive groundwater extraction, distorted river flow regimes and altered groundwater recharge due to changes in land use as the most prominent issues.

Human activities and climate change have also affected such remote areas as Northwest China. Urumqi, the capital of the Xinjiang Uygur Autonomous

K. Fricke, *Analysis and Modelling of Water Supply and Demand Under Climate Change, Land Use Transformation and Socio-Economic Development*, Springer Theses, DOI: 10.1007/978-3-319-01610-8_1,

Province has been founded in a strategically advantageous location at the foot of the Tianshan Mountains and it has developed into the economic and demographic centre of the region. Wang [56] identified three major issues in the water sector for China in general: floods, water shortage and water degradation. Indeed, the strong increase of the population as well as industrial and agricultural production have led to a dramatic land use change and interventions in the hydrological system that are causing severe water scarcity. At the same time, climate change is strongly affecting the local weather patterns and the water supply. The main challenge with regard to water resources is to assess and evaluate these changes and associated consequences for projecting possible developments of the hydrological system into the future.

1.1 Research Goals

In order to adequately react to these challenges, one would need an adapted and integrated water management on the water supply side, but also a decoupling of development and water consumption to mitigate the increasing water demand [19]. For a sustainable use of the water resources they need to be managed within the carrying capacity of the environmental system. Sustainability implies that development has to be compatible with the ecological, economic and social boundary conditions and maintain the environmental resources for the future (cf. [28, 32]). Economic and demographic growth should be within the reproductive and regenerative capacity of the ecosystems and their carrying capacities. However, there is no objective limiting point at which ecosystems start to collapse under an anthropogenic burden. Such limits are difficult to assess, which certainly also applies to any attempt at setting up options of future use [28].

Integrated water resource management (IWRM) tries to consider the entire 'water' ecosystem and to include all parts and sectors of a catchment area, such as the water balance, water quantity and quality, surface and groundwater, waste water, economy, society, technical and institutional disciplines, models, methods, instruments and data [27, 52]. The interconnection between man and nature, complexity of interests and uncertainty of predictions and projections requires an integration of nature and society, of utilisation and consequences. IWRM "promotes the coordinated development and management of water, land and related resources, in order to maximise the resultant economic and social welfare in an equitable manner without compromising the sustainability of vital ecosystems" [27: 8]. Achieving an IWRM preferably requires comprehensive information, investigations, models and scenarios. River catchments and not administrative boundaries serve as planning and implementation units to illustrate and include relationships between up- and downstream catchments as well as the different sectors and stakeholders, in order to develop a holistic approach. The communication needs to be improved between scientists and policy-makers and all available **hydrological knowledge** is required for action and management plans addressing the water challenges [42]. Hence, adaptation to changes and resolving water conflicts will demand more research on the status, distribution, and effects of changes on the water resources [16].

Corresponding to these requirements, **the goal of this research is to evaluate the recent situation of water resources and water demand and the possible changes in Urumqi Region due to the development of population and economy, land use transformation and climate change and to make educated predictions for the future**. Both the water supply and demand side are modelled with a hydrological and a socio-economic model, respectively, which were chosen with regard to the available input data and the desired results. In this study, the recent hydrological situation is modelled, but also possible and reasonable climate and land use scenarios are introduced to simulate the water balance and availability under changing climate and land use conditions. The predicted changes for Urumqi Region are then evaluated. This way, the relation between climate and water resources availability and the response of the hydrological system to natural and human-induced changes can be studied and the relative impact of climate compared to non-climatic factors, mainly land use change, can be assessed [12]. The development of the future water consumption is also estimated and compared to the projected water availability to determine future challenges and possible adaptation strategies.

The results and models are also connected to the task group "water resource efficiency" of the research project "RECAST Urumqi", where they are integrated into a water resources information system as well as for the development of a water conservation strategy. This macroscale and integrated modelling approach corresponds to several of the requirements of integrated hydrological science as stated by Chehbouni et al. [15] and Wagener et al. [54]. Hydrological research and modelling is supposed to document changes and drivers of change and to assess how the hydrological system will respond to these natural and human-induced changes in climate and environment. The dominant processes controlling the overall hydrological functioning of the basins are modelled at the macroscale as required by Chehbouni et al. [15], but not across spatial scales as demanded by Wagener et al. [54]. The latter was not possible due to the demand for strategic suggestion for the larger research area and the lack of data and measurements at smaller scales. However, measures to react to future changes can be provided to enable a sustainable management of hydrological systems.

The case of Urumqi can be regarded as representative for larger populations of the region in various regards: Urumqi Region is one of several dryland cities and regions in Central Asia challenged by a water-limited environment and continuous development. "Water-limited environments include arid, semi-arid, and subhumid regions [...] and occupy approximately 50 % of the global land area" [41: 2]. Additionally, Urumqi's location on the foot of a mountain range and its water supply makes it a **model for numerous cities and areas** inhabited by one sixth of the global population that depend on water resources from nearby mountain watersheds [3, 53]. Urumqi Region also exhibits a large **spatial variation** in climate zones due to the large gradient in altitude over a short horizontal distance, where the variable impacts of climate change on the hydrological system can be observed. Especially, here, it is possible to investigate the importance of snow and ice for the natural storage and water supply and their diminishing role due to climate change (cf. [26]). Hence, one aspect of this research is the transferability of the developed and used methodology and models to other cases.

In summary, the main objective of this study is to **simulate** (i) **the development of water supply and availability under climate and land use change** and (ii) **the development of water demand of the growing population and economy**. For both water supply and demand, the estimated total quantities will be compared. To suggest possible adaptation measures and conservation strategies, the spatial distribution of the water supply and the distribution of water demand among the sectors agriculture, population and industry are evaluated as well. The introduction into the research topic and research area (Chaps. 1 and 2) is followed by an elaboration of the water balance model for the simulation of water demand in Chap. 3. The scenarios of climate change and land use transformation as well as the simulation results are then presented in the next two chapters. For an overview of the scenarios and their abbreviations used in this study please refer to the very last page. The econometric model used to project future water consumption and its driving factors as well as the results are shortly summarised in Chap. 6. In the following chapter, the results are compared and evaluated to lay the foundation of reasonable and optimised adaptation strategies for Urumqi Region. Chapter 8 provides a methodological discussion of the model, its applicability and transferability.

1.2 Current State of Research and Approach

Before describing the current state of research and possible approaches for the research goals introduced above, the main challenges of hydrological modelling and prediction in ungauged basins are stated in the following two sections.

1.2.1 Challenges of Hydrological Modelling

When models are used to predict certain results that cannot be measured completely over a spatial or temporal extent, it is often neglected that these models are simplified representations of reality and the results are only one possible outcome based on the assumptions made in the model. Also, the degree of simplification in turn depends on the (in) completeness of the model and the included variables and factors that actually can be observed and measured. Despite the advantages and benefits of models for the investigation and prediction of hydrological processes, several problems exist that the modeller has to bear in mind when developing and applying any model as described by Beven [4]: the non-linearity of hydrological systems, modelling processes across scales, the uniqueness of watersheds, the equifinality of parameter sets and the uncertainty of parameters and models.

Non-linearity can occur at different scales, where certain thresholds mark a change in the behaviour of output parameter(s) (cf. [50]). Non-linear systems and their initial and boundary conditions should be taken into account in the formulation and application of models [4]. However, they cannot be easily transferred from one scale to another. Subgrid-scale non-linear descriptions for homogeneous plots, such as Richard's equation, are not useful for the model element scale, where the heterogeneity of parameters has to be taken into account [4]. Therefore, the problem of non-linearity of hydrological processes is also connected to the different scales. A process or measurement **scale** can be defined as the extent or duration, spacing or time period, and integral correlation length or integration volume [8]. Conditions and dominant processes often differ in their space or time scale. As Beven [4] remarks, the dominant processes on a certain scale can be represented and measured only to a certain degree of detail with the actually available methods and techniques. Concepts that allow gathering information on one scale are not just valid to make predictions at other scales (scaling problem) [7] and instead of trying to capture all small-scale variability and complexity, the dominant processes for each scale should be acknowledged [50]. For the solution of this problem, two different approaches are being discussed, both starting with the presumption that dominant processes at the different scales need to be identified. Blöschl [9] suggests that a solution of the scale problem is necessary for further advances in hydrology and dominant processes across scales need to be identified, while Beven [6] states that model structures depend on the scale and that scaling theories will prove to be impossible (both cited in Beven [4]). Even the task to describe the dominant processes and to apply hydrological models in particular catchments is challenging due to the **heterogeneity** of and in catchments and because each catchment has its own unique characteristics [4, 50]. This is also expected to be the case in the Urumqi Region as the river catchments extend from mountainous, alpine regions to relatively flat basins with varying climate, vegetation, soil and geology. The dominant processes in the headwaters might differ significantly from the ones of the lower sub-catchments, but should be equally represented by the chosen modelling concept.

The challenge is the development of a general theory of hydrological processes versus taking the **uniqueness** of each place and time into account. However, the unique description of a process is not transferable to other catchments. Also, even in a perfect model, parameters have to be estimated as the current available measurement techniques have limitations in assessing spatial variability [4]. When the parameters in turn are 'only' estimations of the actual situation but not directly measured in place, there may be different estimated or calibrated parameter sets that come to the same model output.

The possibility that there exists more than one optimal parameter set for a model is addressed as **equifinality** [5]. Equifinality is due to limited measurements, unknowability of the subsurface or processes, and heterogeneity over space and time [4]. For instance, remote sensing data has been used to derive knowledge and information about the distributed catchment characteristics and their variability on pixel scale. Unfortunately, the spectral information collected by the sensor does not give direct information about the hydrological variables, but in most cases about the surface characteristic. An interpretative model is necessary for correction and extraction of the desired parameters, which is also subjected to the equifinality in interpretation and uncertainty in prediction [4, 7]. Within this scope of research it was not possible to methodologically test different model structures and many parameter sets, e.g. with a Generalised Likelihood Uncertainty Estimation (GLUE), to define the optimal choices for the research area. Therefore, the problem of equifinality was seen as a question of decidability: other possible models and structures that would lead to the same results are not seen as poorly suitable, but were rejected due to the boundary conditions established for the model and available parameters in the research area.

Equifinality also refers to the **uncertainty** of the optimal model structure and parameter sets to describe a complex natural system (modelling uncertainty). Additionally, other types of uncertainty exist: statistical uncertainty derived from the inadequate description of natural variability, prediction uncertainty due to the limited account of knowledge about effects and processes, and intrinsic uncertainty of parameters that can drastically change over time and space [37]. Modelling and prediction uncertainty are inherent in the limitations to understand real processes and to describe them with models. Only statistical uncertainty (variability, randomness) and intrinsic uncertainty (incertitude) can be reduced by better datasets, research and engineering. Therefore, a sensitivity study was conducted to assess the most important parameters and to improve their accuracy. Additionally, an uncertainty analysis was conducted to account for the variability of the parameters and to assess their effect on the uncertainty of the model and the model outputs (see Sect. 3.2.2).

1.2.2 Modelling and Prediction in Ungauged Basins

The challenges discussed above are inherent to all hydrological models, but they increase when measurement data to revise and calibrate the used concepts and model structures are not available. However, most of the basins on Earth are ungauged and it is necessary to utilize the limited data available to make reliable predictions also in ungauged basins (prediction in ungauged basins, PUB) [25]. Also in the chosen research area, not all catchments are gauged or have their outflow measured. Only for the watershed of one reservoir, more detailed runoff data was available and the model structure and parameters could be defined and calibrated. For the other watersheds, hydrological processes had to be processed as if they were ungauged basins.

Hydrological models for observed processes and relationships, which are calibrated for gauged basins, are usually transferred to ungauged basins using methods such as regression, spatial or physical proximity. Merz and Blöschl [38] simulated the water balance dynamics for 308 catchments in Austria and observed that if available, the average model parameters of the immediate upstream and downstream basins using the method of spatial proximity and regionalisation by kriging are the best ways of transferring model parameters for making PUB. But also mixed performance is reported [22]. Other researchers transferred model parameters from gauged basins to ungauged basins through some of the relationships developed between the model parameters and physical and climatic characteristics [25]. This **'similar basin approach'** is commonly used, where the complete set of model parameters identified at the donor catchment is applied to predict the hydrological responses at ungauged catchments with similar physiographic and climatic characteristics [25]. These approaches use measurements by remote sensing and application of process-based hydrological models where climate inputs are specified or measured [48], which is also employed for the modelling approach of this study. Despite considerable effort, remote sensing has not been useful in the direct extrapolation of hydrological variables; differences in vegetation type or other land surface cover dominate the remotely derived images [7]. From the surface cover data and additional information about subsurface soil characteristics, parameters representing the dominant hydrological processes had to be identified and were calibrated for one gauged basin. Based on the spatial proximity and by using parameters derived from remote sensing and climate data available for the other catchments, the model structure was transferred to the entire research area.

Prediction using hydrological models actually extrapolates the processes and parameters identified for a gauged basin to another time in the future at the same location, for which measurements not yet exist, and applies the concept of PUB [48]. The transferability of hydrological models and the estimation method are subject to limitations when critical but unobservable processes are inadequately represented or incompletely specified, resulting in considerable uncertainties [22]. The uncertainties include input uncertainty due to heterogeneity of input parameters, model structure uncertainty due to process heterogeneity, and parameter uncertainty due to landscape uncertainty in the ungauged basin [48]. Changes to both parameters and processes in space and time can occur, hence making predictions under global change difficult [46]. The prediction uncertainty is very high and even after calibration of the model at the gauged basin's outlet, reasonable predictions at the other ungauged basin's outlet are not guaranteed [25].

1.2.3 Types of Hydrological Models

Modelling in general is used to simulate the relationship or process between the input and output of a system. Within hydrology, the system is a hydrological system and the input data are environmental parameters, while the output is normally related to some kind of water flow. Either input or output data are measured or estimated and compared to the simulated data input or output, respectively. The main challenge is in how much detail the hydrological system has to be modelled when measurements or data needed for this purpose are not available and still reflect the collective knowledge of the hydrological community [4]. According to Beven [4], the working philosophy of most practising environmental scientists is a pragmatic or heuristic realism which is also employed here.

Hydrological modelling can be undertaken to further understand hydrological processes which cannot be measured or observed directly (investigative modelling) or to predict the response of a hydrological system to certain input or output parameters (**predictive modelling**). As the main goal of this research is to estimate the water balance in the future, the hydrological model used here is mainly predictive. Furthermore, hydrological models can be divided according to the method they use to describe hydrological processes and functions. Physical or **process-based** models use formulas closely linked to the actual state, often using semi-empirical models and formulas. Conceptual or **index-based models** describe hydrological processes and phenomena with coefficients and indices. The choice often depends on the available information as data input and for calibration and in how much detail the processes should be modelled [22]. Both model types were employed depending on the available input and calibration data. Models using equations based on fundamental principles of physics or robust empirical methods are also considered as deterministic models. In contrast, models using stochastic methods to find the optimal parameter set for a model are called 'random' models [13]. These random models can also be used to assess the sensitivity of a model to input parameters (see model sensitivity and uncertainty in Sect. 3.2) but were not applied for the modelling itself. Deterministic and process-based models are often expected to simulate more 'realistic' results than models fitted in historical output data, which are then not transferable to other research areas in time and space [4].

Models also differ in the **spatial unit used for modelling**: lumped models treat a catchment as one homogeneous unit with the same processes, input parameters, boundary conditions or hydrological properties and can be calibrated efficiently [11]. The exact opposite is **distributed modelling**, where spatial heterogeneity is especially accounted for and the modelling equations are solved and calculated for each pixel in a grid [13], generating plausible spatial patterns of state variables and processes [11]. Distributed models can take physical data as input data and are mainly used for process-based approaches [2]. Another advantage of distributed modelling is the unproblematic integration of remote sensing data which are normally available as a raster with information for each cell [2]. The distributed approach might better represent the hydrological processes in the watershed [45], but with regard to the modelling results, none of the two model types outperformes the other (cf. [4, 11, 45]). In this research approach, a distributed model was selected to enable the simulation of even small-scale differences and to take the small-scale land use changes into account. Out of practicability, models often employ a combined approach as a lumped framework leads to oversimplification of the spatial variability and processes and a distributed model requires normally much more data and computing requirements. These models, e.g. also the Soil and Water Assessment Tool (SWAT) (see Sect. 3.4.4), are called partially or semi-distributed models.

1.2.4 Estimation of Evapotranspiration

Especially in arid or semi-arid regions, evapotranspiration constitutes the most important water balance component in the hydrological system. Its quantity and spatial distribution is substantial for proper water and land use management [17]. In the best case, in situ measurements of evapotranspiration would be available for different land use and vegetation cover types. As this is not the case in most locations, evapotranspiration rates can be retrieved from satellite data and other ground truth measurements using surface temperature/vegetation index (Ts/VI) methods, physically-based algorithms using, e.g. the Penman–Monteith equation, and numerical or process-based methods, which simulate the transfer of heat and water (e.g. Soil–Vegetation–Atmosphere Transfer models (SVAT) or Land Surface Models (LSMs), cf. Contreras et al. [17]). Surface temperature/vegetation index methods should be applied with calibration and ground truth measurements, which were not possible for the area under investigation. Instead, the physically-based Penman–Monteith equation was employed to estimate potential evapotranspiration. Zhou et al. [60] used a surface energy balance algorithm for land (SEBS/SEBAL) with satellite data as an input to estimate the evapotranspiration for a research area in Xinjiang including Urumqi Region. Although SEBAL is also applying the Penman–Monteith equation, the resolution of the satellite data was insufficient for the scale of the research and not suitable for projections.

1.2.5 Snow Melt Routine and Glacier Modelling

Accumulation and ablation of snow can be calculated based on the radiation budget and other meteorological factors of the snow pack [29, 30]. Alternatively, these processes can be estimated by using the air temperature as indicator of thermodynamic processes (cf. the degree-day-method presented in Sect. 3.1.2 and [44]), which can be expanded with additional factors describing the snow cover (runoff coefficient, proportion of snow cover, recession coefficient) to the snow melt runoff model (SRM) [36]. A distributed temperature-index melt model additionally including potential short-wave radiation has been successfully employed by Huintjes et al. [24] at the eastern branch of the Urumqi Glacier No. 1 to calculate annual mean surface mass balance and the spatial distribution of melt rates.

1.2.6 Econometric Modelling

Predicting the water demand in the future is an equally complex problem as hydrological modelling. The choices and development of the many user groups are as numerous as the modelling approaches. Water use has been forecasted using linear extrapolation of recent water consumption, statistical analysis of water demand time series [61] or more advanced, artificial neural networks [33]. Du et al. [18] already modelled the relationship between water consumption and total population, per-capita GDP, gross industrial output value, and government revenue with polynomial functions to explain the adaptation process between urbanisation and water resources exploitation in Urumqi City. However, the polynomial functions are not suitable for extrapolation in the far future [23], more applicable is an equation developed by Trieb [51] to quantify the growing demand for freshwater in the Middle East and North Africa (MENA) based on population and GDP growth as well as factors of distribution efficiency.

1.2.7 Climate Change and Water Resources in Xinjiang and Urumqi Region

The numerous consequences of climate change have been frequently evaluated for the whole of China [43] or even larger regions such as Central Asia or Asia (cf. [49]). With better data availability, large-scale modelling is obviously a suitable approach. The general findings of models and projections are that the annual runoff over China as a whole will probably increase, but with an uneven distribution with regard to time and space [55]. Even more important is to take a look also at hydrological changes at the regional scale. A research study investigating changes of climate and hydrology closer to Urumqi Region has been published by Shi et al. [47], who found global warming and the enhanced water cycle to effect a continuous increase in precipitation, glacier melt water, river runoff, and air temperature in northwestern China over the last decade. They predict a possible increase in annual runoff by more than 10 %. Cassassa et al. [14] also observed increased runoff that was closely related to temperature rise and enhanced glacier melting under the present warming trends. In contrast, a study by Aizen et al. [1] found climatic and hydrologic changes in the Tianshan and reported that while precipitation and temperatures increased, snow resources decreased and annual runoff has dropped or did not change significantly. Apparently, all research in this region up to now has always revealed some kind of variability as Bolch [10] has shown for the different ways glaciers of continental and maritime type in the Tianshan Mountains react to climatic changes.

Also, the effects of climate change and increasing temperatures on the largest glacier in Urumqi Region, the Urumqi Glacier No. 1, have been investigated in detail by Han et al. [21], Xu et al. [57] and Ye et al. [58]. All studies agree that the glacier has significantly retreated over the last 45 years with accompanying increase in runoff influenced by an increase of annual and summer air temperature. On the other hand, Zhang et al. [59] modelled the monthly runoff at the headwaters of the Urumqi River and confirmed that the river runoff is most sensitive to changes of precipitation and only to a smaller degree to changes of temperature. These examples show that the observed or modelled and predicted effects of climate change on the hydrological system can be quite variable depending on the research area or research scale. All the more important are regional case studies on catchment level to assess and localise future changes.

References

1. Aizen, V. B., Aizen, E. M., Melack, J. M., & Dozier, J. (1997). Climatic and hydrologic changes in the Tien Shan, Central Asia. *Journal of Climate, 10*, 1393–1403.
2. Ajami, N. K., Gupta, H., Wagener, T., & Sorooshian, S. (2004). Calibration of a semi-distributed hydrologic model for streamflow estimation along a river system. *Journal of Hydrology, 298*, 112–135.
3. Barnett, T. P., Adam, J. C., & Lettenmaier, D. P. (2005). Potential impacts of a warming climate on water availability in snow-dominated regions. *Nature, 438*, 303–309. doi:10.1038/nature04141.
4. Beven, K. J. (2001). How far can we go in distributed hydrological modelling? *Hydrology and Earth System Sciences, 5*(1), 1–12.
5. Beven, K. J., & Binley, A. (1992). The future of distributed models: model calibration and uncertainty prediction. *Hydrological Processes, 6*, 279–298.
6. Beven, K. J., (1995). Linking parameters across scales: sub-grid parameterisations and scale dependent hydrological models. *Hydrological Process, 9*, 507–526.
7. Beven, K. J., & Fisher, J. (1996). Remote sensing and scaling in hydrology. In J. B. Stewart, E. T. Engman, R. A. Feddes, & Y. Kerr (Eds.), Scaling up in Hydrology Using Remote Sensing (pp. 1–18). New York: Wiley.
8. Blöschl, P., & Sivapalan, M. (1995). Scale issues in hydrological modelling: a review. *Hydrological Processes, 9*(3–4), 251–290.
9. Blöschl, G., (2001). Scaling in hydrology. *Hydrological Process, 15*, 709–711.
10. Bolch, T. (2007). Climate change and glacier retreat in northern Tien Shan (Kazakhstan/Kyrgyzstan) using remote sensing data: climate change impacts on mountain glaciers and permafrost. *Global and Planetary Change, 56*(1–2), 1–12.
11. Bormann, H., Breuer, L., Giertz, S., Huisman, J. A., & Viney, N. R. (2009). Spatially explicit versus lumped models in catchment hydrology—experiences from two case studies. In P. C. Baveye, M. Laba, & J. Mysiak (Eds.), *Uncertainties in environmental modelling and consequences for policy making* (pp. 3–26). Dordrecht: Springer.
12. Bouwer, L. M., Aerts, J. C., Droogers, P., & Dolman, A. J. (2006). Detecting the long-term impacts from climate variability and increasing water consumption on runoff in the Krishna river basin (India). *Hydrology and Earth System Sciences, 10*, 703–713.
13. Burns, I. S., Scott, S. N., Levick, L. R., Semmens, D. J., Miller, S. N., Hernandez, M., Goodrich, D. C., & Kepner, W. G. (2007). Automated geospatial watershed assessment (AGWA) documentation: version 2.0 (145 p). Tuscon, Arizona.
14. Casassa, G., López, P., Pouyaud, B., & Escobar, F. (2009). Detection of changes in glacial run-off in alpine basins: examples from North America, the Alps, central Asia and the Andes. *Hydrological Processes, 23*, 31–41.

15. Chehbouni, A., Escadafal, R., Duchemin, B., Boulet, G., Simonneaux, V., Dedieu, G., et al. (2008). An integrated modelling and remote sensing approach for hydrological study in arid and semi-arid regions: the SUDMED Programme. *International Journal of Remote Sensing, 29*(17–18), 5161–5181.
16. Committee on Water Resources Activities (2009). Towards a sustainable and secure water future: a leadership role for the U.S. Geological Survey (128 p). Washington, D.C: The National Academies Press.
17. Contreras, S., Jobbágy, E. G., Villagra, P. E., Nosetto, M. D., & Puigdefábregas, J. (2011). Remote sensing estimates of supplementary water consumption by arid ecosystems of central Argentina. *Journal of Hydrology, 397*(1–2), 10–22.
18. Du, H., Zhang, X., & Wang, B. (2006). Co-adaptation between modern oasis urbanisation and water resources exploitation: a case of Urumqi. *Chinese Science Bulletin, 51* (Supp. I) 189–195.
19. Falkenmark, M., Berntell, A., Jägerskog, A., Lundqvits, J., Matz, M., & Tropp, H. (2007). On the verge of a new water scarcity: a call for good governance and human ingenuity (19 p). SIWI, Stockholm, Sweden (SIWI Policy Brief).
20. Grimm, N. B., Grove, J. M., Pickett, S. T., & Redman, C. L. (2000). Integrated approaches to long-term studies of urban ecological systems. *BioScience, 50*(7), 571–584.
21. Han, T., Ding, Y., Ye, B., Liu, S., & Jiao, K. (2006). Mass-balance characteristics of Urumqi Glacier No. 1, Tien Shan, China. *Annals of Glaciology, 43*, 323–328.
22. He, Y., Bárdossy, A., & Zehe, E. (2011). A review of regionalisation for continuous streamflow simulation. *Hydrology and Earth System Sciences, 15*, 3539–3553.
23. Hoffmann, T., & Rödel, R. (2004). Leitfaden für die statistische Auswertung geographischer daten, Greifswald (110 p). (Greifswalder Geographische Arbeiten, 33).
24. Huintjes, E., Li, H., Sauter, T., Li, Z., & Schneider, C. (2010). Degree-day modelling of the surface mass balance of Urumqi Glacier No. 1, Tian Shan, China. *The Cryosphere Discussions, 4*, 207–232.
25. Hunukumbura, P. B., Tachikawa, Y., & Shiiba, M. (2011). Distributed hydrological model transferability across basins with different hydro-climatic characteristics. *Hydrological Processes,* doi:10.1002/hyp.8294.
26. Immerzeel, W. W., Droogers, P., de Jong, S. M., & Bierkens, M. F. (2009). Large-scale monitoring of snow cover and runoff simulation in Himalayan river basins using remote sensing. *Remote Sensing of Environment, 113*, 40–49.
27. Jakeman, A. J., Giupponi, C., Karssenberg, D., Hare, M. P., Fassio, A., & Letcher, R. A. (2006). Integrated management of water resources: concepts, approaches and challenges. In C. Giupponi (Ed.), Sustainable Management of Water Resources. An integrated approach (pp. 3–26). Cheltenham: Elgar.
28. Kahlenborn, W., & Kraemer, R. A. (1999). *Nachhaltige Wasserwirtschaft in Deutschland* (244 p). Berlin, Heidelberg: Springer.
29. Knauf, D. (1976). Die Abflußbildung in Schneebedeckten Einzugsgebieten des Mittelgebirges. Institut für Hydraulik und Hydrologie Darmstadt, Darmstadt (155 p) (Technische Berichte, 17).
30. Kondo, J., & Yamazaki, T. (1990). A prediction model for snowmelt, snow surface temperature and freezing depth using a heat balance method. *Journal of Applied Meteorology, 29*, 375–384.
31. Kreutzmann, H. (2006). Wasser und Entwicklung: Rohstoffverknappung, Marktinteressen und Privatisierung der Versorgung. *Geographische Rundschau, 58*(2), 4–11.
32. Lehn, H., Steiner, M., & Mohr, H. (1996). *Wasser—die elementare Ressource: Leitlinien einer nachhaltigen Nutzung* (368 p). Berlin: Springer.
33. Liu, J., Savenije, H. H., & Xu, J. (2003). Forecast of water demand in Weinan city in China using WDF-ANN model: water resources assessment for catchment management. *Physics and Chemistry of the Earth, Parts A/B/C, 28*(4–5), 219–224.
34. Londong, J., Hillenbrand, T., Otterpohl, R., Peters, I., & Tillman, D. (2004). Vom Sinn des Wassersparens. *KA—Abwasser, Abfall, 51*(12), 1381–1385.

35. Martin, K., & Sauerborn, J. (2006). *Agrarökologie* (297 p). Stuttgart: UTB.
36. Martinec, J., Rango, A., & Roberts, R. (2008). Snowmelt Runoff Model (SRM) User's Manual: Edited by Enrique Gómez-Landesa & Max P. Bleiweiss (178 p). Las Cruces: New Mexico State University.
37. Meinrath, G., & Schneider, P. (2007). *Quality assurance for chemistry and environmental science: metrology from pH measurement to nuclear waste disposal* (326 p). Berlin, Heidelberg: Springer.
38. Merz, R., & Blöschl, G. (2004). Regionalisation of catchment model parameters. *Journal of Hydrology, 287*, 95–123.
39. Meßer, J. (1997). Auswirkungen der Urbanisierung auf die Grundwasserneubildung im Ruhrgebiet unter besonderer Berücksichtigung der Castroper Hochfläche und des Stadtgebietes Herne, Essen. Deutsche Montan Technologie GmbH (DMT-Berichte aus Forschung und Entwicklung, 58).
40. Nagarajan, R. (2006). *Water—conservation, use and management for semi-arid-region* (352 p). New Delhi: Capital Publishing.
41. Newman, B. D., Wilcox, B. P., Archer, S. R., Breshears, D. D., Dahm, C. N., Duffy, C. J., McDowell, N. G., Phillips, F. M., Scanlon, B. R., & Vivoni, E. R. (2006). Ecohydrology of water-limited environments: a scientific vision. *Water Resources Research,* 42 (W06302).
42. Oki, T., & Kanae, S. (2006). Global hydrological cycles and world water resources. *Science, 313*(5790), 1068–1072.
43. Piao, S., Ciais, P., Huang, Y., Shen, Z., Peng, S., Li, J. Z., et al. (2010). The impacts of climate change on water resources and agriculture in China. *Nature, 467*, 43–51.
44. Rango, A., & Martinec, J. (1995). Revisiting the degree-day method for snowmelt computations. *Water Resources Bulletin, 31*(4), 657–669.
45. Ruelland, D., Larrat, V., & Guinot, V., (2010) A comparison of two conceptual models for the simulation of hydro-climatic variability over 50 years in a large Sudano-Sahelian catchment. In E. Servat & S. Demuth (Eds.), Global change. Facing risks and threats to water resources. *Proceeding of the 6th World FRIEND Conference*, Fez, Morocco, 25–29 October, 2010 (pp. 668–678). Wallingford: IAHS Press.
46. Schaefli, B., Harman, C. J., Sivapalan, M., & Schymanski, S. J. (2010). Hydrologic predictions in a changing environment: behavioral modelling. *Hydrology and Earth System Sciences Discussions, 7*, 7779–7808.
47. Shi, Y., Shen, Y., Kang, E., Li, D., Ding, Y., Zhang, G., et al. (2007). Recent and future climate change in Northwest China. *Climatic Change, 80*, 379–393.
48. Sivapalan, M., Takeuchi, K., Franks, S. W., Gupta, V. K., Karambiri, H., Lakshmi, V., et al. (2003). IAHS Decade on predictions in ungauged basins (PUB), 2003–2012: Shaping an exciting future for the hydrological sciences. *Hydrological Sciences Journal/Journal Sciences Hydrologiques, 48*(6), 857–880.
49. Solomon, S., Qin, D., Manning, M., Alley, R. B., Berntsen, T., Bindoff, N. L., Chen, Z., Chidthaisong, A., Gregory, J. M., Hegerl, G. C., Heimann, M., Hewitson, B., Hoskins, B. J., Joos, F., Jouzel, J., Kattsov, V., Lohmann, U., Matsuno, T., Molina, M., Nicholls, N., Overpeck, J., Raga, G., Ramaswamy, V., Ren, J., Rusticucci, M., Somerville, R., Stocker, T. F., Whetton, P., Wood, R. A., & Wratt, D. (2007). Technical summary. In S. Solomon, D. Qin, M. Manning, Z. Chen, M. Marquis, K. B. Averyt, & H. L. Miller (Eds.). *Climate change 2007: the physical science basis. Contribution of working group I to the fourth assessment report of the Intergovernmental Panel on Climate Change* (pp. 19–91). Cambridge: Cambridge University Press.
50. Tetzlaff, D., Carey, S. K., Laudon, H., & McGuire, K. (2010). Catchment processes and heterogeneity at multiple scales—benchmarking observations, conceptualisation and prediction. *Hydrological Processes, 24*, 2203–2208.
51. Trieb, F. (2007). AQUA-CSP: Concentrating solar power for seawater desalination (279 p). DLR (Final report, Bundesministerium für Umwelt, Naturschutz und Reaktorsicherheit (BMU)).

52. van Edig, A., & van Edig, H. (2005). Integriertes Wasserressourcen-Management: Schlüssel zur Nachhaltigen Entwicklung. In S. Neubert (Ed.). Integriertes Wasserressourcen-Management (IWRM). Ein Konzept in die Praxis Überführen (135–157). Baden–Baden: Nomos Verl.-Ges.
53. Viviroli, D., Dürr, H. H., Messerli, B., Meybeck, M., & Weingartner, R. (2007). Mountains of the world, water towers for humanity: Typology, mapping, and global significance. *Water Resources Research, 43*, W07447. doi:10.1029/2006WR005653.
54. Wagener, T., Sivapalan, M., Troch, P. A., McGlynn, B. L., Harman, C. J., Gupta, H. V., Kumar, P., Rao, P. S., Basu, N. B., & Wilson, J. S. (2010). The future of hydrology: an evolving science for a changing world. *Water Resources Research*, *46* (W05301).
55. Wang, W., Yang, X., & Yao, T. (2012). Evaluation of ASTER GDEM and SRTM and their suitability in hydraulic modelling of a glacial lake outburst flood in southeast Tibet. *Hydrological Processes, 26*(2), 213–225.
56. Wang, S. C. (2006). Resource-oriented water management. Towards harmonious coexistence between man and nature (218 p). Singapore: World Scientific.
57. Xu, X., Pan, B., Hu, E., Li, Y., & Liang, Y. (2011). Responses of two branches of Glacier No. 1 to climate change from 1993 to 2005, Tianshan, China. *Quaternary International, 236*(1–2) 143–150.
58. Ye, B., Yang, D, Jiao, K., Han, T., Jin, Z., Yang, H., & Li, Z. (2005). The Urumqi river source Glacier No. 1, Tianshan, China. Changes over the past 45 years. *Geophysical Research Letters, 32*, doi:10.1029/2005GL024178.
59. Zhang, W., Ogawa, K., Ye, B., & Yamaguchi, Y. (2000). A monthly stream flow model for estimating the potential changes of river runoff on the projected global warming. *Hydrological Processes, 14*(10), 1851–1868.
60. Zhou, Y., Nonner, J. C., Li, W., & et al. (2007). Strategies and techniques for groundwater resources development in Northwest China (338 p). Beijing: China Land Press.
61. Zhou, S. L., McMahon, T. A., Walton, A., & Lewis, J. (2002). Forecasting operational demand for an urban water supply zone. *Journal of Hydrology, 259*(1–4), 189–202.

Chapter 2
Research Area

The research area is the region around Urumqi City (Wūlǔmùqí Shì), located in Northwest China. Urumqi is the provincial capital of Xinjiang Uygur Autonomous Region (Xīnjiāng Wéiwú'ěr Zìzhìqū), the most northwestern province of PR China and close to Central Asia. Xinjiang belongs to the dry temperate zone and is characterised by two large basins, the Tarim and Junggar Basin with the Taklamakan and Gurbantünggüt Desert, respectively, which are separated and surrounded by the Altai, Tianshan and Kunlun Mountains from North to South. The largest area is taken by semi-arid steppe or deserts [1]. Urumqi itself is located at the southern margin of the Junggar Basin and the northern slope of the Tianshan Mountains. The city is a former oasis settlement, which developed between the semi-desert in the North and the mountain ranges in the South.

Agricultural and human development is only possible on a relatively small green grassland corridor between the extreme environments of mountains and deserts, "making it a highly sensitive ecology" [2: 451]. **Oases** make up only 4–5 % of the total area of the region, but harbour over 90 % of the population and over 95 % of the social wealth (Han 2001 cited in [3]).

From the basin at 400–500 m a.s.l. to the mountains with over 4,000 m a.s.l., **a gradient of more than 3,500 m difference in altitude over a distance of 55 km exists**, while Urumqi City is located at 600–1,100 m. South of Urumqi a rift divides the Tianshan Mountain range, the Dabancheng Corridor, which acts as a wind channel [4] and as pass to the Tarim Basin south of the mountain range. The rivers in Urumqi Region flow mainly from the Tianshan Mountains in the South to the Junggar Basin in the North, passing by the urban area of Urumqi City as there an opening for the Urumqi River valley is located between the two mountain ridges of the Tianshan ([5], see Fig. 2.1). The administrative area of Urumqi City exceeds the urban area and covers large parts of the adjacent watersheds in the Tianshan Mountains draining towards the city, and also areas draining to the Turpan Basin in the South and a part of the Grubantünggüt Desert in the North. The latter areas are of little interest for the hydrology of Urumqi Region as they are either not connected to the hydrological system of the city area or their relationship is unsure. Thus, Urumqi Region in this study refers to the watershed of the rivers

K. Fricke, *Analysis and Modelling of Water Supply and Demand Under Climate Change, Land Use Transformation and Socio-Economic Development*, Springer Theses, DOI: 10.1007/978-3-319-01610-8_2,

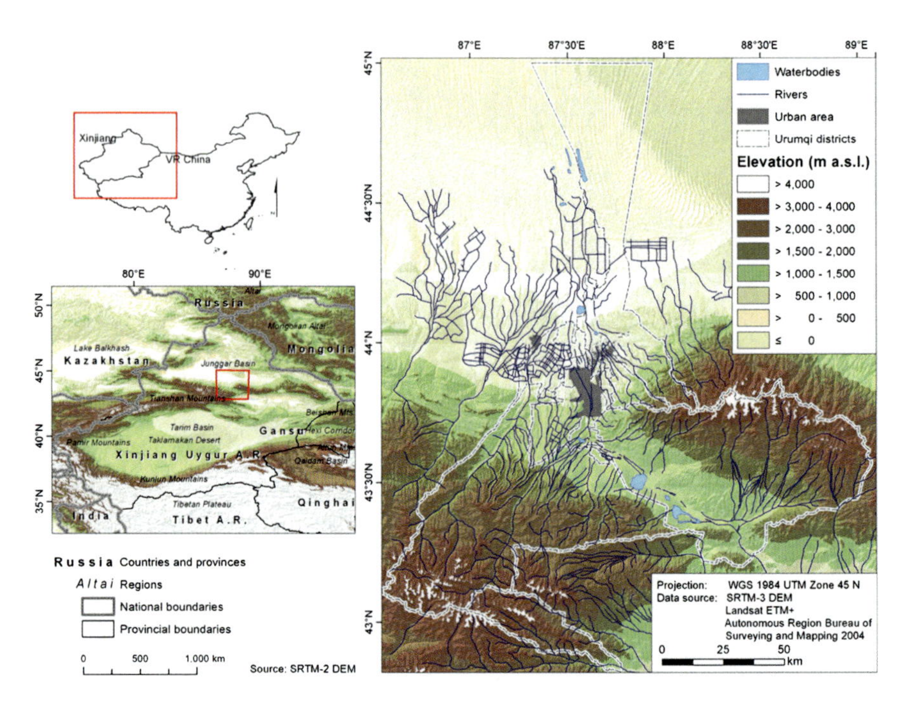

Fig. 2.1 The research area Urumqi Region within its regional context and with the administrative area of Urumqi city, elevation, rivers and water bodies (*Data source* SRTM-2, SRTM-3 [80], Landsat ETM+ [6, 81])

flowing through the urban area, namely Urumqi River, Toutun River, and Shimogou River, along with several smaller creeks flowing parallel. The **climate is continental** with a mean annual air temperature of 7.3 °C (1975–2010, see Fig. 2.2), including a large temperature amplitude of mean monthly air temperatures from −12.4 °C in January to 24.1 °C in July. The mean annual preciptation is 358 mm at Wulumuqi station, but can vary from below 300 mm in the basin to 600 mm in the high mountains [6]. The **Tianshan mountains serve as a meteorological divide** and intercept humid air masses with relatively high precipitation values, leading to convective precipitation and summer maxima (May and July) [7]. The Tianshan mountains are also a **water reservoir**, where precipitation in winter is stored in snow and ice, and provide water for irrigation agriculture and extensive pasture farming on the steppe belt adjacent to the mountain range [2]. Most of the rivers in Urumqi Region are 'mountain rivers' that are fed by snowfall and snow melt of 50–80 %. Rivers originating in areas wir over 3,000 m altitude are also supplied by glacier melt water [8].

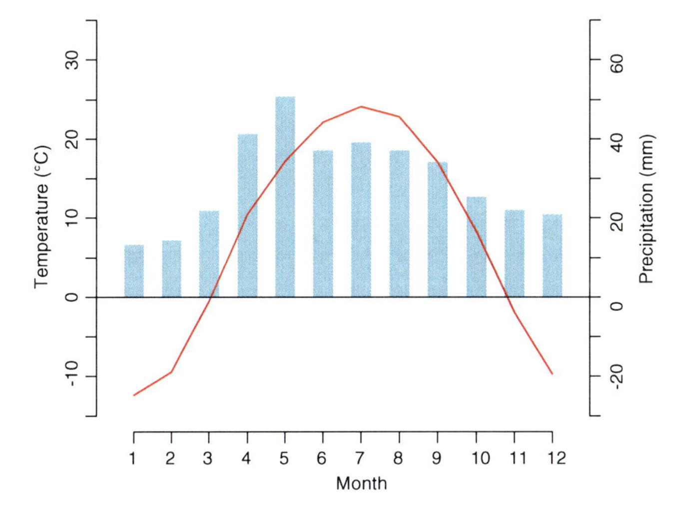

Fig. 2.2 Climate diagram at Wulumuqi station with average monthly air temperature (mean annual air temperature 7.3 °C) and precipitation sums (annual sum 347 mm) for the period 1975–2010 (*Data source* [62])

2.1 Landscape Units

The research area is referred to as a 'region', implying a common characteristic, which is the watershed boundary. Several types of landscape can be found within this area. The landscape type was **classified according to slope steepness and relief amplitude** as well as dominant **lithology** (cf. [9]) **and hydrogeology** (cf. [5]). They are described along the vertical gradient taking its course from South to North: In the centre of the Tianshan range, **high mountains** are dominant, with more than 1,000 m difference in elevation compared to the surrounding areas and mainly consisting of bedrock from the Carboniferous, Permian and Triassic Period. The runoff is generated here, but cannot be stored as the sedimentary and metamorphoic rocks of the Tianshan have a low permeability; they are permeable only at joints, faults and open contacts. They are followed by **lower mountains** with less relief amplitude at the base of the mountain range and rocks from the Permian, Triassic, Jurassic and Cretaceous Period. Adjacent are the slightly sloped accumulation and erosion zones in the foreland of the mountain ranges, the **piedmont**. Loess deposits can be found here, also on the bedrock on the Northern Tianshan slopes, in the northwestern parts of the Dabancheng Corridor or in the old alluvial fans and terraces [10]. The fine sediments have been blown here from the semi-desert basin in the North [8]. Then the **alluvial plain** slopes down in the Dabancheng Corridor and the alluvial fans towards the Junggar Basin build of depositions of coarse Quarternary sediments and formed by fluvial processes. Groundwater recharge from rivers with very good water quality infiltrates into sandy deposits alternating with clays, leading to a multiple aquifer system with shallow aquifers with high salinity and deep aquifers with good water quality.

Especially in the Chaiwopu Basin, a single layer of coarse sand and gravel offers several 100 m of a permeable, good groundwater reservoir [5]. Towards Urumqi Valley, the thickness is changing and due to complicated river bed geology, groundwater storage is limited (Fig. 2.3).

In contrast, the Quarternary deposits in the **accumulation basin** are finer sediments, sand, sandy loam, fine soil, gravel and also loess in lower lying areas, influenced by aeolian transportation processes if not covered by semi-desert vegetation. The aquifers here consist of fine sands and store groundwater with a high salinity (see Table 2.1). In the past, the groundwater was recharged by the rivers flowing from the Tianshan Mountains into the basin and groundwater spring flow occurred in the Urumqi Valley. In the present state, the water is led through artificial channels and reservoirs, the leakage through the river bed is not predominant but instead water infiltrates from delivery- and field channels and irrigation losses. Additionally, groundwater is extracted by pumping [5].

The **soil types** in Urumqi Region cover a large variety from chernozems in the higher mountains, castanozems in the lower mountain areas and foothills with forest and steppe vegetation, changing into sierozems in the foothills and the piedmont, solonchaks and solonetz in the basin and xerosol-yermosol soil groups

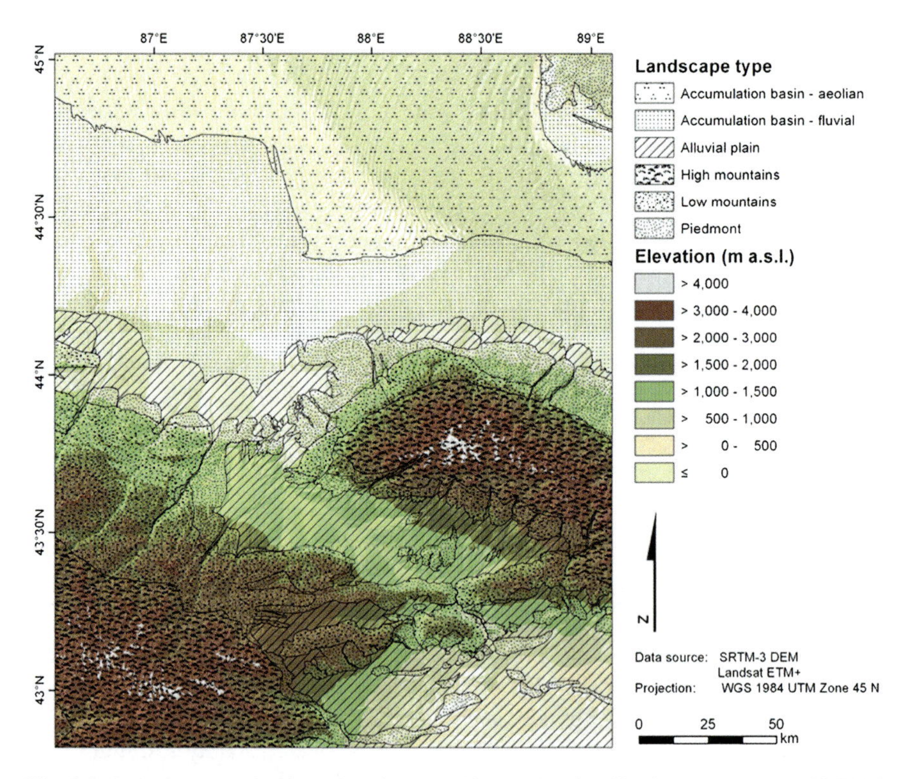

Fig. 2.3 Lanscape types in Urumqi region according to the classification (*Data source* SRTM-3 [80], Landsat ETM+ [81], for classification see Table 2.1)

Table 2.1 Landscape type in Urumqi region with description and dominant lithology

Landscape type	Description (slope)	Simplified lithology
High mountains	With more than 1,000 m relief energy compared to the surrounding areas, bedrock ($>50°$), low permeability	Carboniferous, Permian, Triassic
Low mountains/ foothills	Hills at the base of the mountain range, transition zone between mountain range and plain basin (10–50°), low permeability	Permian, Triassic, Jurassic, Cretaceous, partly quartenary sediments
Piedmont	Slightly sloped accumulation and erosion zones in the foreland of the mountain range, alluvial–colluvial with loess deposits (3–10°), low permeability	
Alluvial plain and cones	Accumulation zone in the mountain foreland, flooded periodically, deposition of coarse sediments with occasional loess deposits ($<3°$), high permeability	Quarternary sediments
Accumulation basin	Part of the Junggar Basin, accumulation of quarternary deposits, finer sediments than the alluvial plain, with loess deposits and sand dunes (slope $<3°$), high permeability	Quarternary sediments

Source own classification, [9, 64–66]

in the semi-desert areas. Locally, other soil types with a high variability can occur due to changes of rocktype, relief, land use and age of land surfaces [10]. For example the oasis soil near human settlements is influenced by anthropogenic reclamation and cultivation and in the fluvial terraces and irrigation areas Gleysols and Fluvisols are abound [8, 10].

The **catchments** located within the administrative area and chosen as research area are Urumqi River basin (3,126 km^2), Toutun River basin (1,069 km^2) and the smaller catchments in Midong District. Midong District does not have one clearly identifiable outlet; therefore the results for three sub-catchments covering the district area were summed up to Midong catchment (1,918 km^2). The combined catchments of Urumqi Region cover about 7,552 km^2, therefore the hydrological processes are modelled on the catchment level and the **spatial macro scale** according to Blöschl (1996 cited in [11]). The precipitation-runoff processes cannot be described with a conceptual model on the catchment scale but it has to be divided in several smaller areas or raster cells as in this study [12]. The characteristics **of the catchment areas** are slightly different: Urumqi River basin includes the catchment of Urumqi Mountain and Wulapo Reservoir, which extend over mountains, foothills and alluvial areas with agriculture, and the urban and sealed area of Urumqi City. The largest part of the catchment of Toutun River is covered by mountainous and foothill areas. Midong catchment includes a small mountainous part, but mainly semi-vegetated foothills and alluvial areas with agriculture and little sealed area of Midong District. The upstream watersheds are nested in the downstream catchment outlets (e.g. Urumqi mountain catchment is included in Wulapo catchment which is included in the Urumqi River catchment), except the catchment Urumqi City, which was calculated also separately to assess specifically the characteristics of the urban area (Fig. 2.4).

2.2 Recent Development in Urumqi

Urumqi is an old Central Asian oasis city, formerly a trade centre on one branch of the Silk Road. Not until the twentieth century the Chinese government established their power and sovereign rights permanently. In 1948, the 'urban area' had only 88,000 inhabitants [13: 64], in 2004, the population of Urumqi City exceeded the 2 million mark and in 2009, the population counted 2.41 million inhabitants [14]. The central Chinese government began intensified **reclamation of farmland** in the northwestern basins **in the 1950s** to improve the food self-sufficiency of Northwest China [5], release the population pressure from densely populated provinces in the East and to support its dominant position in the Central Asian area. For this purpose, state farms were established and farmland was reclaimed and cultivated with the construction of irrigation channels [15, 16]. This strategy is still supported by subsidised low water prices [5]. The development of Xinjiang and also Urumqi was mainly based on the strategy 'one black, one white', meaning oil production and cultivation of cotton or more generally resource extraction and agriculture

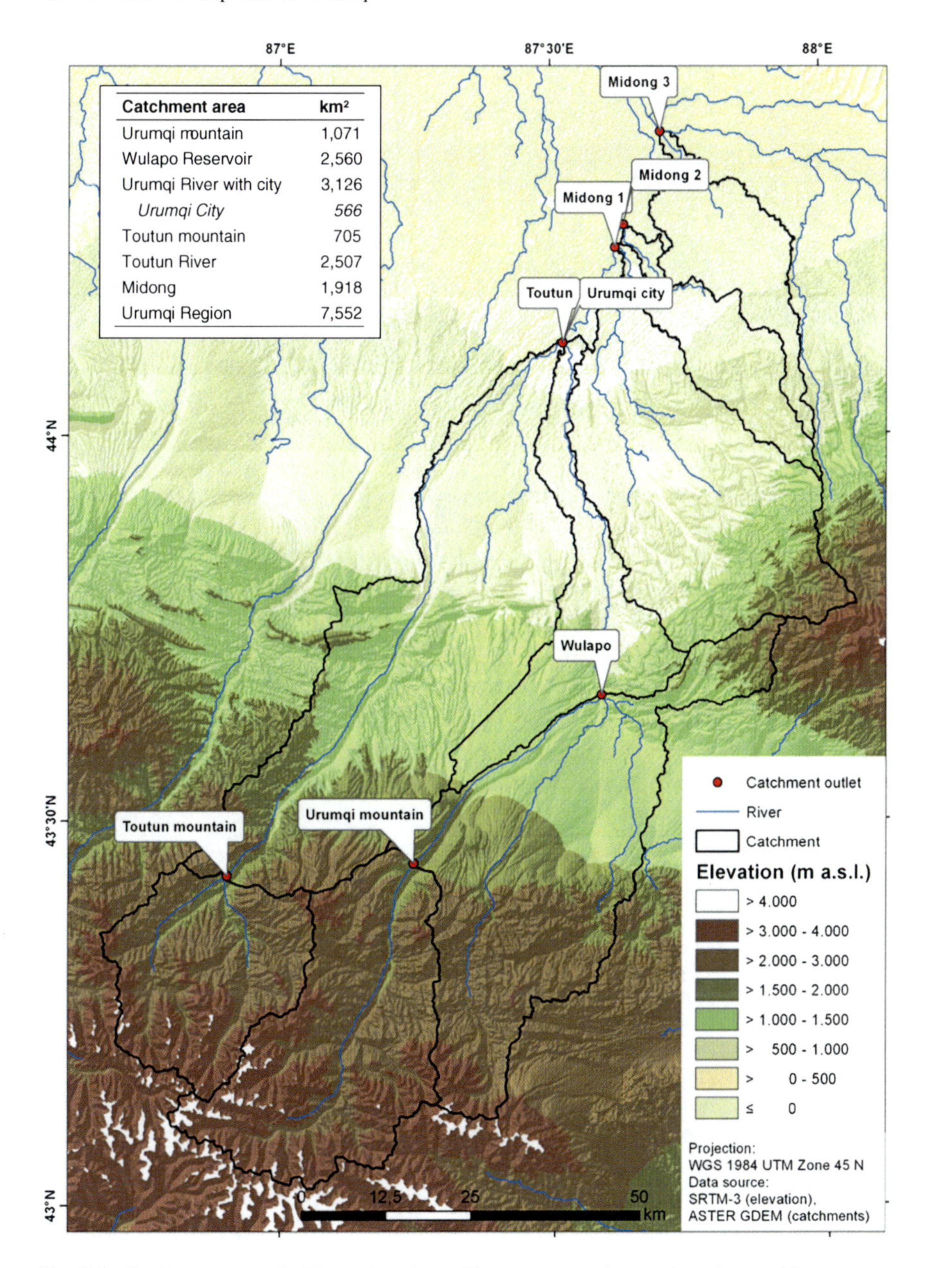

Fig. 2.4 Catchment areas in Urumqi region with streams, outlets and total area (*Data source* SRTM-3 (elevation) [80], ASTER GDEM (catchments) [82])

[17]. The expansion of arable land was accompanied by mechanisation of agriculture, which was only possible in the almost flat basins, and an enormous immigration by Han Chinese workers and farmers [18]. The largest area of

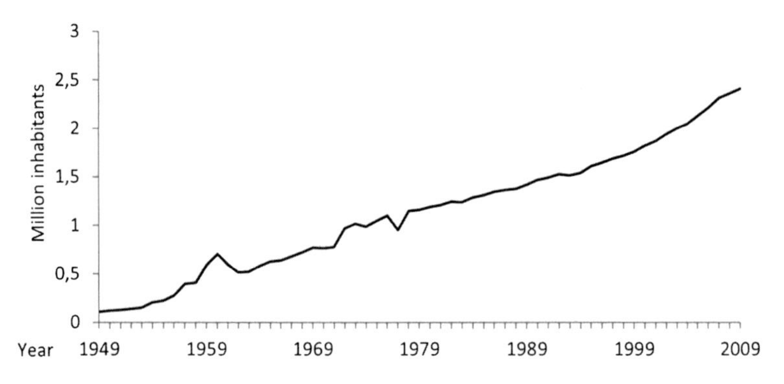

Fig. 2.5 Population development of Urumqi city 1949–2009 (*Data source* [14])

regional development was located in the oasis belt and expanded from there into the basins (cf. also [3]).

With the agricultural expansion, industry developed based on coal deposits and oil resources as well as natural gas, salt and valuable ore deposits in the circumference of Urumqi Region, but since the 1980s, the economic output mainly of the tertiary sector increased. During the last 60 years, Urumqi has undergone **the transformation from a region based on agriculture to a regional centre for trading and industry**. Urumqi is part of the Tianshan Northern Piedmont Plain Industrial Belt, where the economic development of Northwestern China is concentrated. The last two decades, Urumqi City had an average annual GDP growth rate of 10 % [19].

The economic development is accompanied by **population growth and encroachment of the settlement area** (Fig. 2.5). Brohmeyer [20] investigated the land use change in Urumqi Region using Landsat satellite images and reported an enlargement from 84.57 km^2 in 1975 to 765.68 km^2 in 2007, which corresponds to a factor of 9. The urban area is mainly growing to the North occupying favourable arable land on the alluvial fans as the mountains restrict development to the South. Dong et al. [21] reported an expansion of sealed areas in the administrative area of Urumqi City from about 77.48 km^2 in 1975 to 173.26 km^2 in 2004, still corresponding to an impressive factor of 2.23 from 1975 to 2004. The expansion also mainly changed cultivated land, grassland and unused lands into sealed areas.

One example of the recent urban development in Urumqi is the new Midong District northeast of the city centre. Two former settlements, Miquan and Dongshan were merged and the foundation was laid for a new district with the largest industrial park in Western China. The sealed area is planned to increase in two steps from 77.52 km^2 in 2007 to 130 km^2 in 2020 and 343 km^2 in 2050. This planned land use development has been investigated by Fricke [22] and described in more detail. Its consequences for the hydrological system are evaluated in Sect. 5.3.

2.3 Water Supply and Consumption

Urumqi City has a comparatively **favourable situation with regard to water resources as it is supplied with water from the Tianshan Mountains**, an area with high water yield. The rivers from the Tianshan Mountains are supplied by precipitation, mountain springs and melted snow and glaciers [5]. The Urumqi and Toutun River provide the largest part of surface water resources for the city. Additionally, several smaller river, for example originating in the Bogdan Shan, contribute to the water supply. The water is diverted through a widely ramified canal system and either used directly in the agricultural and urban areas, or stored in the reservoirs above and below the city [2]. However, about **70 %** of the measured precipitation is **lost to evapotranspiration** (see Fig. 2.6). On the piedmont plains and the alluvial fan again 60 % of the **surface runoff infiltrate** into the gravel sediments, feeding a deep groundwater systems with large storage capacity in and around the Tianshan Mountains as stated by Aizen et al. [7].

The groundwater then emerges in spring clusters on the lower part of the alluvial fan north of the urban area (cf. [5, 8, 23]). The **exchange between surface and groundwater is intensive** and it is difficult to define clear boundaries between the water bodies. Nonetheless, the city and the oasis ecosystem benefits from the lateral water supplies [24]. Another advantage is the storage of winter precipitation as snow and ice in the higher altitudes and the melt water flow right at the beginning of the vegetation period in spring and early summer [25, 26]. In the Urumqi River, 70 % of the runoff occurs during the flood season from June to August [5].

The **water consumption of the different sectors** agriculture, industry and population in 2007 is shown in Fig. 2.7. Although the external water supply is large, water consumption is high and the overall consumption rate of the available water resources is 74 %. The amount of surface and groundwater consumed is almost the same, but due to the different supply, the consumption rate of the more easily accessible surface water is 89 % and significantly higher than the one of groundwater (62 %). The largest water consumer is agriculture with 65 %, followed by households with 23 % and industry with only 12 % despite the high

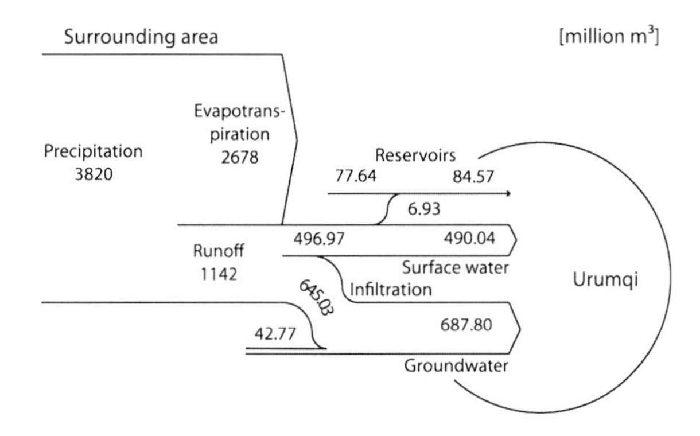

Fig. 2.6 Water resources in Urumqi city according to the Water Report 2007, proportion of snow fall and melt water not specified (*Source* [2], *data source* [52])

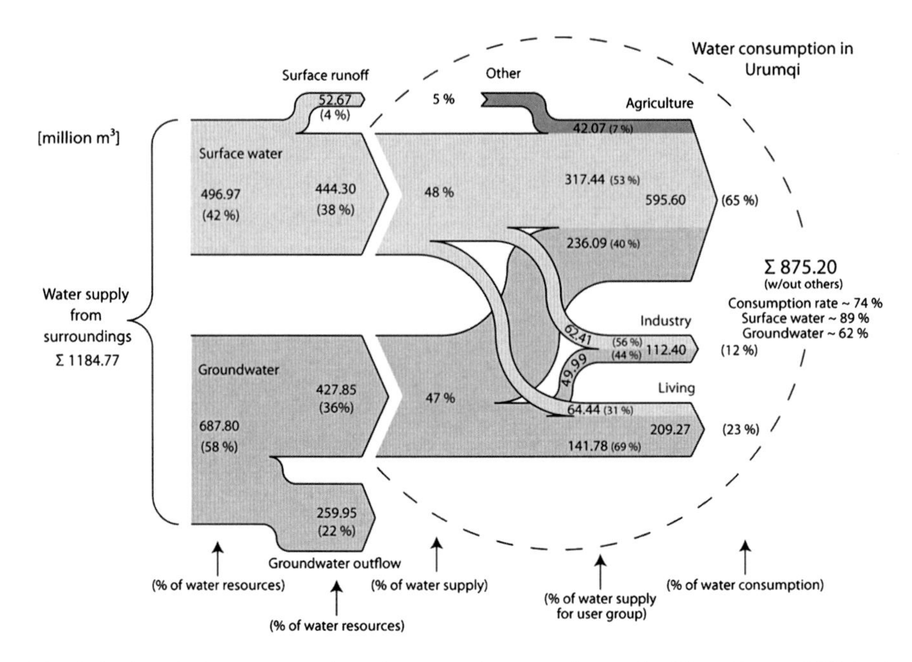

Fig. 2.7 Urumqi city water supply and consumption according to the Water Report 2007 (*unit* $10^6 m^3$)(*Source* [2], *data source* [52])

degree of industrialisation in Urumqi. Most of the surface water supply is consumed by agriculture, but also 55 % of the groundwater supply. For the water use of households, groundwater is even more important as it provides almost 70 % of the water used [2].

The water demand differs importantly on district level and depending on the population density in the districts (Fig. 2.8): the highest population density can be seen as expected in the city centre in the middle of the administrative area and it is lower in the periurban areas Dabancheng District, Midong District and Urumqi County. In these areas, the percentage of agricultural water consumption is also high, while the city districts are characterised by higher water consumption of households and industry. Dabancheng District and Urumqi County also have very high water consumption per capita as water consumption for agriculture is high and population density low.

2.4 Climate Change in Urumqi Region

The effects of the global climate change can be observed in Urumqi Region. Shi et al. [27] have shown that Urumqi lies within a region where the climate notably changed from warm-dry to warm-wet. Based on the meteorological data and time series from Wulumuqi station, the local trend is evaluated and compared to

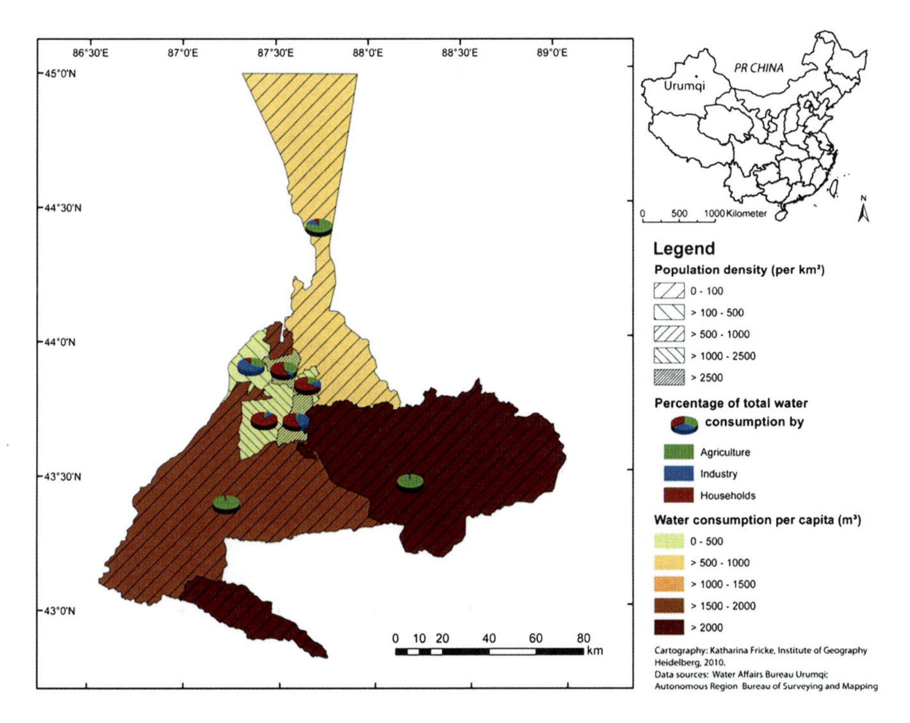

Fig. 2.8 Water consumption per population and sector in Urumqi 2007 (*Data source* [6, 52] *source* [63])

regional and global trends. In the following, the expected consequences are discussed, especially for glacier and melt water.

2.4.1 Effects of Climate Change on Temperature and Precipitation

In this section the observed changes of temperature and precipitation are assessed and possible future effects of climate change are determined based on the climate data from Wulumuqi station (cf. [28]) and the findings of the IPCC Report [29].

The trend analysis of the mean annual **air temperature** of Wulumuqi station conducted by Barth shows an increasing mean annual temperature with a trend value of 1.74 °C 1975–2010[1] [28]. The standard deviation over the time period is at 0.88 °C and the plausibility calculated with a T/s-value[2] showed a significance level

[1] A linear regression function was fitted into the station data and calculated value for 1975 and 2010. The trend value is the difference between the modelled value at the beginning and at the end of the modelled period.

[2] T/s value is the trend value T divided by the standard deviation s and can be used to determine the level of significance.

Table 2.2 Linear trends in land-surface air temperatures in the Northern Hemisphere

Northern hemisphere: land surface dataset	Temperature trend (°C per decade)		
	1850–2005	1901–2005	1979–2005
CRU (Brohan et al. 2006)	0.063 ± 0.015	0.089 ± 0.025	0.328 ± 0.087
NCDC (Smith and Reynolds 2005)		0.072 ± 0.026	0.344 ± 0.096
GISS (Hansen et al. 2001)		0.083 ± 0.025	0.294 ± 0.074
Lugina et al. (2006)		0.079 ± 0.029	0.301 ± 0.075

Source [31]

of 90 %. According to the classification of probability, the occurrence of this trend is very likely [30]. The mean decadal trend was 0.48 °C per decade which exceeds the linear trends in land-surface air temperatures calculated for the Northern Hemisphere land surface of 0.294–0.344 °C per decade published by [31] for the period 1979–2005 (see Table 2.2). Solomon et al. [32] also stated that temperature trends of certain land regions can be greater than the global average rate.

But the calculated local trend is confirmed by the modelled regional temperature change of +1.0–2.0 °C 1970–2004 from the IPCC Synthesis Report 2007 [29] or +0.35–0.45 °C per decade in Northwest China [31]. Trend values from other references differ slightly, but are calculated for other periods: Aizen et al. [7] found that during the period 1940–1991 over the entire Tianshan the mean annual air temperature rose by an average rate of 0.18 °C per decade. This trend is relatively low compared to the other trend values from [28, 31] for Northwest China, but it was calculated for an earlier time period when the temperature trend was not as pronounced as for recent periods (see Table 2.2). Aizen et al. [7] also reported slightly different trends for the Central, Northern and Western Tianshan.

Piao et al. [33] stated that the annual mean temperature rose about 0.25–0.55 °C for the four different seasons per decade from 1960 to 2006 in Northwest China, confirming the trend values cited before. The local trend for Wulumuqi station near Urumqi was reflected by analogous temperature increases at other stations in Xinjiang. Barth [28] also found out that the largest increase of mean monthly temperature occurred in November, March, October, April and December (in that order) while the trend was negative in July or small in August, February, January and September. Important to note is that temperatures increased especially in the months when snowfall and snow melt are probably most sensitive to temperature changes.

The trend analysis of mean annual **precipitation** showed an increase: For Wulumuqi station the trend value was absolute 92 mm or relative 26 % over the whole period (25.5 mm or 7 % decadal trend value) [28]. However, the variation coefficient was very high at 35 % of mean annual precipitation and therefore the trend was observed only at 80 % level of significance, which means the trend is only likely to occur [28, 30]. It is important to mention that Wulumuqi station experienced a trend quite different from most of the stations in a 500 km radius with a pronounced increase in annual precipitation, while most of the other stations showed a decrease. The annual precipitation seems to have become more stable and less variable over the last 36 years. The values for mean annual precipitation and standard deviation showed that the mean annual precipitation increased or remained stable, while the standard deviation decreased especially over the last 12 years (Table 2.3). Also, the monthly trend values showed mostly positive trend values. Especially the summer months July and August benefit from the increasing precipitation values, but also January through May [28]. This corresponds to the statements of the IPCC [29] that "precipitation increased significantly in eastern parts of North and South America, northern Europe and northern and central Asia" [29, pp. 30]. A more specific evaluation of the calculated trend value is difficult as available references differ widely probably depending on the stations included. The regional trend in annual precipitation 1979–2005 calculated with GHCN precipitation data set from NCDC was 3–15 % per decade [31]. Piao et al. [33] also reported for Northwest China from 1960 to 2006 a larger increase in precipitation of 16 % per decade.

Aizen et al. [7] stated that during the period 1940–1991, the annual precipitation increased by 12–14 %. But they also observed general patterns of variability in annual precipitation throughout the Tianshan and also different trends for the warm and the cold season as well as for altitudes above and below 2,000 m. The increase of **heavy precipitation events** observed by [32] could not be validated as the climate data available for Wulumuqi Station has not been evaluated for daily or shorter time intervals. A higher amount of precipitation will fall as rainfall due to the rising temperatures and cause a more concentrated and fast runoff regime [34]. Already since the 1980s, the risk of droughts and short-term flooding rose [35].

Trenberth et al. [31] also reported that mid-latitude westerly **winds** have generally increased in both hemispheres. In Urumqi, the main wind direction is North-Northwest which is mainly controlled by the local topography (cf. [8, 36]. As the wind direction was not documented by the climate records available, the trend in wind speed and direction was not investigated further due to the meager data basis and few references compared to other climate variables. For the station **pressure** changes could be expected as well due to the changing temperature patterns.

The **correlation between industrial and socio-economic development and warming trends** has often been cited as a confirmation of the connection between urbanisation, land surface changes and the warming observed (see also [37]). However, the connection is due to large-scale coherences and warming patterns and "the correlation of warming with industrial and socioeconomic development ceases to be statistically significant" [31: 244]. These findings are also confirmed by [28], who found no statistical significant correlation between the location of the

Table 2.3 Mean annual precipitation sum and standard deviation of three 12-year periods 1975–2010 (*unit* mm)

Climate station	1975–1986	1987–1998	1999–2010
Wulumuqi	310 ± 117	381 ± 154	381 ± 85

Source [28]

climate station in relation to urban areas and the warming trend in Northern Xinjiang. Individual stations might suffer biases and require treatment on a case-by-case basis due to local geography and climate. However, these adjustment algorithms may not be applicable to other stations and parts of the world [38].

2.4.2 Consequences for Glacier and Melt Water

Snow and glacier melt water are important for the water supply of Urumqi City due to its location, but climate change will also influence these two parameters of the hydrological system. Variations in temperature and precipitation affect the snow cover by determining snow fall and snow melt, albedo and surface energy budget or indirect feedbacks (e.g. summer moisture) [39]. For extratropical glaciers, glacier mass balance reacts to changes in temperature and precipitation in several ways: the location of the equilibrium line between ice accumulation and ablation is mainly controlled by temperature, while the rate of accumulation and the rate of ablation depend on snowfall as well as temperature and ice albedo, respectively.

With regard to **snow fall and snow cover**, [40] investigated the hydrological processes in subalpine forests during winter and modelled the system's response to climatic change. According to their model, the rise in air temperature led to a higher proportion of rainfall in winter and the annually average snow depth was reduced by 25 % $°C^{-1}$. Aizen et al. [7] also state that an increase in air tempertature is accompanied by more liquid precipitation in the Tianshan Mountains from the middle of the twentieth century to the present lead to a decrease in snow resources. In contrast, according to Qin et al. [41] no trends in snow depth or snow cover were detected in Western China since 1957 (cited in [39]). The effects on freezing depths were ambiguous, thinner snow depths increased the freezing depth under less isolating snow cover and rising temperatures decreased the freezing under not snow covered areas. The decrease of snow cover due to increasing temperatures would be compensated by increasing precipitation values. Whether the increase of precipitation would also increase runoff could not be evaluated as the simulated changes of winter precipitation were too low [40].

On a global level, **glaciers** reportedly react to the rising temperatures due to climate change. Meehl et al. [42] calculated a global average glacier and ice caps (all land ice except for the ice sheets of Greenland and Antarctica) surface mass balance sensitivity of –0.32–0.41 m $year^{-1}$ $°C^{-1}$ for a geographically and seasonally uniform rise in global temperature. However, these global average sensitivities cannot be used for local projections, which have to consider regional and seasonal temperature changes [37, 42]. Generally, precipitation can outweigh the loss of glacier mass due to increases in surface air temperature and ice melt: Meehl et al. [42] cite that an increase in precipitation of 20–50 % $°C^{-1}$ is required to balance increased ablation representing a variety of climatic regimes. Whether this assumption can be applied depends on many local factors (local climate change,

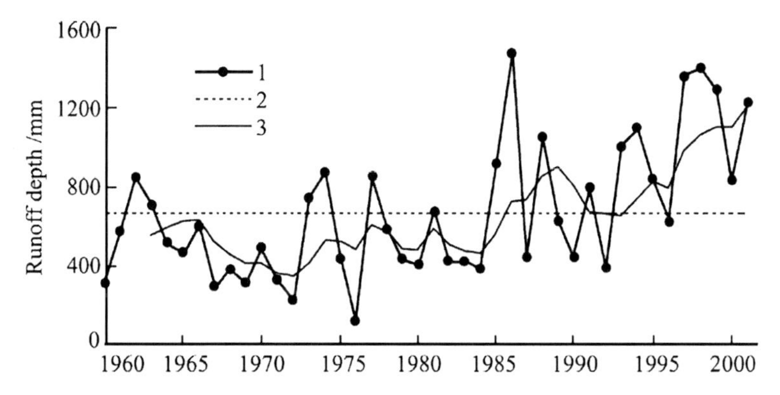

Fig. 2.9 Variation of melt water runoff 1960–2001 of Glacier No. 1 (42°30 N, 86°26 E) at the source area of Urumqi river, Tianshan (*1* Runoff depth, *2* Mean annual runoff depth, *3* 5-year moving average of runoff depth) (*Source* Li et al. 2003 in [27])

reaction of glacier to climate). Aizen et al. [7] reported for different regions in the Tianshan different longterm trends of the key hydroclimatic variables such as surface air temperature, precipitation, runoff, glacier mass, and snow thickness. In general, surface air temperature increased as well as precipitation, but also a decrease in snow resources and glaciers was observed. The annual runoff corresponds with the type of precipitation (liquid or solid) and glacier behaviour: "Over the last few decades, periods of glacier decline have coincided with declining river runoff" [7, pp. 1,393].

A survey of the glaciered areas in Urumqi Region within a Bachelor's thesis based on remote sensing images revealed a significant decrease in glaciered or perennial snow cover area [36]. The reduction of glaciated area of 66 % (−61 km^2) from 1975 to 2010 or 37 % (−34 km^2) from 1989 to 2010 (31 km^2) is comparable to other studies in the Tianshan [37, 43]. The comparison with information about the Glacier No. 1 collected by World Glacier Monitoring 1988–2009 confirms the extent and decrease of glacier area (cf. 70–79). The Urumqi Glacier No. 1, one of the best documented glaciers in the region and located only 100 km from Urumqi City at the headwater of Urumqi River has experienced a dramatic decrease in volume and length in the last decades [36, 44]. Piao et al. [33] also give an annual rate of −0.5–0.3 % $year^{-1}$ change in glacier area for glaciers near Urumqi. During the same period, melt water runoff from glaciered areas increased reportedly (see Fig. 2.9). Since the beginning of the 1980s the runoff from glacier melt water to Urumqi River has increased [45], but it is unsure how long the additional runoff will last.

The average proportion of **glacier melt water** in the total water flow was estimated using several related information. The glaciated area in the catchment area of Urumqi River was 38 km^2 in 1990 according to [46] or 40.5 km^2 in 1989 according to [36]. For the Glacier No. 1 with a glaciated area of 1.95 km^2 on average 1988/1989–2000/2001 (according to World Glacier Monitoring Service, compiled by [36]), the average annual runoff of glacial melt water from 1985 to 2001 accounts for 936.6 mm a^{-1} (Li et al. 2003 in [27]). When transferring this melt water rate to the total glaciered area in the Urumqi River catchment, the average annual runoff volume of glacier melt water can be calculated (see Table 2.4). Accordingly, glacier melt water was recently responsible for almost 10 % of the total river runoff. The calculated volume of annual runoff from glacier melt in the Urumqi River catchment (35.5–38·106 m^3) was significantly higher than the volume of 23.6· 106 m^3 published by [47] for an unknown measurement period, which was probably based on the long-term average glacier runoff from 1960 to 2001, 680 mm a^{-1} (Li et al. 2003 in [27]). The comparison of these numbers illustrates the significant changes in glacier melt and river runoff that have taken place the last decades: the average annual runoff from glacier melt has increased by 413 mm during the period 1980–2003 [48]. For almost the same time, a negative annual glacier mass balance of −404 mm a^{-1} (1980–2003 [48]) or −425.6 mm a^{-1} (1985–2002 [44]) has been reported, indicating that the increase in runoff was mostly due to glacier loss [48]. An increase in runoff of about 400 mm a^{-1} as observed for the period 1980–2000 or of 250 mm a^{-1} as projected by [48] due to rising air temperature could lead to additionally 2–4 % of total river runoff (see Table 2.4). Accumulation and ablation of the Urumqi Glacier No. 1 take place mainly during the summer when more precipitation is available ([49, 50]). A correlation analysis by [44, 48] reveal that annual mass balance is negatively correlated to mean air temperatures in summer. Regression

Table 2.4 Glaciated area, average annual runoff of glacial melt water and proportion of the total surface runoff in Urumqi Region catchments

Area	Glaciated area (reference year)	Average annual runoff from glacial melt water [mm][a]	Average annual runoff from glacial melt water [10° m][e]	Average annual surface runoff for several years [10° m][b]	% of total average runoff[e]
Glacier No. 1	1.95[c]	936.6	1.83		
	(1988–2001)	+400	0.78		
		+250	0.49		
Urumqi River Catchment	38[d] (1990)	936.6	35.59	427.31	8.33
		+400	15.20		3.56
		+250	9.50		2.22
	40.5[c] (1989)	936.6	37.93	427.31	8.88
		+400	16.20		3.79
		+250	10.13		2.37
Urumqi and Toutun River Catchment	57.7[c] (1989)	936.6	54.04	578.66	9.34
		+400	23.08		3.99
		+250	14.43		2.49

The average annual runoff from glacier melt water at Urumqi Glacier No. 1 of 936.9 mm was taken as baseline value. 40 and 250 mm were suggested as possible maximum and minimum increase in glacier melt runoff due to climate change (*Source* [a] [27, 48], [b] [59], [c] [36], [d] [46], [e] own calculations) results of the relationship between mean air temperature, glacier mass balance and melt water runoff "suggest that the 1 °C summer temperature change leads to 486 mm glacier mass loss and 250 mm runoff change" ([48]: Climate Change and Glacier Response).

2.5 Water Problems

Despite the advantageous location on the foot of the Tianshan Mountains, Urumqi experiences several problems related to water supply and consumption. They have already been illustrated by [2], so the most important ones will be presented in this subsection. Due to the continental climate and high temperatures, a large amount of precipitation is lost to transpiration and especially evaporation as soon as it is put into use, during transport and storage in the open canal and reservoir system or common flood irrigation [51]. This leads to reduced water flow and an increasing concentration of minerals and pollutants further away from the headwater of the stream. The continuous dynamic economic development of the region is the cause for other hydrological problems [52]. The growth of economy and population leads to a growing direct (drinking, hygiene, production) and indirect water demand (water-intensive food production or production of needed goods). If the economy and population are growing too fast, water demand will rise despite adaptation of demand and a general reduction of the per capita water consumption which has been documented for Urumqi City [53]. Simultaneously, water consumption rises due to the (also increasing) standard of living, the distribution of water, new requirements for water supply and discharge of pollutants and waste water will become one of the largest challenges for urban infrastructure and municipal governments [53].

The increased water demand during the last 20–30 years led to a pronounced overexploitation of surface and groundwater resources [53] and a **demand-driven water scarcity** which is measured by the use-to-availability indicator or consumption ratio as introduced before. In 2007, the consumption ratio in Urumqi Region was on average 74 %, 89 % for the surface and 62 % for the groundwater resources [54]. According to [55], this indicates that the extraction exceeds the appropriate level and is threatening the environmental flow necessary for the aquatic ecosystems. The population-driven water shortage is measured by the number of people per 10^6 m^3 available water resources per year, also called water crowding [55]. The value was 1,953 inhabitants per 10^6 m^3 available water resources per year for Urumqi in [54] and indicates real and severe water shortage.

A **decrease of river flows to the downstream basins** could be observed, which led to a continuous deterioration of natural vegetation and large scale land desertification (cf. [5, 23]). The streams coming from the Tianshan Mountains and flowing through Urumqi City and the green belt are increasingly used, dry up faster and sooner than before [56]. Only 4 % of the original surface water flow is available to the environment, further increasing the risk of degradation and desertification [57]. The water diverted for irrigation evapotranspirates instead of sustaining river flow or infiltration into the groundwater. Falling groundwater tables below the city indicate a reduced recharge and unsustainable consumption rate (cf. [54, 58, 59]). Zhou et al. [5] hold the irrigation water demand, sealed man-made reservoirs, canals, and diverted river flows accountable for the reduction of groundwater recharge in the Gobi zone. Due to the lower groundwater tables, percolation and springs outflows out of the alluvial fan joining the surface water flow to the terminal lakes have been reduced as well.

High **evapotranspiration rates and insufficient irrigation and drainage management** lead to the accumulation of pollutants and salts on the agricultural fields and soil salinisation on the middle sections of the downward slope from the mountains to the semi-arid basin [5]. Overirrigation can also cause rising groundwater tables in certain areas and secondary salinisation when the capillary forces transports water to the surface where it is evaporated and leaves the solid substances behind. Wang and Cheng [60] reported that almost 20 % of the cultivated area in Urumqi was classified as salinised by the end of the 1990s. The displacement and expansion of agricultural areas to the north by the enlarging urban area led to more agriculture in areas with less advantegous pedological and hydrological prerequisites. The pastoral area used by cattle husbandry is also diminished and threatening the livelihood of stock breeders in this area mostly coming from other ethnic groups than the Han majority in Urumqi. Thus, the expansion could also increase the social and ethnic conflict potential [61]. Growing water withdrawal for the newly reclaimed areas and less groundwater infiltration on the sealed areas aggravates this situation. Another problem connected to the hydrological system is the air pollution over Urumqi City. With the strong Northwest winds, the accumulated pollutants are blown out of the dried up downstream basins and agricultural fields and transported to Urumqi urban area in the South as illustrated in Fig. 2.10.

The **quality of surface water** is generally poor due to the high consumption rate and pollution through sewage discharge and transportation of waste. One of the most polluted rivers in Xinjiang due to the high density of larger industries and smaller enterprises along its water side is the Shuimo River, which flows parallel to the Urumqi River [56]. Urumqi has an annual total wastewater discharge of $144.9 \cdot 10^6$ m^3 and an average amount of waste water treated of $74.96 \cdot 10^6$ m^3 per year (1996–2007), i.e. the average treatment rate was 52.37 % [62]. Since 2009 and the operation of a new treatment plant the estimated treatment capacity of all treatment plants at a national discharge standard of at least class 'II' (GB18918-2002) was $0.7353 \cdot 10^6$ m^3 per day. Despite the large capacity, the average amount of treated waste water in Urumqi City was only $0.1912 \cdot 10^6$ m^3 per day, averaging 25 % of the treatment capacity. Both treatment rate and degree of capacity utilisation need to be improved. A remaining problem is furthermore the untreated sewage in the past that led to an accumulation of pollutants and particles in the reservoirs below the city [2]. Because large amounts of pollutants are disseminated into the hydrological system and less water is available for their dilution, groundwater bodies become more and more polluted. Major problems stated by

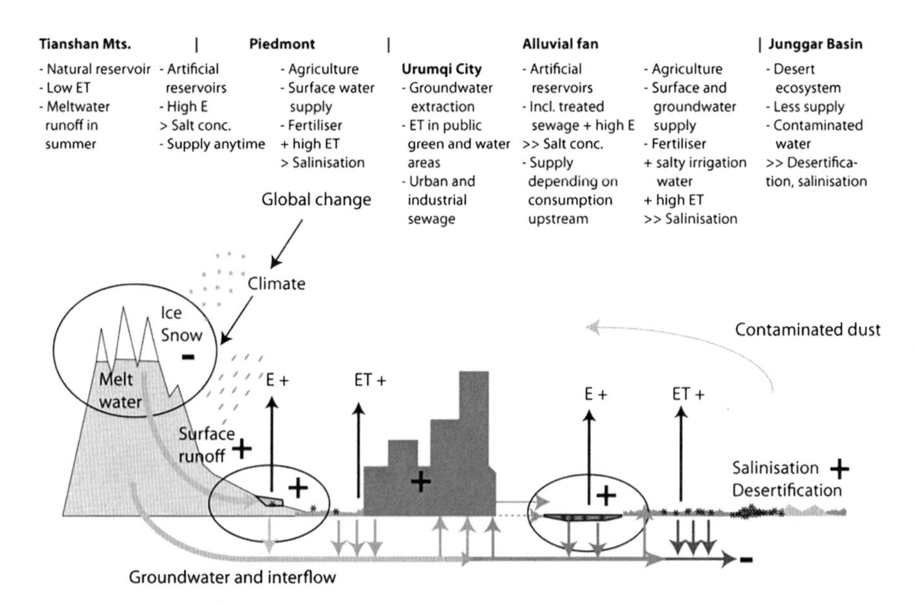

Fig. 2.10 Hydrological system, future changes and consequences for Urumqi and its surroundings, *E* Evaporation, *ET* Evapotranspiration, + Increase in the future, – Decrease in the future (*Source* [2])

[63] such as serious water pollution of the middle and small rivers and underground water nearby cities, increase of heavy metal contents and chemical fertilizers possibly leading to eutrophication, overextraction of groundwater, and declination of groundwater table can also be observed in Urumqi. Especially the users downstream of the main pollution area, the Urumqi City Centre, suffer from the decreasing groundwater quality. In Midong District located northeast of the city centre, the first large aquifer (80 m) is not usable anymore due to contamination and the next aquifer (150 m) had to be tapped [64]. The deterioration of the water quality causes a quality-driven water scarcity as parts of the water resources become unsuitable for certain uses [55], but this was not investigated here because only few reliable data is available for water quality.

On the **water management** side, several shortfalls can be observed as well. The connection between water supply and water consumption is insufficient as water fees are less than the variable costs of water production and treatment [22, 62]. Rising fees, however, are very difficult to enforce because the water supply of all inhabitants is seen as part of the national public services. The difficulties of proper irrigation and drainage management have already mentioned before, but [5] also stated a shortage of professional staff trained in practical groundwater investigation techniques and a poor information system for the public and decision-makers as a major problem for a sustainable water management in Urumqi Region.

In a nutshell, the demand-driven water scarcity (see also Sects. 2.2 and 2.3) is prone to worsen due to the socio-economic development despite the quite advantegeous situation at the Tianshan Mountains. At the same time, the effects of climate change on the hydrological system and water availability in the future are unsure and spatially variable (see Sects. 1.2.7 and 2.4) potentially inducing a climate-driven water scarcity [55]. To predict the consequences for Urumqi Region, on the one hand a water balance model simulating the water availability will be presented in the next chapters, on the other the water demand will be projected in Chap. 6.

References

1. Schultz, J. (2002). *Die Ökozonen der Erde: 17 Tabellen, 5 Kästen*. Ulmer, Stuttgart (UTB Geowissenschaften, Ökologie, Agrarwissenschaften, 1514), p. 320.
2. Fricke, K., Sterr, T., Bubenzer, O., & Eitel, B. (2009). The oasis as a mega city: Urumqi's fast urbanisation in a semi-arid environment. *Die Erde, 140*(4), 449–463.
3. Jia, B., Zhang, Z., Ci, L., Ren, Y., Pan, B., & Zhang, Z. (2004). Oasis land-use dynamics and its influence on the oasis environment in Xinjiang China. *Journal of Arid Environments, 56*(1), 11–26.
4. Dowamat, T. (1993). *Xinjiang—My beloved home* (452 p). Urumqi: Xinjiang People's Publishing House.
5. Zhou, Y., Nonner, J. C., Li, W., et al. (2007). *Strategies and techniques for groundwater resources development in northwest China*. Beijing: China Land Press, 338 p.
6. Autonomous Region Bureau of Surveying and Mapping (2004). *Xinjiang-Weiyu'er-Zizhiqu-dituji: Xinjiang Uygur Autonomous Region Atlas* (307 p). Beijing: Zhongguo ditu chubanshe.
7. Aizen, V. B., Aizen, E. M., Melack, J. M., & Dozier, J. (1997). Climatic and Hydrologic Changes in the Tien Shan, Central Asia. *Journal of Climate, 10*, 1393–1403.
8. Roberts, B. (1987). *Die ökologischen Risiken der Stadtentwicklung und Landnutzung in Urumqi, Xinjiang/China*. Univ. Studiengang Geographie Fachber. 8, Bremen (Bremer Beiträge zur Geographie und Raumplanung, 12), 277 p.
9. Bocco, G., Mendoza, M., & Velázquez, A. (2001). Remote sensing and GIS-based regional geomorphological mapping: A tool for land use planning in developing countries. *Geomorphology, 39*, 211–219.
10. Yuan, G., Lichtenfeld, A., & Stahr, K. (1988). Soils of the Manas River Area in Northern Xinjiang, People's Republic of China. *Zeitschrift für Pflanzenernährung und Bodenkunde, 151*(3), 152–163.
11. Uhlenbrook, S. (1999). *Untersuchung und Modellierung der Abflussbildung in einem mesoskaligen Einzugsgebiet* (201 p). Institut für Hydrologie der Universität Freiburg i. Br., Freiburg (Freiburger Schriften zur Hydrologie, 10).
12. Plate, E. J., Zehe, E., & Maurer, Th. (2008). Einführung. In E. J. Plate & E. Zehe (Eds.), *Hydrologie und Stoffdynamik kleiner Einzugsgebiete* (pp. 1–28). Stuttgart: Schweizerbart.
13. Roberts, B. R. (1993). *Water management in desert environments: A comparative analysis* (337 p). Berlin: Springer (Lecture notes in earth sciences, 48).
14. Statistics Bureau of Urumqi (2010). *Urumqi Statistical Yearbook* 2010 (456 p). Beijing: China Statistics Press.
15. Gruschke, A. (1991). *Neulanderschließung in Trockengebieten der VR China und ihre Bedeutung für die Nahrungsversorgung der chinesischen Bevölkerung* (227 p). Hamburg: IfA.
16. Kolb, A. (1986). Xinjiang als Naturraum und ökologisches Problemgebiet. *Geoökodynamik, 7*, 29–40.
17. Becquelin, N. (2000). Xinjiang in the Nineties. *The China Journal, 44*, 65–90.

18. Li, R. M. (1989). Migration to China's Northern Frontier, 1953–1982. *Population and Development Review, 15*(3), 503–538.
19. Statistics Bureau of Urumqi (2009). *Urumqi Statistical Yearbook* 2009 (506 p). Beijing: China Statistics Press.
20. Brohmeyer, F. (2011). *Klassifikation der Landnutzung und ihrer Veränderung von 1975 bis 2007 anhand von Satellitenbildern in der Region Urumqi (Nordwest-China)* (53 p). Unpublished bachelor thesis, Ruprecht-Karls-Universität Heidelberg.
21. Dong, W., Zhang, X., Wang, B., & Duan, Z. (2007). Expansion of Urumqi urban area and its spatial differentiation. *Science China Series D: Earth Sciences, 50*(Supp. I), 159–168.
22. Fricke, K. (2008). *The development of Midong New District, Urumqi, PR China: Ecological and historical context and environmental consequences* (165 p). Heidelberg University: Unpublished diploma thesis.
23. Chen, M., & Cai, Z. (2000). Groundwater resources and hydro-environmental problems in China. In M. Chen & Z. Cai (Eds.), *Groundwater resources and the related environ-hydrogeologic problems in China* (pp. 38–44). Beijing: Seismological Press.
24. Contreras, S., Jobbágy, E. G., Villagra, P. E., Nosetto, M. D., & Puigdefábregas, J. (2011). Remote sensing estimates of supplementary water consumption by arid ecosystems of central Argentina. *Journal of Hydrology, 397*(1–2), 10–22.
25. Berkner, A. (1993). Wasserressourcen und ihre Bewirtschaftung in der Volksrepublik China. *Petermanns Geographische Mitteilungen, 137*(2), 103–126.
26. Zhu, Y., Wu, Y., & Drake, S. (2004). A survey: obstacles and strategies for the development of ground-water resources in arid inland river basins of Western China. *Journal of Arid Environments, 59*, 351–367.
27. Shi, Y., Shen, Y., Kang, E., Li, D., Ding, Y., Zhang, G., et al. (2007). Recent and future climate change in Northwest China. *Climatic Change, 80*, 379–393.
28. Barth, N. C. (2011). *Auswertung der Temperatur- und Niederschlagsdaten von 15 Klimastationen im Umkreis von Urumqi (AR Xinjiang, China)* (65 p). Unpublished bachelor thesis, Ruprecht-Karls-Universität Heidelberg.
29. Intergovernmental Panel on Climate Change (IPCC) (2007). *Climate change 2007: Synthesis report* (73 p). Retrieved 11 Nov, 2011, from http://www.ipcc.ch/pdf/assessment-report/ar4/syr/ar4_syr.pdf.
30. Le Treut, H., Somerville, R., Cubasch, U., Ding, Y., Mauritzen, C., Mokssit, A., et al. (2007). Historical overview of climate change. In S. Solomon, D. Qin, M. Manning, Z. Chen, M. Marquis, K. B. Averyt et al. (Eds.), *Climate change 2007: The physical science basis. Contribution of working group I to the fourth assessment report of the Intergovernmental Panel on Climate Change* (pp. 93–127). Cambridge: Cambridge University Press.
31. Trenberth, K. E., Jones, P. D., Ambenje, P., Bojariu, R., Easterling, D., Klein Tank, A., et al. (2007). Surface and atmospheric climate change. In S. Solomon, D. Qin, M. Manning, Z. Chen, M. Marquis, K. B. Averyt et al. (Eds.), *Climate change 2007: The physical science basis. Contribution of working group I to the fourth assessment report of the Intergovernmental Panel on Climate Change* (pp. 235–336). Cambridge: Cambridge University Press.
32. Solomon, S., Qin, D., Manning, M., Alley, R. B., Berntsen, T., Bindoff, N. L., et al. (2007). Technical summary. In S. Solomon, D. Qin, M. Manning, Z. Chen, M. Marquis, K. B. Averyt et al. (Eds.), *Climate change 2007: The physical science basis. Contribution of working group I to the fourth assessment report of the Intergovernmental Panel on Climate Change* (pp. 19–91). Cambridge: Cambridge University Press.
33. Piao, S., Ciais, P., Huang, Y., Shen, Z., Peng, S., Li, J. Z., et al. (2010). The impacts of climate change on water resources and agriculture in China. *Nature, 467*, 43–51.
34. Halike, Y., Eitel, B., & Küchler, J. (2008). Wasserverknappung und Wasserkonflikte in der zentralasiatischen Wüstenmetropole Urumqi/NW China. *TU International, 61*, 12–14.
35. Jiang, F., Zhu, C., Mu, G., & Hu, R. (2005). Magnification of flood disasters and its relation to regional precipitation and local human activities since the 1980s in Xinjiang, Northwestern China. *Natural Hazards, 36*, 307–330.

36. Fuchs, J. (2011). *Multitemporale Detektion der Gletscherveränderung im östlichen Tian Shan (AR Xinjiang, China) im Kontext des Klimawandels: Untersuchungen am Beispiel der Flusseinzugsgebiete von Toutun, Shuixi und Urumqi* (69 p). Heidelberg: Unpublished bachelor thesis, Ruprecht-Karls-Universität Heidelberg.
37. Bolch, T. (2007). Climate change and glacier retreat in northern Tien Shan (Kazakhstan/ Kyrgyzstan) using remote sensing data: Climate change impacts on mountain glaciers and permafrost. *Global and Planetary Change, 56*(1-2), 1–12.
38. Trenberth, K. E., Jones, P. D., Ambenje, P., Bojariu, R., Easterling, D., Klein Tank, A., et al. (2007). Observations: Surface and atmospheric climate change. In S. Solomon, D. Qin, M. Manning, Z. Chen, M. Marquis, K. B. Averyt et al (Eds.), *Climate change 2007: The physical science basis. Contribution of working group I to the fourth assessment report of the Intergovernmental Panel on Climate Change* (SM.3-1-SM.3-11). Cambridge: Cambridge University Press.
39. Lemke, P., Ren, J., Alley, R. B., Allison, I., Carrasco, J., Flato, G., et al. (2007). Observations: Changes in snow, ice and frozen ground. In S. Solomon, D. Qin, M. Manning, Z. Chen, M. Marquis, K. B. Averyt et al. (Eds.), *Climate change 2007: The physical science basis. Contribution of working group I to the fourth assessment report of the Intergovernmental Panel on Climate Change* (pp. 337–383). Cambridge: Cambridge University Press.
40. Stadler, D., Bründl, M., Schneebeli, M., Meyer-Grass, M., & Flühler, H. (1998). *Hydrologische Prozesse im subalpinen Wald im Winter* (145 p). vdf Hochsch.-Verl. an der ETH, Zürich.
41. Qin, D., S. Liu, & P. Li (2006). Snow cover distribution, variability, and response to climate change in Western China. *Journal of Climate, 19*, 1820–1833.
42. Meehl, G. A., Stocker, T. F., Collins, W. D., Friedlingstein, P. G., Gregory, J. M., Kitoh, A., et al. (2007). Global climate projections. In S. Solomon, D. Qin, M. Manning, Z. Chen, M. Marquis, K. B. Averyt et al. (Eds.), *Climate change 2007: The physical science basis. Contribution of working group I to the fourth assessment report of the Intergovernmental panel on Climate Change* (pp. 747–845). Cambridge: Cambridge University Press.
43. Niederer, P., Bilenko, V., Ershova, N., Hurni, H., Yerokhin, S., & Maselli, D. (2008). Tracing glacier wastage in the Northern Tien Shan (Kyrgyzstan/Central Asia) over the last 40 years. *Climatic Change, 86*, 227–234.
44. Han, T., Ding, Y., Ye, B., Liu, S., & Jiao, K. (2006). Mass-balance characteristics of Urumqi Glacier No. 1, Tien Shan China. *Annals of Glaciology, 43*, 323–328.
45. Ye, B., & Chen, K. (1997). A model simulating the processes response of Glacier and runoff to climatic change: A case study of Glacier No 1. In the Urumqi River China. *Chinese Geographical Science, 7*(3), 243–250.
46. Zhang, G., Sang, S., & Wang, X. (1990). Simulation of daily runoff in the Urumqi River basin with the improved tank model. In H. Lang & A. Musy (Eds.), *Hydrology in mountainous regions* (pp. 693–700). IAHS: Wallingford.
47. Zhou, Y. (1999). *River hydrology and water resources in Xinjiang* (445 p). Urumqi: Xinjiang Science and Technology Publishing House.
48. Ye, B., Yang, D, Jiao, K., Han, T., Jin, Z., Yang, H., & Li, Z. (2005). The Urumqi river source Glacier No. 1, Tianshan, China. changes over the past 45 years. *Geophysical Research Letters, 32*. doi:10.1029/2005GL024178.
49. Huintjes, E., Li, H., Sauter, T., Li, Z., & Schneider, C. (2010). Degree-day modelling of the surface mass balance of Urumqi Glacier No. 1, Tian Shan, China. *The Cryosphere Discussions, 4*, 207–232.
50. Xu, X., Pan, B., Hu, E., Li, Y., & Liang, Y. (2011). Responses of two branches of Glacier No. 1 to climate change from 1993 to 2005, Tianshan, China. *Quaternary International, 236*(1–2), 143–150.
51. Tao, S., Fu, C., Zeng, Z., & Zhang, Q. (1997). *Two long-term instrumental climatic data bases of the People's Republic of China: ORNL:CDIAC-47, NDP039*. Oak Ridge National Laboratory, TN: Carbon Dioxide Information Analysis Center. Retrieved 25 March, 2009, from http://cdiac.ornl.gov/epubs/ndp/ndp039/ndp039.html.

52. Dong, W., & Zhang, X. (2011). Urumqi. *Cities, 28*(1), 115–125.
53. Du, H., Zhang, X., & Wang, B. (2006). Co-adaptation between modern oasis urbanisation and water resources exploitation: A case of Urumqi. *Chinese Science Bulletin, 51* (Supp. I), 189–195.
54. Water Affairs Bureau Urumqi (2007). *Water report 2007* (24 p). Urumqi: Water Affairs Bureau Urumqi City.
55. Falkenmark, M., Berntell, A., Jägerskog, A., Lundqvits, J., Matz, M., & Tropp, H. (2007). *On the verge of a new water scarcity: A call for good governance and human ingenuity* (19 p). Stockholm: SIWI.
56. Hao, Y. (1997). Water environment and sustainable development along the belt of Xinjiang section of the new Eurasian continental bridge. *Chinese Geographical Science, 7*(3), 251–258.
57. Babaev, A. G. (1999). *Desert problems and desertification in Central Asia: The researches of the desert institute* (293 p). Berlin: Springer.
58. Water Affairs Bureau Urumqi (2004). *Water report 2004* (27 p). Urumqi: Water Affairs Bureau Urumqi City.
59. Water Affairs Bureau Urumqi (2005). Water Report 2005 (27 p). Urumqi: Water Affairs Bureau Urumqi City.
60. Wang, G., & Cheng, G. (1999). The ecological features and significance of hydrology within arid inland river basins of China. *Environmental Geology, 37*(3), 218–222.
61. Hamann, B. (2007). *Ökologische und sozioökonomische Entwicklungen am Südrand des Dsungarischen Beckens/AR Xinjiang: vor dem Hintergrund des chinesischen Transformationsprozesses in den 90er Jahren des 20. Jahrhunderts* (255 p). Dissertation, Technische Universität Berlin.
62. Yao, Y. (2011). Water reuse: A case study of Urumqi, China. In IWA (Ed.), *1st Central Asian Regional Young and Senior Water Professionals Conference 22–23 Sept, 2011*. Almaty: CD-ROM.
63. Deng, W., Bai, J., & Yan, M. (2002). Problems and countermeasures of water resources for sustainable utilisation in China. *Chinese Geographical Science, 12*(4), 289–293.
64. Fricke, K. (2007). Protocol of the excursion to Midong New District October 28th 2007 (4 p), not published.
65. National Oceanic and Atmospheric Administration of the U.S. Department of Commerce National Climatic Data Centre (NOAA NCDC) (2011). *Global summary of the day Wulumuqi station, station number 514630. 22.08.1956–31.12.2010*. Retrieved 16 May, 2011, from http://www7.ncdc.noaa.gov/CDO/cdo.
66. Fricke, K., & Bubenzer, O. (2011). Available water resources and water use efficiency in Urumqi, PR China. In: German Academic Exchange Service (Ed.), *Future megacities in balance young researchers' symposium in Essen 9–10 Oct, 2010. DAAD Dok and Mat, 66*, 134–140.
67. Bruns, B. R., Ringler, C., & Meinzen-Dick, R. (2005). *Water rights reform—Lessons for institutional design* (360 p). Washington, DC: International Food Policy Research Institute.
68. Brunotte, E., Martin, C., Gebhardt, H., Meurer, M., Meusburger, P., Nipper, J., et al. (Eds.) (2002). *Gast bis Ökol* (426 p). Heidelberg, Berlin Spektrum Akad. Verl. (Lexikon der Geographie, 2).
69. Zhang, X., Zhou, K., & Ahati, J. (2011). *Natural resources of arid Metropolitan Urumqi* (223 p). Urumqi: Xinjiang Art and Photography Publishing House.
70. WGMS (1991). *Glacier mass balance bulletin No. 1 (1988–1989)* (70 p). Zürich: IAHS (ICSI)/UNEP/UNESCO, World Glacier Monitoring Service.
71. WGMS (1993). *Glacier mass balance bulletin No. 2 (1990–1991)* (74 p). Zürich: IAHS (ICSI) / UNEP / UNESCO, World Glacier Monitoring Service.
72. WGMS (1994). *Glacier mass balance bulletin No. 3 (1992–1993)* (80 p). Zürich: IAHS (ICSI) / UNEP / UNESCO, World Glacier Monitoring Service.
73. WGMS (1996): *Glacier mass balance bulletin No. 4 (1994–1995)* (90 p). Zürich: IAHS (ICSI) / UNEP / UNESCO, World Glacier Monitoring Service.

74. WGMS (1999): *Glacier mass balance bulletin No. 5 (1996–1997)* (96 p). Zürich: IAHS (ICSI) / UNEP / UNESCO. World Glacier Monitoring Service.
75. WGMS (2001): *Glacier mass balance bulletin No. 6 (1998–1999)* (93 p). Zürich: IAHS (ICSI) / UNEP / UNESCO / WMO, World Glacier Monitoring Service.
76. WGMS (2003): *Glacier mass balance bulletin No. 7 (2000–2001)* (87 p). Zürich: IAHS (ICSI) / UNEP / UNESCO / WMO, World Glacier Monitoring Service.
77. WGMS (2005): *Glacier mass balance bulletin No. 8 (2002–2003)* (100 p). Zürich: IUGG (CCS) / UNEP / UNESCO / WMO, World Glacier Monitoring Service.
78. WGMS (2007): *Glacier mass balance bulletin No. 9 (2004–2005)* (100 p). Zürich: IUGG (IACS) / UNEP / UNESCO / WMO, World Glacier Monitoring Service.
79. WGMS (2009): *Glacier mass balance bulletin No. 10 (2006–2007)* (96 p). Zürich: ICSU (WDS) / IUGG (IACS) / UNEP / UNESCO / WMO, World Glacier Monitoring Service.
80. U.S. Geological Survey (2005). Shuttle Radar Topography Mission, Version 2, 3-arc second resolution. Global Land Cover Facility, University of Maryland, College Park, Maryland. Retrieved 31 Jan, 2008, from http://glcf.umiacs.umd.edu/data/srtm/index.shtml.
81. NASA Landsat Program (2010). Landsat ETM+ scenes L1T, USGS, Sioux Falls. Retrieved from http://glovis.usgs.gov/.
82. U.S. Geological Survey & Japan ASTER Programme (2009). ASTER GDEM Version 1. NASA Land Processes Distributed Active Archive Center, Sioux Falls. Retrieved from 15 Jul, 2009. http://earthexplorer.usgs.gov/.

Chapter 3
Water Balance Model

The water balance model, which was used to simulate the water balance and water distribution, consists of several modules, each describing one part of the hydrological system: Bare soil evaporation, transpiration by vegetation, processes related to ice and snow (snow accumulation and snow melt), surface runoff, groundwater recharge, and finally river runoff at the catchment outlet. These are described in the next sections, then the model sensitivity and uncertainty is assessed to identify the most important parameters for which the according input dataset are described. Before applying the water balance model, its accuracy is evaluated with different measures and against other datasets and model results.

3.1 Modelling Concept

The model type used for the research was chosen based on the model scale, the available input, calibration and validation data and the required model output. The research area with the chosen watersheds extends over an area of over 7,000 km^2, thus suggesting modelling at catchment and macro scale and not for partial areas and hill slopes (meso scale) or soil profiles (micro scale) (cf. [1, 2], see also Sect. 2.1). The focus of this study was not on the development of new methods for the scaling of process representations, but also in the **application of models and practical predictions of the effects of climate and land use change** as suggested by Beven [3]. Only limited data were available for model calibration, namely monthly runoff data for one catchment and the amount of surface and groundwater of Urumqi Region for several years (for a comprehensive overview of the data used refer to Table A.24). Therefore, the model should not require large amounts of data to calibrate process parameters for the research area, but be easily transferable to the ungauged basins within the catchment (Fig. 3.1).

This prerequisite excluded models designed to simulate runoff-regimes which require an extensive calibration of parameters and coefficients [4–6]. Some models are not suitable due to the different input data required for simulation, for example the retention parameters in soil and the catchment areas necessary for LARSIM

K. Fricke, *Analysis and Modelling of Water Supply and Demand Under Climate Change, Land Use Transformation and Socio-Economic Development*, Springer Theses, DOI: 10.1007/978-3-319-01610-8_3,

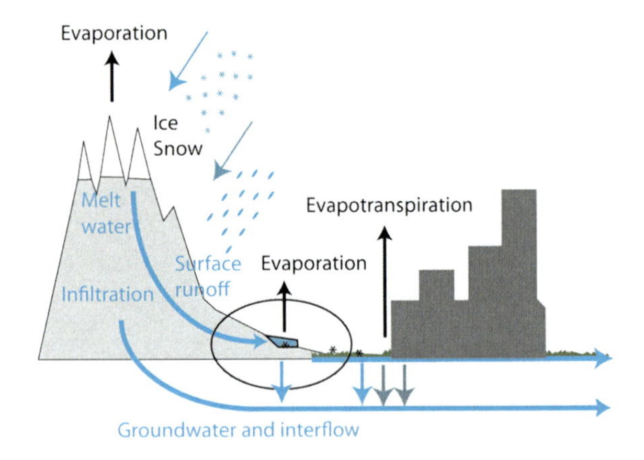

Fig. 3.1 Overview of the hydrological system (*Source* own design)

could not be obtained (cf. [1]). As the model results were supposed to be available on the pixel scale, lumped or semi-distributed models such as SWAT (Soil and Water Assessment Tool) were also not suitable. However, the simulation of micro-scale processes was neither desired nor was the necessary input data available, thus models simulating hydrological processes of the micro-scale were also not employed.

The **available input data** existed as point measurement at one climate station or spatially distributed in the raster or shapefile format such as a global soil database, land use and land cover (LULC), spatial distribution of temperature and precipitation, and elevation including all derivatives such as slope and catchment areas. The resolution of the available input data varied from 15 m × 15 m for the elevation data (ASTER GDEM) to 1° × 1° for the precipitation data (TRMM). For the hydrological modelling, pixels corresponding to the resolution of the processed data are used as process boundaries. The pixel results are later summarised as a weighted flow accumulation, heterogeneity could not be accounted for below the pixel scale. The requirements of the model output were the annual water balance at pixel scale to enable the simulation of even small scale land use changes, but also to simulate the effects of climate change on the long-term water balance.

The **hydrological model used is based on the water balance model ABIMO** which was developed for modelling surface runoff formation in Germany [7]. It was further developed by Meßer [8, 9] for the calculation of long-term groundwater recharge rates also in urban areas including a runoff model by Schroeder and Wyrich (1990 cited in [10]). The model is a conceptual model and designed for macro-scale modelling on a raster dataset. It is based on the water balance equation [11]:

$$Q_n = P - ET_a \tag{3.1}$$

with total flow Q_n, precipitation P and actual evapotranspiration ET_a (all abbreviations and physical quantities are also summarised in the abbreviations section). The original module integrated in ABIMO for the calculation of effective (actual) evapotranspiration is based on the Bagrov relationship between actual evapotranspiration, maximum evapotranspiration and water availability (Rachimov 1996 in Glugla et al. [12]). The efficiency parameters applied in the model take into account the modification of the water storage capacity in the evaporative zone due to land use, degree of imperviousness and soil properties. However, these parameters were quantified and calibrated for locations and evapotranspiration values in Europe which are not necessarily representative for the semi-arid climate of Urumqi and could not be confirmed or corrected in a field survey. Therefore, **potential maximal evapotranspiration** was substituted by applying the method of Penman-Monteith which is a process-based model and uses physical and climatic input parameters available for the research area. The Penman-Monteith equation is not suitable for the description of evapotranspiration processes at the micro-scale. It has been widely used for modelling potential evapotranspiration (ET_p) at the meso- or macro-scale for simulating soil evaporation and crop evapotranspiration under different land covers and irrigation schemes in Turkey [13] and Morocco [14], for estimating irrigation needs in the Heihe River Basin in Nortwest China [15] or for the calculation of groundwater recharge rates in the state of Baden-Württemberg [16]. As ET_p was mainly calculated with input from climate variables averaged over time, the results were considered to adequately represent also long-term (ET_p) averages as required by the water balance model. ET_a was then derived based on precipitation and ET_p:

$$ET_a = \begin{cases} ET_p, & \text{if } P > ET_p \\ P, & \text{if } P \leq ET_p \end{cases}. \tag{3.2}$$

Evaporation from other, secondary water resources such as accumulated snow, soil water and surface water was not considered because additional intermediate storages would have to been necessary. Unfortunately, the data base was not sufficient for this expansion. Groundwater fed evaporation was also not specifically taken into account. At the end of the alluvial plain and at the transition to the Junggar basin, groundwater tables come close to the land surface. In these areas, the water is mostly drained into man-made channels and reservoirs to avoid water logging and salinisation. Evaporation from groundwater was considered less important than evaporation from open water bodies such as reservoirs or flooded fields, that could clearly be recognised from the land use classification. Hence evaporation from open water bodies was taken into account. The **net precipitation** P_n available for snow accumulation and runoff is calculated as follows:

$$P_n = Q_n = P - ET_a. \tag{3.3}$$

Other processes considered include snow accumulation, snow water melt and runoff formation, they were simulated by using index models as presented in Sects. 3.1.2 and 3.1.3. These index models do not exactly describe the physical or hydrological processes, instead they use empirically derived parameters and coefficients representing physical characteristics or functions. **Snow accumulation** was calculated based on the proportion of snow fall of precipitation as suggested by Merz [17] and Huintjes et al. [18]. **Snow melt** was then estimated using the degree-day-method which calculates the amount of snow melt in mm per day and degree temperature. This is an empirical approach for calculating snow melt and runoff from mountain basins which has been widely used due to its applicability and effectiveness [19]. The procedure is explained in Sect. 3.1.2. The water balance equation was subsequently expanded to:

$$Q_n = P - ET_a - P_s + Q_m, \tag{3.4}$$

with the amount of precipitation that falls as snow Ps and melt water flow Q_m from snow melt. **Surface and groundwater flows** (Q_{sf} and Q_{gw}) were calculated from total water flow Q_n with the runoff coefficient p_{runoff} developed by Meßer [8, 9] who modified an approach of Schroeder and Wyrich (1990) (see also [10]):

$$Q_n = Q_{sf} + Q_{gw} = \left(Q_n \cdot p_{runoff}\right) + \left(Q_n \cdot \left(1 - p_{runoff}\right)\right). \tag{3.5}$$

The derivation of the runoff parameter from the land use classes is presented in Sect. 3.1.3. For an overview of the model structure, refer also to Fig. 3.5. ET_p, ET_a, P, Ps, Q_m, Q_n, Q_{sf} and Q_{gw} were calculated as water equivalent in mm and only later converted into m^3 for the evaluation of the model results.

3.1.1 Calculation of Evapotranspiration

The Penman-Monteith equation was chosen for calculating evapotranspiration on land surfaces (see Eqs. (3.6) and (3.7)). It is physically based, can be applied globally at different time scales from hours to months depending on the input data and offers a practicable approach for estimating evapotranspiration [20]. Additionally, it is well documented and has been calibrated and validated by other research projects and publications (e.g. [21, 22]). The necessary ground-based climate data (air temperature, wind speed, dew point temperature, station pressure) was available from the climate station Wulumuqi (see Sect. 3.3.2) and could be combined with satellite data to estimate the distribution of solar radiation, precipitation and temperature over the research area. The classified Landsat satellite data was used to provide additional information about parameters and coefficients influencing the evapotranspiration, namely albedo and resistance terms. Over water bodies, the Penman-Monteith equation was used with the albedo adapted to the seasonal albedo of water bodies, but basically with the same input climate data as for the reference evapotranspiration over land surface. Unfortunately, the calculated evapotranspiration could not be locally calibrated due to the lack of ground measurements.

The Penman-Monteith equation was used as following to calculate the **potential evapotranspiration** ET_{PM} [mm day^{-1}] [23]:

$$ET_{PM} = \frac{1}{\lambda} \cdot \frac{\Delta \cdot (R_n - G) + 86400 \cdot \frac{\gamma \cdot \varepsilon \cdot \lambda}{T_{Kv} \cdot R \cdot r_a} \cdot (e_s - e_a)}{\Delta + \gamma \cdot \left(1 + \frac{r_s}{r_a}\right)} \tag{3.6}$$

$$= \frac{\frac{\Delta}{\lambda} \cdot (R_n - G) + 86400 \cdot \frac{\gamma \cdot 0.622}{1.01 \cdot (T + 273) \cdot 0.287 \cdot r_a} \cdot (e_s - e_a)}{\Delta + \gamma \cdot (1 + \frac{r_s}{r_a})} \tag{3.7}$$

with the latent heat of vaporisation λ [2.45 MJ kg^{-1}], the slope of the vapour pressure curve Δ [kPa °C^{-1}], net radiation at the surface R_n [MJ m^{-2} day^{-1}], soil heat flux density G [MJ m^{-2} day^{-1}], the psychrometric constant γ [kPa °C^{-1}], the air temperature T measured at 2 m height [°C], the saturation vapour pressure deficit e_s–e_a [kPa], (bulk) surface resistance r_s [s m^{-1}], and aerodynamic resistance r_a [s m^{-1}]. For a complete overview of the equations used refer to Fig. 3.3.

To calculate the slope of the **saturation vapour pressure curve**Δ, the mean air temperature (daily or monthly, depending on the time step) is used in the Penman-Monteith equation "as the effect of temperature variations on the values of the climatic parameter is small in these cases" [23: 33]:

$$\Delta = \frac{4098 \cdot \left[0.6108 \cdot exp\left(\frac{17.27 \cdot T_{mean}}{T_{mean} + 237.3}\right)\right]}{(T_{mean} + 237.3)^2} \tag{3.8}$$

with slope vapour pressure curve Δ [kPa °C^{-1}] and **mean air temperature** T_{mean} [°C]. The calculation of T_{mean} based on the daily maximum and minimum temperatures T_{max} and T_{min} [°C] is standardised by Allen et al. [23] and preferred over the mean daily air temperature derived from the summary of several observations for the day:

$$T_{mean} = \frac{T_{max} + T_{min}}{2}. \tag{3.9}$$

The **net radiation** at the crop surface R_n [MJ m^{-2} day^{-1}] in Eq. (3.7) is calculated as the difference between incoming net solar or shortwave radiation R_{ns} [MJ m^{-2} day^{-1}] and net outgoing longwave radiation R_{nl} [MJ m^{-2} day^{-1}]:

$$R_n = R_{ns} - R_{nl}. \tag{3.10}$$

R_{ns} in turn depends on the albedo α of the surface which accounts for the amount of solar radiation which is not reflected by the land surface and the solar radiation R_s [MJ m^{-2} day^{-1}] at the ground surface:

$$R_{ns} = (1 - \alpha) \cdot R_s. \tag{3.11}$$

The albedo α or canopy/surface reflection coefficient for different land cover types in the research area was taken from several references (cf. [24–26]) (see Table 3.1). The calculation of ET_p over water bodies was very similar to the calculation of ET_p over land surfaces, however with a seasonal albedo for water bodies α_w taken from [24] (see Table 3.2). The seasonal albedo of water is given

Table 3.1 Land use and land cover class and albedo

Class	Albedo (−)
Agriculture	0.26[a]
Agriculture dry	0.20[a]
Sparse vegetation	0.25[a, b]
Grass	0.25
Soil	0.20[a]
Coniferous forest	0.12[a, b]
Rock	0.18[a]
Sealed areas	0.18[a]
Ice	0.90[a]
Snow	0.85[a]
Snow old	0.35[a]
Irrigation	0.06–0.12[c]
Water	0.06–0.12[c]
Dark areas	0.15[a]

Source[a] Iqbal [25], [b] Schulla [26], [c] DVWK [24]

Table 3.2 Seasonal albedo for water bodies, frozen water bodies are classified as ice (*Source* DVWK [24])

Month	α_w (−)
1	0.11
2	0.09
3	0.07
4	0.06
5	0.06
6	0.06
7	0.06
8	0.06
9	0.07
10	0.09
11	0.11
12	0.12

for areas in Central Europe. Urumqi Region is located only slightly more south and experiences distinct seasonal changes as well. Therefore, the albedo values are expected to be transferable to Urumqi Region and hence were used for the calculation of a specific net radiation over water bodies R_{nw} and potential evapotranspiration ET_{pw}.

Global solar radiation is the amount of solar radiation reaching a given location and actually depends on the time of year, latitude and surface geometry as well as cloud cover and turbidity of the clean air. Direct radiation, diffuse radiation from the sky, direct and diffuse radiation reflected by nearby terrain are summed up to global solar radiation [25]. To simplify the calculation of ET_p only the solar radiation depending on the time of year, latitude and surface geometry was necessary as input (see Eq. (3.12)) and the cloud cover and the turbidity of the clean air are taken into account by empirical regression constants.

The **solar radiation** R_s [MJ m^{-2} day^{-1}] **at the ground surface** is calculated including the regression constants a_s and b_s and the ratio of actual duration of sunshine n [hours] and maximum possible duration of sunshine N [hours]:

$$R_s = \left(a_s + \left(b_s \cdot \frac{n}{N}\right)\right) \cdot R_0. \quad (3.12)$$

To account for the effects of the atmosphere on the **incoming solar radiation**R_0 [MJ m^{-2} day^{-1}] **at the top of the atmosphere**, a_s = 0.25 and b_s = 0.50 are suggested by Doorenbos and Pruitt (1977 cited in [23, 27]). The **net outgoing longwave radiation** R_{nl} [MJ m^2 day^{-1}] is calculated as follows:

$$R_{nl} = \sigma\left[\frac{(T_{max} + 273.16)^4 + (T_{min} + 273.16)^4}{2}\right](0.34 - (0.14\sqrt{e_a}))\left(\left(1.35\frac{R_s}{R_{so}}\right) - 0.35\right) \quad (3.13)$$

with the Stefan-Boltzmann constant σ (4.903·10^{-9} MJ K^{-4} m^{-2} day^{-1}), actual vapour pressure e_a [kPa] (see Eq. (3.22)), solar radiation R_s as explained above and the **clear-sky radiation** R_{so} [MJ m^{-2} day^{-1}]. The latter can be obtained from empirical coefficients and the station's elevation above sea level z_{st} [m] (for Wulumuqi station 947 m):

$$R_{so} = (0.75 + (0.00002 \cdot z_{st})) \cdot R_0. \quad (3.14)$$

The **soil heat flux density**G [MJ m^{-2} day^{-1}] in Eq. (3.7) depends on the volumetric soil heat capacity c_v [MJ m^{-3} °C^{-1}], the air temperature difference ΔT over the time interval Δt [°C], the length of the time interval Δt (normally a day) and the effective soil depth Δz [m] (see also appendix A.1):

$$G = c_v \frac{\Delta T}{\Delta t} \Delta z_s. \quad (3.15)$$

For monthly periods, constant c_s of 2.1 MJ m^{-3} °C^{-1} is assumed according to Allen [23]. An appropriate soil depth can be derived according to appendix A.1 and the equation for soil heat flux density can be simplified (cf. [23]). Assuming that the change in temperature over time is linear and Δt is one month the following equation is applied:

$$\Delta T = T_i - T_{i-1} = \frac{T_{i+1} - T_{i-1}}{2} \quad (3.16)$$

with average air temperature of the i-th month T_i [°C], the previous month T_{i-1} [°C], and the following month T_{i+1} [°C]. The average daily soil heat flux density G_i of month i is then calculated with the parameters in Table 3.3:

$$G_i = c_v \frac{T_{i+1} - T_{i-1}}{\Delta t \cdot 2} \Delta z. \quad (3.17)$$

The **psychrometric constant**γ [kPa °C^{-1}] in Eq. (3.7) can be calculated from the atmospheric pressure P [kPa] measured at the climatological station, the

Table 3.3 Typical values for mass density, specific heat capacity, volumetric heat capacity, thermal conductivity and effective depth of heat flux for 1 and 60 days

Material	Condition	Mass density ρ (kg m^{-3}·10^{3})	Specific heat capacity c_s (J kg^{-1} K^{-1}·10^{3})	Volumetric heat capacity c_v (MJ m^{-3} K^{-1})	Thermal conductivity K (MJ m^{-1} K^{-1} day1)	Effective depth Δz for $\Delta t = 1$ day (m)	Effective depth Δz for $\Delta t = 60$ days (m)
Air[a]	20 °C, still	0.0012	1.01	0.0012	0.002	1.33	10.3
Water[b]	20 °C, still	1.00	4.2	4.2	0.05	0.11	0.8
Ice[b]	0 °C, pure	0.91	2.11	1.92	0.19	0.31	2.4
Snow[a]	Fresh	0.10	2.09	0.21	0.007	0.18	1.5
Snow[a]	Old	0.48	2.09	1.00	0.036	0.19	1.5
Soil[c]	–	–	–	2.1	0.14	0.26	2.0
Rock[a]	Solid	2.7	0.75	2.03	0.25	0.35	2.72

Source[a] http://apollo.lsc.vsc.edu/classes/met455/notes/section6/2.html, accessed 12.11.2011, [b] Roth [157], [c] Allen [23], for the calculation of effective depth see appendix A.1)

specific heat of air c_p ($1.013 \cdot 10^{-3}$ MJ kg^{-1} °C^{-1}), the ratio of the molecular weight of water vapour and dry air ε (0.622), and the latent heat of vapourisation λ [2.45 MJ kg^{-1}]:

$$\gamma = \frac{c_p \cdot P}{\varepsilon \cdot \lambda} \tag{3.18}$$

$$\gamma = 0.665 \cdot 10^{-3} \cdot P. \tag{3.19}$$

The **saturation vapour pressure deficit** is used to indicate the air humidity as the climate data does not provide direct measurements. The saturation vapour pressure deficit $e_s - e_a$ [kPa] is calculated subtracting the actual vapour pressure e_a [kPa] from the saturation vapour pressure e_s [kPa]. e_s is calculated "as the mean between the vapour pressure at the daily maximum and minimum air temperature of that period" assuming non-linearity of humidity [23: 32]:

$$e_s = \frac{e^0(T_{max}) + e^0(T_{min})}{2}. \tag{3.20}$$

For the computation of saturation vapour pressure, the vapour pressure e^0 [kPa] is calculated for the maximum T_{max} and the minimum air temperature T_{min} [°C]:

$$e^0(T) = 0.6108 \, exp\left[\frac{17.27 \cdot T}{T + 237.3}\right]. \tag{3.21}$$

The actual vapour pressure e_a [kPa] is the vapour pressure calculated with the dewpoint temperature T_{dew} [°C] from the climatological station:

$$e_a = e^0(T_{dew}) = 0.6108 \, exp\left[\frac{17.27 \cdot T_{dew}}{T_{dew} + 237.3}\right]. \tag{3.22}$$

The relationships of **aerodynamic and canopy resistance** in the evaporative process were introduced into the Penman combination equation by Monteith [28] with two resistance terms instead of the wind function developed by Penman [29]: the surface resistance r_s and the aerodynamic resistance r_a, both in s m^{-1}. The surface resistance parameter r_s combines the resistance of vapour flow through stomata openings, total leaf area and soil surface [27]. It is only applicable to densely vegetated areas as it neglects the soil evaporation and the derivation of surface resistance terms. For only partly vegetated areas, it has to be adapted accordingly [16]:

$$r_s = \frac{r_l}{LAI_{active}}. \tag{3.23}$$

The average bulk stomatal resistance of an individual leaf r_l is crop and management specific, but due to limited studies and data available for this parameter, ~ 100 s m^{-1} for a single leaf under well-watered conditions was assumed (see also [23]). LAI_{active} is the active leaf area index of the area for which ET_p is computed and is commonly estimated as half of the total LAI.

$$LAI_{active} = 0.5 \cdot LAI. \tag{3.24}$$

The leaf area index, a dimensionless quantity, is the leaf area (upper side only) per unit area of soil below it and here estimated as suggested by Allen [23]:

$$LAI = 24 \cdot h. \tag{3.25}$$

The aerodynamic resistance r_a determines the transfer of heat and water vapour from the evaporating surface into the air above the canopy. It decreases with growing crop height and increasing wind speed (see Fig. 3.2) and is calculated from the following equation:

$$r_a = \frac{ln\left(\frac{z_m - d}{z_{om}}\right) \cdot ln\left(\frac{z_h - d}{z_{oh}}\right)}{k^2 u_2} \tag{3.26}$$

based on the height of wind measurements z_m (2 m) and humidity measurements z_h (2 m), the Karman's constant k (0.41), the windspeed u_2 measured at height 2 m [m s^{-1}], the zero plane displacement height d of the surface [m], the roughness length z_{om} [m] and z_{oh} [m] governing momentum transfer and the transfer of heat and vapour, respectively. The last three parameters d, z_{om} and z_{oh} can be estimated with empirical equations from the crop height h [m] when they cannot be measured [23, 30, 31]:

$$d = \frac{2}{3} \cdot h \; z_{om} = 0.123 \cdot h \; z_{oh} = 0.1 \cdot z_{om} = 0.0123 \cdot h. \tag{3.27}$$

Under neutral conditions or the condition of atmospheric stability, "i.e., where temperature, atmospheric pressure, and wind velocity distributions follow nearly adiabatic condition" [23: 21] in the planetary boundary layer, the typical vertical profile of wind speed and a logarithmic wind profile is established as described in the equation above (see also [33, 34]) (Fig. 3.2).

Although both the surface and the aerodynamic resistance terms are only an approximation of a more complex resistance of the land surface, they are relatively simple to obtain and show good correlation between measured and estimated ET_p compared to other methods estimating ET_p [27]. However, their values have to be carefully chosen as they have a large influence on the quality of the chosen approach to estimate evapotranspiration and as ET_p proved to be very sensitive to r_s and r_a (see sensitivity analysis in Sect. 3.2.1, [16, 23]. The concept of bulk stomatal resistance r_s has been developed and tested for crops and vegetation describing actual physiological processes. It is not defined for other land use classes such as barren land and soil, rock, sealed areas, snow and water. Thus, r_s was calculated based on the Eqs. (3.23)–(3.25) only for agricultural and irrigated areas and taken from the hydrological model LARSIM for the other land use classes [1]. For the evaporation and sublimation from land covers other than vegetation, for example snow and ice, the surface resistance r_s has been set to zero. The aerodynamic resistance term r_a is calculated as a function of the surface type (see Eqs. (3.26)–(3.27a, b and c)) and can vary depending on local wind profiles

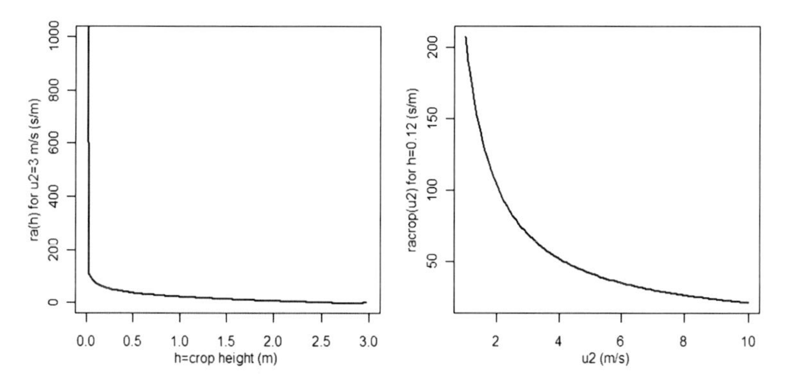

Fig. 3.2 Illustration of the relationships between crop height and aerodynamic resistance as well as between wind speed and aerodynamic resistance according to Eqs. (3.26) and (3.27a, b and c)

[12]. Therefore it was tried to calculate the aerodynamic resistance based on the average surface or roughness height published from different research studies (see Table 3.4). For the aerodynamic resistance r_a of snow and ice, alternative equations have been suggested by Thom and Oliver [33]. However, their term led to lower r_a values and a higher evaporation than compared to the equation suggested by Monteith [28] or to reported reference measurements for a study in Northern Mongolia [35]. Thus, the equation for r_a suggested by Monteith was used for all LULC classes only with adapted values for h.

Based on the equations presented above, the evapotranspiration potential is determined and the **actual evapotranspiration** ET_a can be calculated taking the available precipitation into account as given in Eq. (3.2). ET_a is limited by the available water through precipitation, but not by the retention capacities of soil and vegetation. The net precipitation P_n was then calculated as precipitation minus ET_a as shown in Eq. (3.3). If P was lower than ET_p, P_n was set to zero. Therefore, negative values of P_n were ignored and both lateral flow of soil water and capillary rise of groundwater were neglected (Fig. 3.5).

3.1.2 Snow Accumulation and Snow Melt

Due to the continental climate and cold winters in Urumqi Region, about one third of the precipitation falls as snow. Depending on the location and elevation, the snow persists until spring and early summer and then melts or feeds glaciers at higher altitudes. The change in glacier ice was neglected in the water balance model (see also Eq. (3.4)) as (i) it is not detectable at a significant level over the course of 1 year and could not be calibrated with reliable runoff data, (ii) it is of less importance for the water supply of the research area (see Sect. 2.4.2), and (iii) additional data about the projection of the long-term development was not available. Intermediate storage of water in snow was believed to play a significant

Table 3.4 Surface height, surface and aerodynamic resistance for LULC classes

LULC	h (m)	r_s (s m^{-1})	Sources
Agri	0.4	20	*h*: modified according to Foken [158] (grain), r_s: Bremicker et al. [1]
Agriculture dry	0.1	40	*h*: estimated, r_s: Bremicker et al. [1]
Sparse vegetation	0.2	70	*h*: estimated, r_s: Bremicker et al. [1]
Soil	0.05	100	*h*: estimated acc. to Foken [158], r_s: Bremicker et al. [1]
Coniferous forest	8	70	*h*: Kelliher et al. [159], Nakai et al. [160] (mean values), Hurtalová et al. [30], r_s: Bremicker et al. [1] (coniferous forest)
Grass	0.1	65	*h*: estimated (see agriculture dry), r_s: Bremicker et al. [1] (grassland)
Rock	1	500	*h*: estimated, r_s: Bremicker et al. [1] (sealed)
Sealed	5	500	*h*: estimated, r_s: Bremicker et al. [1] (sealed)
Ice	0.001	0	*h*: Foken [158] (ice), r_s: Wimmer et al. [35]
Snow	0.01	0	*h*: Foken [158] (snow), r_s : Wimmer et al. [35]
Snow old	0.01	0	*h*: Foken [158] (snow), r_s: Wimmer et al. [35]
Irrigation	0.15	65	*h*: estimated, r_s: Bremicker et al. [1] (wet lands)
Water	0.05	0	*h*: mean between Thom and Oliver [33], Penman [29] and Foken [158] (open sea, calm), r_s: Bremicker et al. [1], Thom and Oliver [33]
Dark areas	5	500	*h*: Foken [158] (air field), r_s: Bremicker et al. [1] (sealed)

role in the water balance and seasonal discharge which was to be evaluated. Hence, the module snow accumulation and snow melt was calculated additionally to the calculation of evapotranspiration and precipitation.

3.1.2.1 Snow Accumulation

The snow module is based on the **concept of a snow water equivalent**, which measures the amount of snow added to the snow pack in millimeter precipitation that has fallen as snow. For the modelling it was assumed that at temperatures below 0 °C the documented precipitation would fall as snow and is then accumulated until it melts. For the modelling of snow melt, several methods are available. However, methods based on the modelling of heat transfer in snow packs and other physical concepts require input variables such as absorption coefficients, snow pack temperature and vapour pressure, precipitation intensity and a regression coefficient describing the decrease of runoff during the time without snow melt or precipitation [36]. Since most of these variables were unknown and could only be roughly estimated, these methods were dismissed in favour of a simpler temperature index method using the degree day method [37]. The degree-day-method and degree-day-factors have been documented and measured for a large number of land surfaces and seasons (cf. [18, 37–41]) and have been used by other hydrological models such as SWAT [41] and other studies [43].

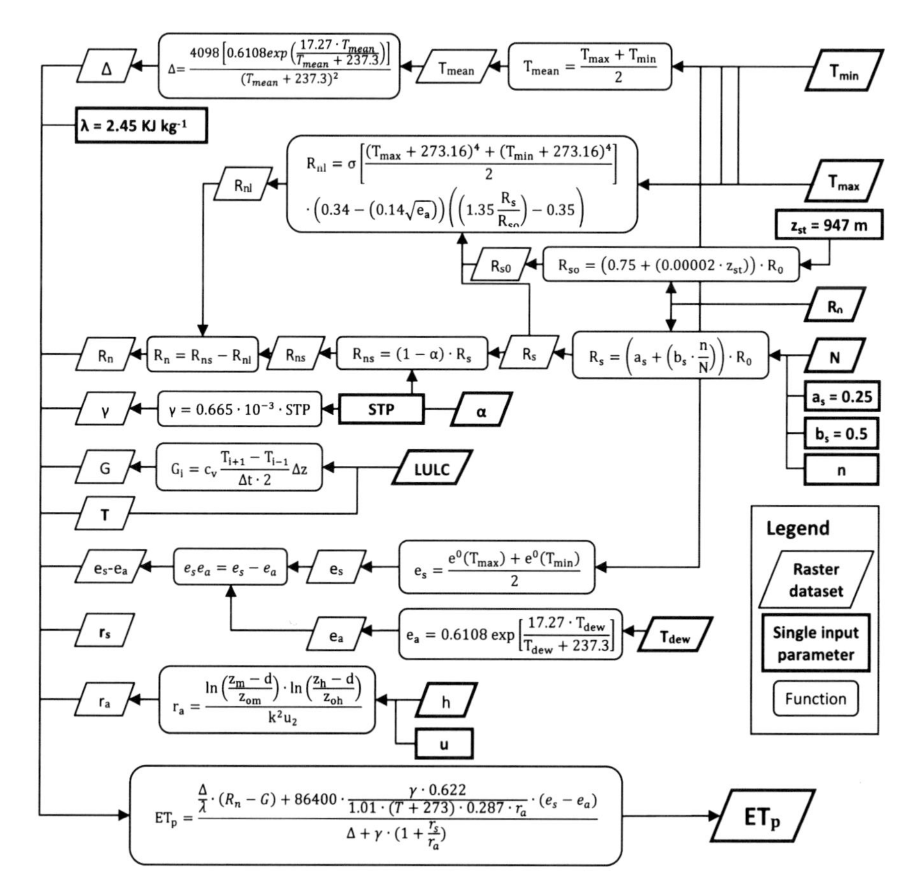

Fig. 3.3 Overview over the modelling and raster calculation of ET_p, input and final output data sets in *bold*

The precipitation falls as snow once the temperature falls below a certain threshold and is accumulated in a snowpack, whose amount is measured with the snow water equivalent (SWE) [mm m^{-2}] (cf. [43]). This process is based on the following equations:

$$P_n = P_s + P_r \tag{3.28}$$

$$P_s = x \cdot P_n \quad x = \begin{cases} 0, \text{if } T \geq T_r \\ 1, \text{if } T \leq T_s \\ \frac{T_r - T}{T_r - T_s}, \text{if } T_s < T < T_r \end{cases} \tag{3.29}$$

$$P_r = (1 - x) \cdot P_n, \tag{3.30}$$

where P_n is the daily value of catchment average net precipitation [mm], P_s and P_r are snow and rainfall in mm, respectively, T is the mean air temperature in °C and T_s and T_r are the lower and upper transition from snow to rain fall, set at 0 and

2 °C [18]. The ratio of P_s and P_r is assumed to change linearly between the threshold temperatures. The accumulated *SWE* of month i (SWE_i) is derived from *SWE* of the last month (SWE_{i-1}), the precipitation fallen as snow during month i, P_s, and the snow melt Q_m during month i in mm:

$$SWE_i = SWE_{i-1} + P_s - Q_m. \tag{3.31}$$

3.1.2.2 Snow Melt

Snow melt starts as soon as the temperature reaches the threshold T_m [°C]. The melting process is described by the temperature index method, which is a simplification of the physical processes occurring in the snow. Over each land type a certain amount of snow is melting per °C temperature and day, specified by the degree-day-factor *DDF* [mm day^{-1} °C^{-1}]:

$$Q_m = (T - T_m) \cdot DDF, \text{if} \quad T > T_m \text{ and } SWE > 0, \tag{3.32}$$

$$Q_m = 0, \text{otherwise.} \tag{3.33}$$

It is assumed that the snow pack has homogeneous characteristics regarding temperature and heat transfer, and the air temperature approximately represents the amount of heat available to melt the snowpack. Other processes such as internal refreezing and change of the snow pack structure as described by Kondo and Yamazaki [44] are not accounted for (Fig. 3.4). As the research area exhibits a quite large spatial and temporal diversity with regard to land use and snow types, the threshold temperature T_m for the beginning of the snow melt should have been measured for every location and land cover type. However, T_m **= 0 °C** is mostly used and was chosen as **average threshold temperature** (cf. [38]). Differences of the snow melt due to exposition and varying solar insolation in the mountains were taken into account through the spatial distribution of temperature values. Snow melt largely depends on the albedo, which varies from 0.9 for freshly fallen snow to about 0.4 for mature snow [38, 44] as well as the land cover type the snow has fallen on. "The difference between the incident solar energy and that reflected back to the atmosphere is absorbed in the snowpack and this portion of the energy is largely responsible for the melting of snow" [38: 500]. Measured degree-day-factors are by far more heterogenic than the generalised values for the land and snow types suggest [37, 38].

This can be seen in the **variety of degree-day-factors** from different publications for the same land cover types in Table 3.5. Degree-day-factors also vary over the course of the year, which can be modelled as a function of time of year, directly by employing physical surface properties (e.g. snow density or albedo) or indirectly by adding more input variables from physical properties (e.g. radiation components). The latter would lead to a simplified energy balance model requiring additional input variables and parameters, which are not easily available at the

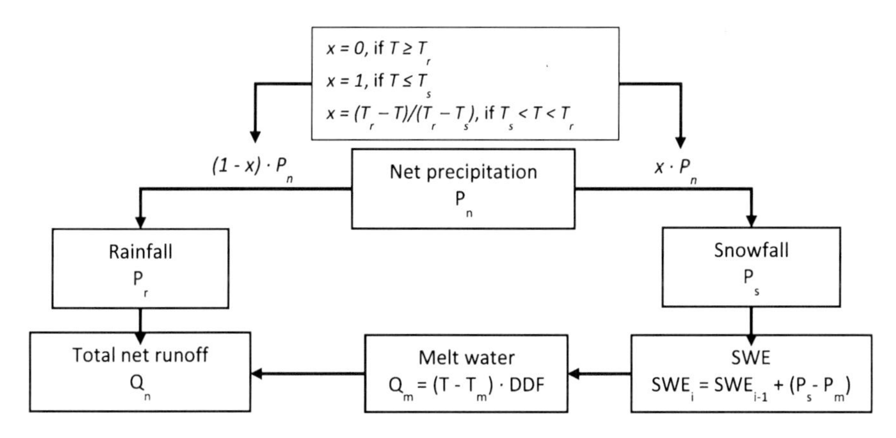

Fig. 3.4 Snow accumulation and melt water module, × is the factor determining the proportion of snow fall of the net precipitation

catchment scale [45]. Hence, the classical degree-day-method was kept as it is and considered to be adequate for 'average conditions' when the analysis is spatially and temporally restricted to the catchment scale and to periods exceeding a couple of days (Lang and Braun 1990 in Hock [45]). The degree-day-factors were not treated as constant, but adjusted to the seasonal changing land surface characteristics based on the values documented by several publications as suggested by Hock [45] (see Table 3.6). The additional integration of a snow correction factor to account for the catch deficit of precipitation gauges during snowfall as recommended by Merz and Blöschl [43] could not be applied.

The **total runoff** Q_n is then calculated as the sum of the rainfall fraction P_r and the runoff from snow melt Q_m (all in mm):

$$Q_n = P_r + Q_m. \tag{3.34}$$

When Eq. (3.34) is expanded by Eqs. (3.3) and (3.28), Q_n is calculated as follows:

$$Q_n = P - ET_a - P_s + Q_m. \tag{3.35}$$

The average contribution of the glaciated areas to melt water runoff cannot be neglected (see Sect. 2.4.2 and Table 3.2, 3.3, and 3.4). It has been taken into account by simulating the annual snow accumulation and snow melt over the glaciered areas with degree-day-factors for ice and snow that were adapted from a study modelling the mass balance of Urumqi Glacier No. 1 [18]. Glacier movement or long-term glacier melt and resulting runoff have not been considered because the potential long-term changes were estimated to account only for 2–4 % of the total surface water runoff. This is estimated to be smaller than the natural runoff variation and uncertainty of the water balance model (see the uncertainty analysis in Sect. 3.2.2 and the results of the climate scenarios in Sect. 5.2). The integration of glacier melt would also have required an additional module and

Table 3.5 Degree-day-factors from various reference

Characteristics		DDF in mm $°C^{-1}d^{-1}$
Glaciers at the end of summer		>6[a]
Ice		5.6[b]
Snow		2.8[b]
Snow (calibrated)		3.319–4.825[c]
Snow (calibrated)		1–2[d]
Deciduous forest with few coniferous trees		3–4.3[a]
Coniferous forest or dense deciduous forest		1.5–2.3[a]
Forest		2.25[e]
Dense coniferous		1.4–1.5[f]
Average density coniferous and deciduous forest		1.7–1.8[f]
Low density coniferous and deciduous forest		3–4[f]
Forest		1.75–3.36[g]
Moderate forest cover:	April	2[h]
	May	3[h]
	June	4[h]
Partial forest cover:	April	3[h]
	May	4[h]
	June	6[h]
Forest	Beginning of April	1.85[i]
	Beginning of June	3.7
Open area		2.7[e]
Open area		2.82–4.94[g]
No Forest:	April	4[h]
	May	6[h]
	June	7[h]
Open area	Beginning of April	3.7[i]
	Beginning of June	7.4[i]

Source[a] Martinec et al. [40], [b] Huintjes et al. [18], [c] Rahman [161], [d] Shin et al. [41], [e] US Corps of Engineers in Bagchi [38], [f] WMO in Bagchi [38], [g] Kuusisto [39], [h] WMO 1964 in Rango and Martinec [37], [i] Weiss and Wilson 1958 in Rango and Martinec [37]

model for the glacier mass balance for which sufficient data and observations were not available.

3.1.3 Calculation of Surface Runoff and Recharge

Based on the available water Q_n as calculated with Eq. (3.32), surface runoff and groundwater infiltration are calculated. **The processes leading to surface or near surface runoff are in general and at the catchment scale very complex**. When water is meeting the land surface, the processes and factors controlling surface runoff formation and infiltration happen on the hydrological microscale [2]:

Table 3.6 Degree day factors for the land use and cover classes used

LULC class	Time period	October–November	December–February	March–April	May–July	August–October	November–January
Agriculture		4	–	4	8	4	–
Agriculture dry		4	–	–	8	4	–
Sparse vegetation		4	4	4	8	4	4
Soil		4	4	4	8	4	4
Coniferous forest		3	3	3	4	3	3
Grass		4	–	–	8	4	–
Rock		4	4	4	–	4	4
Sealed areas		4	4	4	8	4	4
Ice		–	–	–	6	–	–
Snow		4	3	4	5	6	3
Snow old		3	3	3	4	3	3
Irrigation		–	–	–	–	4	–
Water		4	4	4	8	4	4
Dark		4	–	4	8	4	–

'–' indicates that the LULC class did not exist in the specific classification; *Source* taken from references in Table 3.5, own design

precipitation intensity and infiltration rate depending on soil type and land surface cover decide how precipitation that impinges on the land surface is partitioned into surface flow, interflow, base flow or groundwater flow. Surface runoff alone can be additionally differentiated into Hortonian overland flow, saturation flow from saturated surfaces or return flow of infiltrated water resurfacing, depending on the precipitation characteristics, antecedent (soil) water content and microtopography. Underground pressure transfer leading to groundwater ridging and the piston-flow effect can occur, as well as lateral macro-pore flow [2]. When working on the macro-scale however, these processes have to be aggregated to an impulse-response function 'available precipitation-runoff' [46]. Within the water balance model, processes and intermediate storages in soil and subsurface have not been modelled separately.

Surface and groundwater runoff in mm (Q_{sf} and Q_{gw}) **are calculated with the infiltration coefficient** p_{runoff} according to the concept of the ABIMO and Meßer models (cf. [8, 10]) with the following equations:

$$Q_{sf} = Q_n \cdot \frac{p_{runoff}}{100} \tag{3.36}$$

$$Q_{gw} = Q_n - Q_{sf} \tag{3.37}$$

$$Q_{gw} = Q_n - \left(Q_n \cdot \frac{p_{runoff}}{100}\right). \tag{3.38}$$

The proportion of surface and groundwater flow, p_{runoff}, can yield values between 0 and 100 and is derived depending on the soil type, grain size and field capacity, land use and degree of imperviousness, groundwater level and slope gradient (cf. [8–10]). As most of the soils in the research area have a high silt and loam content according to the Soil and Terrain Database for China [47] and soil parameter estimates presented by Batjes [48], only the p_{runoff} values for loamy soils were assigned to the LULC class 'soil' depending on the slope [8]. For the other LULC classes, the p_{runoff} values were assigned depending on their slope and imperviousness (see Table 3.7). As there were only small areas with shallow groundwater tables, a differentiation according to groundwater depths has been omitted.

The equations and concepts used for the calculation of the water balance model are summarised in Fig. 3.5. As the snow accumulates already before the calendar year starting in January, the water balance for the months October to December preceding the year chosen for calculation were modelled as well, but not included into the evaluation. All modules were implemented in a *Python* script, which accessed *ArcGIS* via the *ArcPy* package, and calculated for a raster data set based on the cell size and the extent of the LULC data sets. Using the *Hydrology toolbox* in *ArcGIS*, the catchment areas, streams and outlet points were calculated. For the catchments as presented in Sect. 2.1 and Figs. 3.2, 3.3, and 3.4, the monthly output of the water balance model, including ET_p, ET_a, P, P_s, Q_m, Q_n, Q_{sf} and Q_{gw}, was summarised with a *weighted flow direction* grid (Fig. 3.5).

3.2 Evaluation of Model Sensitivity and Uncertainty

Before simulating the water balance in Urumqi Region and analysing the model results, the model sensitivity and uncertainty was be evaluated, because "it is inevitable that both model and data input will present some uncertainty" as stated by Buttafuoco et al. [20: 2319]. The error will propagate towards the calculated

Table 3.7 Degree of imperviousness and proportion of baseflow p_{runoff} for the land use classes

Class	Imperviousness	Proportion of baseflow p_{runoff} over loamy soil (%)[c]			
	(%)	0–2 %	2–4 %	4–10 %	>10 %
Agriculture	0[a]	10	30	55	80
Agriculture dry	0[a]	10	30	55	80
Sparse vegetation	0	5	25	50	70
Soil	0[a]	10	30	55	80
Coniferous forest	0	0	20	45	60
Grass	0	5	25	50	70
Rock	–	75	80	85	90
Sealed areas	Ø 65[a]	85	90	95	100
Ice	100	100	100	100	100
Snow[b]	–	30	35	40	45
Snow old[b]	–	60	65	70	75
Irrigation	0	0	0	0	0
Water	0	0	0	0	0
Dark	40–60[*]	100	100	100	100

Source modified according to [a] Haase [10], [b] Johnsson and Lundin [162], [c] Meßer [8]

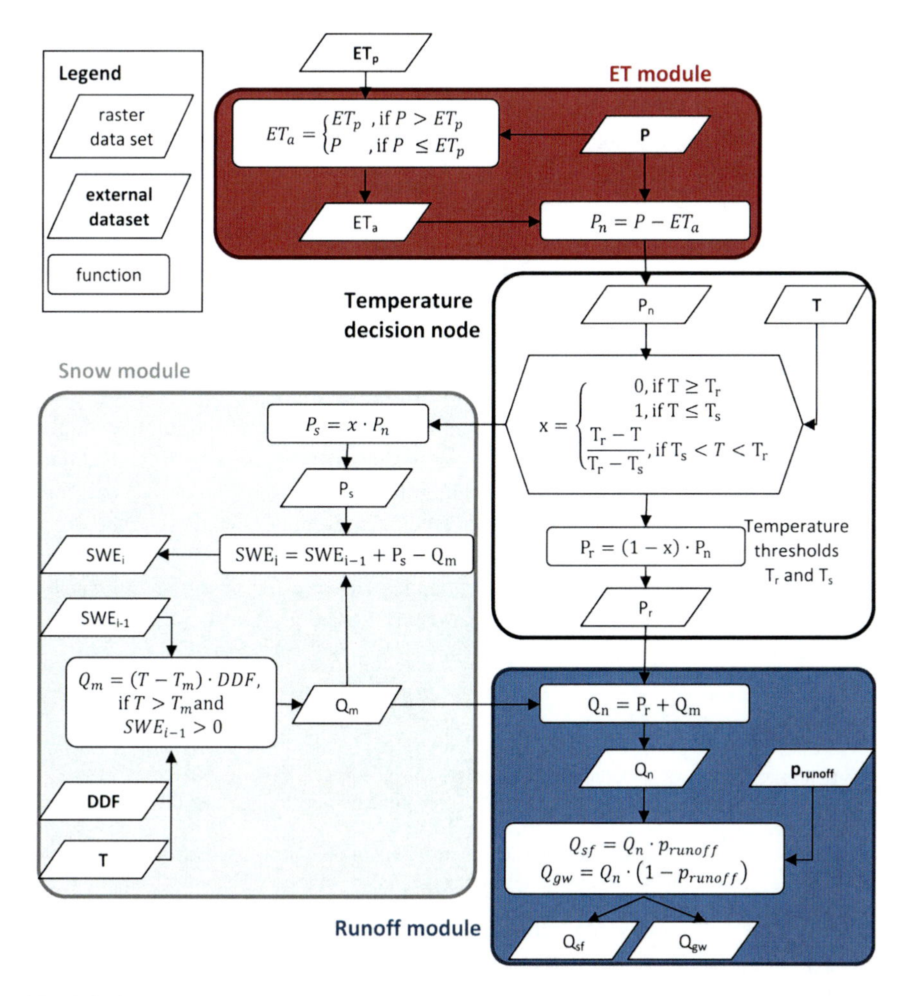

Fig. 3.5 Overview over the modelling and raster calculation of the water balance model, input data sets in *bold*

model output and when neglected will lead to improper evaluation and decision making. There are several sources of error and uncertainties based on the quality of the data and the model as well as the parameter interactions. Buttafuoco et al. [20] identify two main causes of model uncertainty, the **limitations in the models** used to simulate physical systems, and the errors introduced by the **uncertainty of input parameters** and data. Modelling uncertainty due to a simplified description of physical processes is inherent to human understanding of the natural systems and can only be quantified in retrospective analysis when comparing the model results with observed measurements [20, 49]. This error will be evaluated in Sect. 3.4 where the modelled results are compared with the observed data.

Data uncertainty on the other hand is caused by measurement errors, incomplete knowledge of spatial and temporal variations and heterogeneities [20]. Meinrath and Schneider [49] further divide the data uncertainty and the incomplete knowledge of input parameters into (i) **statistical uncertainty**, which is caused by the inadequate description of natural variability, randomness and heterogeneity on the temporal and spatial scale, and (ii) the **systematic intrinsic uncertainty**, which is based on the variability of the physical parameters. Statistical and intrinsic data uncertainty can be reduced by improving the accuracy and precision of measurement techniques. Even if only limited information is available for some properties, the uncertainty must be reduced as much as possible [49]. The statistical uncertainty was not assessed in this work, but was reduced by taking the spatial heterogeneity of the most sensitive input factors into account and using spatially distributed input datasets. The intrinsic uncertainty of the input variables and parameters was assessed by calculating the sensitivity index with the standard deviation of the input factors which takes the measurement error into account and determines the importance of a parameter for the model results as shown in Sect. 3.2.1.

'Sensitivity' describes the extent of changes for one factor depending on the changes of another factor. Hamby [50] states that there are two ways a model can be sensitive to input parameters:

> The consensus among authors is that models are indeed sensitive to input parameters in two distinct ways:
>
> (1) the variability, or uncertainty, associated with a sensitive input parameter is propagated through the model resulting in a large contribution to the overall output variability, and
> (2) model results can be highly correlated with an input parameter so that small changes in the input value result in significant changes in the output [50: 137].

The former describes the parameter importance for which an uncertainty analysis is applied, while the latter refers to the parameter sensitivity in the model which is analysed via a sensitivity analysis [50]. A parameter may be sensitive because the model is sensitive to the input, but this same parameter may not be important because it is known precisely and adds little variability or uncertainty to the model. However, an important parameter is always sensitive, since parameter uncertainty would not appear in the model output if the model was not sensitive to it. In a sensitivity analysis, the reaction of the model result y_i to changes of input parameter x_i is estimated. It can be used locally for the screening of one factor or for determining the local impact of the factors (x) on the model and its results (y), but also as a global sensitivity analysis (GSA) for assessing the output uncertainty depending on several model factors [51]. Using a GSA, the focus is on the expected value and variance of the output variable which are estimated, also called **uncertainty analysis** [51].

Sensitivity analysis can be used to optimise model calibrations, assess relationships between model factors, quantify model uncertainties and understand model behaviour [5]. It is not only a compulsory tool to compare and analyse 'black box' modelling output, but according to Meinrath and Schneider [49: 166] also an element of "good modelling practice". As defined by Campolongo et al. [51], any input included in the sensitivity study is a factor. Input variables refer to factors of the model which are directly observable in the corresponding real system, in this study for example climate variables such as temperature and precipitation. Parameters are factors which are not directly observable in the corresponding real system, but they may be estimated (e.g. runoff coefficient, degree-day-factor).

The **techniques available for sensitivity analysis** include screening designs to isolate the most important factors (one-at-a-time (OAT) design), differential analysis (mostly locally), Monte Carlo (MC) analysis (global), measures of importance (including FAST, global), and response surface methodology (global) [52]. As the aim of the analysis was to assess the most important parameters and variables and to determine the uncertainty of the model output, only factor screenings to generate sensitivity indices and Monte-Carlo analysis to assess the relationships between input and output parameters and the model output uncertainty were used.

In the first subsection, the parameter sensitivity and importance are assessed. For the local sensitivity analysis of the input variables and parameters, a OAT design was applied and sensitivity indices were calculated. In the second part, the general relationship between input and output and the uncertainty of the model results was estimated based on a MC analysis. Within the MC approach, multiple evaluations of the model with different input variables and parameters in given ranges were performed and assessed.

3.2.1 Analysis of Parameter Sensitivity and Importance

In this section, the significance of every parameter for the model results is investigated, which is also referred to as sensitivity. The importance of the model parameters affecting the resulting factors evapotranspiration, runoff and infiltration will be assessed and decided which parameter should be given as accurate as possible.

The influence of each variable or parameter on the output values is assessed using the **differential sensitivity analysis** [50]. Mathematically, this dependence is expressed by the partial derivative $\partial y/\partial x$ with x being the model input and y being the model output [53]. To numerically approximate this expression, non-linearities are neglected and a finite difference is used: An initial parameter value x_0 is varied by $\pm\Delta x$ resulting in $x_1 = x_0 - \Delta x$ and $x_2 = x_0 + \Delta x$ with corresponding values y_1 and y_2. Δx is determined in Sect. 3.2.1.1. The sensitivity index according to Lenhart et al. [53: 646] is then "calculated as the ratio between the relative change of model output and the relative change of a parameter":

$$I = \frac{(y_2 - y_1)}{2 * \Delta x}. \quad (3.39)$$

This sensitivity index I is normalised to achieve a dimensionless form I':

$$I' = \frac{(y_2 - y_1)/y_0}{2 * \Delta x / x_0}. \quad (3.40)$$

The normalised sensitivity index I' is calculated for each factor using an OAT design where the value of one factor is changed at a time and its impact is evaluated. The importance of the factors can be assessed by ranking them according to the magnitude of their output. The sign of the index shows in which direction the input parameter influences the output [53].

3.2.1.1 Input Values and Output Classification for the Sensitivity Analysis

The input values for the sensitivity analysis were derived with slightly different methods for point measurements and raster datasets. The $\pm \Delta x$ of the climate variables were based on the distribution of the measurements from 1975 to 2010, errors due to spatial interpolation were not taken into account. For the raster datasets, the summary statistics were used to account for the heterogeneity of land surface parameters and include spatial uncertainties. Buttafuoco et al. [20] point out the necessity to assess spatial correlations of input variables to obtain realistic values of the model output uncertainty via applying joint stochastic simulations. However, this stochastic simulations method is computational intensive, despite being applied for a relatively simple model for evapotranspiration in Buttafuoco et al. [20]. Due to computation power limitations this was not possible in this work.

The subjective decision made by the analyst when determining the parameter boundaries is a general challenge of sensitivity and uncertainty analyses, as it may distort the parameter sensitivities. To avoid this problem, standardised characteristics of the frequency distribution of all parameters are used, e.g. the standard deviation after testing or assuming a normal distribution [5]. However, using a fixed percentage of the initial value for the parameter changes could lead to two problems: If the model is non-linear, the magnitude of changes in the output depends on the chosen initial value. If the initial parameter value is located close to the upper or lower possible parameter range, the variation by a fixed percentage could lead to inadmissible input values [53]. Lenhart et al. [53] suggest two possible ways to calculate Δx: Δx_1 is 25 % of the total parameter range and Δx_2 is 10 % of the mean parameter value, leading to two different sensitivity indices I_1 and I_2. In their study, both approaches lead to similar results for the sensitivity index and the overall ranking of the parameters were also the same [53]. To analyse the consequences of possible individual non-linear relationships between input and output values which are expected due to the model equations (this chapter), both were applied to the input parameters. Additionally, the sensitivity index I_3 using $\Delta x_3 = \sigma$ was calculated. For most of the parameters, the standard deviation σ also represents the error or uncertainty of the parameter [20]. In general, physically based and distributed models have a quite low uncertainty with regard to model assumptions. However, due to the large parameter sets they are prone to uncertainties of the input parameters [5]. With $\Delta x3$, the effect of parameter uncertainty and importance for the output values can also be estimated [50]. The summary statistics of the input parameter for the sensitivity analysis were calculated directly from the input data used in the modelling (see Table 3.8). They were then used to calculate Δx_1, Δx_2 and Δx_3 and the corresponding sensitivity indices for the relative importance in the input factors (Table 3.10, [52]). As the normalisation reduces the impact of the different range of input factors on the sensitivity index, the normalised index values can be compared directly or classified according to Table 3.9.

The main advantage of this approach to evaluate the parameter sensitivity are the low computational costs compared to a global sensitivity analysis, where all parameter values are changed, or the sensitivity approach according to Meinrath and Schneider [49], where only an individual entry in the complete set of input data is fixed while all other entries change randomly.

Also, the evaluation of parameter sensitivity depends strongly on the parameter's value range [5]. Hence, thresholds that would change parameter behavior are eventually not reached and represented. Such a "local sensitivity approach is practicable when the variation around the midpoint of the input factor is small" [52: 16] and the input–output relationship is assumed to be generally linear. Most evaluation measures for model sensitivity and uncertainty have problems when dealing with non-linear and non-monotonicity relationships and variance based measures that could overcome "the difficulties posed by model non-monotonicity" are relatively time consuming to apply [52: 32–33]. The problem was approached by calculating three sensitivity indices with different value ranges for the input parameters to assess possible non-linear or non-monotonic behavior and thresholds inside the model's ranges. A correlation analysis was also conducted with a global approach where the model was run with 5,000 different random values for each parameter. For each parameter and result combination, Pearson's correlation coefficient and Kendall's and Spearman's rank correlation coefficient was calculated. The correlation coefficients did not add new information to the sensitivity analysis and are therefore not presented here.

3.2.1.2 Results of the Sensitivity Analysis

When evaluating the sensitivity of the ET_p to the input factors, several variables and parameters exhibit stronger influence on the output. The most important parameter is the canopy height h with a very high sensitivity index (class IV) because it is determining aerodynamic and stomatal resistance (r_a and r_s) of vegetation and land cover in the water balance model. In the second highest class III, maximum daily

Table 3.8 Summary statistics and Δx of the input parameters for the sensitivity analysis

Summary statistics	T_{mi} (°C)	T_{mi-1} (°C)	T_{mi+1} (°C)	T_{min} (°C)	T_{max} (°C)	T_{dew} (°C)	n (h/day)	N (h/day)	R_0 (daily MJ/m^2)	α (%)	STP (kPa)	u (m/s)	h (m)	P (mm)	SWE (mm)	DDF (mm l/°C day)	p_{runoff} (%)
Max	34.5	34.5	34.5	19.2	41.0	18.5	9.10	14.70	30.40	0.85	94.42	18.7	8.00	224.54	207.01	8.00	1
Min	–29.2	–29.2	–29.2	–24.0	–23.7	–32.3	3.10	0.00	0.59	0.06	89.58	0.0	0.01	0.00	0.00	2.00	0
Mean	7.4	7.4	7.4	2.5	12.4	–2.5	7.00	10.19	12.41	0.40	91.32	2.4	0.71	29.80	25.66	4.52	0.47
Std	14.0	14.0	14.0	12.7	14.5	9.1	2.60	1.36	1.57	0.18	1.30	1.1	2.04	28.10	6.93	0.61	0.25
x_0 = mean	7.4	7.4	7.4	2.5	12.4	–2.5	7.00	10.19	12.41	0.40	91.32	2.4	0.71	29.80	25.66	4.52	0.47
Δx_1 = range*25 %	15.9	15.9	15.9	10.8	16.2	12.7	1.50	3.67	7.45	0.20	1.21	1.9	0.80	22.45	20.70	1.50	0.25
Δx_2 = mean*10 %	0.7	0.7	0.7	0.3	1.2	–0.3	0.70	1.02	1.24	0.04	9.13	0.2	0.07	2.98	2.57	0.45	0.5
Δx_3 = std	14.0	14.0	14.0	12.7	14.5	9.1	2.60	1.36	1.57	0.18	1.30	1.1	2.04	28.10	6.93	0.61	0.25

Data source [82] (T_{mi}, T_{mi-1}, T_{mi+1}, T_{min}, T_{max}, T_{dew}, *STP*, *u*, *P*), [89] (*n*), [76] Area solar radiation tool (*N*, R_0), various references (α, *h*, *DDF*, p_{runoff}), own calculations (*SWE*)

Table 3.9 Magnitude of the sensitivity index according to Lenhart et al. [53]

Class	Index	Sensitivity
I	$0.00 \leq \mid I \mid < 0.05$	Small to negligible
II	$0.05 \leq \mid I \mid < 0.20$	Medium
III	$0.20 \leq \mid I \mid < 1.00$	High
IV	$\mid I \mid \geq 1.00$	Very high

temperature, wind speed and station pressure show still a strong influence and in class II, minimum daily temperature, dewpoint temperature and solar radiation R_0 show a medium influence. For the actual evapotranspiration ET_a, precipitation is of high importance while all other parameters show less influence. The effective runoff Q_n is highly sensitive to the degree-day-factor, average daily temperature T_{mi} and *SWE*. This might be surprising, as one should expect that the precipitation value is also of high importance for possible runoff quantity. However, the amount of effective runoff at a certain date is obviously regulated by the available melt water which is in turn controlled by the available snow water equivalent, temperature und melting factor. For the surface runoff Q_{sf}, the runoff coefficient p_{runoff} is the most important input parameter to the ones already described for Q_n, as it is also the only one used when calculating Q_{sf} from Q_n.

Almost all of the input factors with high magnitudes for I_1 and I_2, the parameter uncertainty and importance for the model output I_3 are high as well. The three different sensitivity indices also show the effect of different boundary limits: For most of the factors, they all three show very similar values. But I_2' seems to assign less importance to the factors T_{max}, T_{dew}, u, h, and P than the other sensitivity indices probably because Δx_2 is significantly smaller than the other Δx shown in Table 3.8. Similarily, sensitivity indices for T_{mi}, u, z, P, SWE, DDF and p_{runoff} vary for Q_n and Q_{sf} (see Table 3.10). All these factors are important for the respective output results, which is also the reason for the sensitivity to the changing input ranks. Especially the canopy height h leads to very different reaction of the modelling results and sensitivity indices as it is used to calculate r_a with a non-linear function (see equation in Sect. 3.1.1 and Fig. 3.6).

A similar approach for sensitivity analysis based on the partial derivative $\partial y/\partial x$ is presented by Campolongo et al. [52]. They approximate the expression by **calculating a standardised regression coefficient** to quantify the effects of a variable in an OAT design. The advantage is that the model coefficient of determination R^2 can be calculated to assess how well the regression model can explain the variance of the output value and predict the accuracy of the model with the predicted error sum of squares. This approach was tested by calculating a GSA using a Monte Carlo design, plotting scatter plots and applying a regression analysis. The output variable y_i is plotted against x_{ij} for the selected input factors x_j to reveal relationships between model inputs and model predictions, such as non-linear relationships and thresholds (Helton 1993 in Campolongo et al. [52]). The findings of this analysis were **used to introduce limits to some of the input parameters** of the water balance model in order to avoid unreasonable results. For

Table 3.10 Sensitivity indices for input parameter T_{mi}, T_{mi-1}, T_{mi+1}, T_{min}, T_{max}, T_{dew}, n, N, R_0, α, STP, u, h, P, SWE, DDF, and p_{runoff} as well as the results ET_p, ET_a, Q_n and Q_{sf}

Result	ET_p									ET_a								
Input parameter	I_1	I'_1		I_2	I'_2		I_3	I'_3		I_1	I'_1		I_2	I'_2		I_3	I'_3	
T_{mi} (°C)	–0.13	–0.23	III	–0.13	–0.24	III	–0.13	–0.24	III	–0.13	–0.23	III	–0.13	–0.24	III	–0.13	–0.24	III
T_{mi-1} (°C)	0.01	0.02	I	0.01	0.02	I	0.01	0.02	I	0.01	0.02	I	0.01	0.02	I	0.01	0.02	I
T_{mi+1} (°C)	–0.01	–0.02	I	–0.01	–0.02	I	–0.01	–0.02	I	–0.01	–0.02	I	–0.01	–0.02	I	–0.01	–0.02	I
T_{min} (°C)	0.20	0.12	II	0.22	0.13	II	0.20	0.12	II	0.20	0.12	II	0.22	0.13	II	0.20	0.12	II
T_{max} (°C)	0.33	1.01	IV	0.33	0.99	III	0.37	1.13	IV	0.33	1.01	IV	0.33	0.99	III	0.37	1.13	IV
T_{dew} (°C)	–0.20	0.12	II	0.00	0.00	I	–0.27	0.17	II	–0.20	0.12	II	0.00	0.00	I	–0.27	0.17	II
n (h day^{-1})	–0.05	–0.09	II	–0.05	–0.09	II	–0.05	–0.09	II	–0.05	–0.09	II	–0.05	–0.09	II	–0.05	–0.09	II
N (h day^{-1})	0.04	0.10	II	0.04	0.09	II	0.04	0.09	II	0.04	0.10	II	0.04	0.09	II	0.04	0.09	II
R_0 (MJ m^{-2} day^{-1})	0.07	0.21	III	0.07	0.21	III	0.07	0.21	III	0.07	0.21	III	0.07	0.21	III	0.07	0.21	III
α (%)	–1.42	–0.14	II	–1.41	–0.14	II	–1.41	–0.14	II	–1.42	–0.14	II	–1.41	–0.14	II	–1.41	–0.14	II
STP (kPa)	0.02	0.50	III	0.02	0.50	III	0.02	0.50	III	0.02	0.50	III	0.02	0.50	III	0.02	0.50	III
u (m s^{-1})	1.56	0.92	III	1.28	0.75	III	1.54	0.90	III	1.56	0.92	III	1.28	0.75	III	1.54	0.90	III
h (m)	–117.8	–20.4	IV	4.39	0.76	III	–31.22	–5.41	IV	–117.8	–20.4	IV	4.39	0.76	III	–31.22	–5.41	IV
P (mm)										0.00	0.00	I	0.00	0.00	I	0.04	0.31	III
Result	Q_n									Q_{sf}								
Input parameter	I_1	I'_1		I_2	I'_2		I_3	I'_3		I_1	I'_1		I_2	I'_2		I_3	I'_3	
T_{mi} (°C)	1.67	0.24	III	0.13	0.02	I	1.90	0.27	III	0.79	0.24	III	0.06	0.02	I	0.89	0.27	III
T_{mi-1} (°C)	–0.01	0.00	I	–0.01	0.00	I	–0.01	0.00	I	–0.01	0.00	I	–0.01	0.00	I	–0.01	0.00	I
T_{mi+1} (°C)	0.01	0.00	I	0.01	0.00	I	0.01	0.00	I	0.01	0.00	I	0.01	0.00	I	0.01	0.00	I
T_{min} (°C)	–0.20	–0.01	I	–0.22	–0.01	I	–0.20	–0.01	I	–0.09	–0.01	I	–0.10	–0.01	I	–0.09	–0.01	I

(continued)

Table 3.10 (continued)

Result	Q_n									Q_{sf}								
Input parameter	I_1	I'_1		I_2	I'_2		I_3	I'_3		I_1	I'_1		I_2	I'_2		I_3	I'_3	
T_{max} (°C)	–0.33	–0.08	II	–0.33	–0.08	II	–0.37	–0.09	II	–0.16	–0.08	II	–0.15	–0.08	II	–0.17	–0.09	II
T_{dew} (°C)	0.20	–0.01	I	0.00	0.00	I	0.27	–0.01	I	0.09	–0.01	I	0.00	0.00	I	0.13	–0.01	I
n (h day^{-1})	0.05	0.01	I	0.05	0.01	I	0.05	0.01	I	0.02	0.01	I	0.02	0.01	I	0.02	0.01	I
N (h day^{-1})	–0.04	–0.01	I	–0.04	–0.01	I	–0.04	–0.01	I	–0.02	–0.01	I	–0.02	–0.01	I	–0.02	–0.01	I
R_0 (MJ m^{-2} day^{-1})	–0.07	–0.02	I	–0.07	–0.02	I	–0.07	–0.02	I	–0.03	–0.02	I	–0.03	–0.02	I	–0.03	–0.02	I
α (%)	1.42	0.01	I	1.41	0.01	I	1.41	0.01	I	0.67	0.01	I	0.66	0.01	I	0.66	0.01	I
STP (kPa)	–0.02	–0.04	I	–0.02	–0.04	I	–0.02	–0.04	I	–0.01	–0.04	I	–0.01	–0.04	I	–0.01	–0.04	I
u (m s^{-1})	–1.56	–0.07	II	–1.28	–0.06	II	–1.54	–0.07	II	–0.73	–0.07	II	–0.60	–0.06	II	–0.72	–0.07	II
h (m)	117.77	1.62	IV	–4.39	–0.06	II	31.22	0.43	III	55.35	1.62	IV	–2.06	–0.06	II	14.67	0.43	III
P (mm)	1.00	0.58	III	1.00	0.58	III	0.96	0.56	III	0.47	0.58	III	0.47	0.58	III	0.45	0.56	III
SWE (mm)	0.69	0.34	III	1.00	0.50	III	1.00	0.50	III	0.32	0.34	III	0.47	0.50	III	0.47	0.50	III
DDF (mm $°C^{-1}$ day^{-1})	1.10	0.10	II	0.00	0.00	I	0.00	0.00	I	0.52	0.10	II	0.00	0.00	I	0.00	0.00	I
P_{runoff} (%)										0.41	0.80	III	0.55	1.06	IV	0.52	1.00	IV

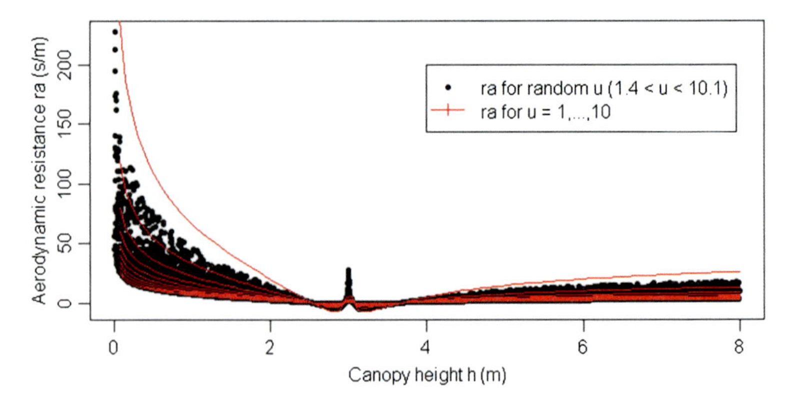

Fig. 3.6 Relationship between canopy height and aerodynamic resistance calculated for random wind speed (1.4 <u <10.1) (*black dots*) and for $u = 1, \ldots, 10$ (*red lines*). For $h = 3$ m $r_a = \ln(0)$ is not a real number

example the values for crop or surface height h had to be restricted due to the logarithmic relationship to the aerodynamic resistance r_a (see Fig. 3.6).

3.2.2 Analysis of Result Uncertainty and Error

So far, the relationship between the uncertainty of single factors and model results was evaluated in the sensitivity analysis. This uncertainty is also referred to as intrinsic parameter uncertainty. However, Harlin and Kung [54] found that **parameter uncertainty is also connected to combinations of parameters** and not only to the absolute values of each. Hence, a GSA using a Monte-Carlo approach was performed (cf. [20, 49]). In other cases, uncertainty analysis is used to determine the best parameter sets for the model, but here the aim was to assess the uncertainty of the whole model including the effects of relationships between parameters and the predictive uncertainty of the model, which could not be assessed separately. This time, the uncertainty analysis included a whole modelling period of 15 months to take into account connections between the monthly timesteps, i.e. carrying the accumulated SWE from one month to another. Actually, the uncertainty would have to be calculated either for a single cell or for a group of cells, for example a (sub-)catchment, and then compared to the cell or catchment output as the water balance model estimates the water flow for each raster cell separately. But measurements for evaluating the uncertainty were only available for a larger catchment, for which the uncertainty analysis would have demanded too much computing time. Therefore, the uncertainty of the result was calculated for one exemplary data set based on the summary statistics of the input data and evaluated independently to analyse the model uncertainty and dimension of the possible error. The parameter values were sampled randomly in fixed ranges

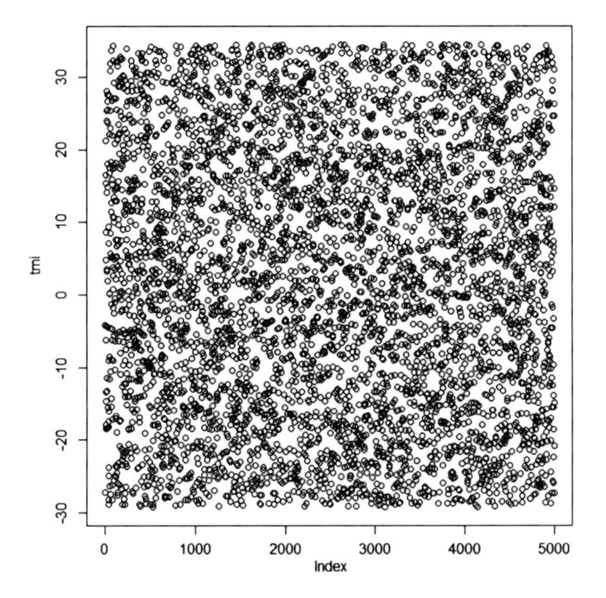

Fig. 3.7 Example for random variables with number of runs $k = 5{,}000$ on x-axis and random values on y-axis

determining the probability that model runs are located inside a certain interval. The parameter range was set as

$$x_{min/max} = \overline{x} \pm \sigma, \tag{3.41}$$

where $x_{min/max}$ is the upper or lower parameter boundary, $\overline{x}$ is the parameter mean value and σ is the standard deviation of the measurements (see Table 3.8). For the parameters assigned to land use classes, the standard deviation of the spatial variation of parameter values in the whole research area was used as the parameter error could not be estimated otherwise. The range for the input parameter was also considered from minimum to maximum of the monthly values. But such extreme values rarely occur at the same time and caused improbable output values in a test run. Thus the standard deviation of the input data was selected to represent the systematic error of the measurements or parameter estimation.

Inside the chosen boundaries derived from the summary statistics of variables and parameters (see Table 3.8), the input values were generated for a number of runs k = 10,000 equally distributed inside the set boundaries and each of them changed for every run (Fig. 3.7). This random sampling is also referred to as pseudo-random. Since the program uses an algorithm to generate the numbers, they are generated by a deterministic process and can be predicted and reproduced. Their statistical advantage is that they produce unbiased estimates of the mean and the variance of y [51, 55].[1]

[1] If the computing time was limited, a factorial set up could have been introduced, where only a given number of 'levels' are chosen for each parameter or possible interaction among factors are examined (cf. [50, 52]). However, this was not necessary as the uncertainty was only run for one point and not the raster data set.

Table 3.11 Input variables and parameters and intermediate and final results

Input parameters and variables	Intermediate results	Results
$T_{mi} = T_{mi-1} = T_{mi+1}$	G_{mi}	ET_p
T_{min}, T_{max}, T_{dew}	$e_s e_a$	ET_a
n	sv	P_n
N	R_s	P_s
R_0	R_{s0}	Q_m
α	R_{nL}	SWE
STP	R_n	Q_n
u	γ	Q_{sf}
h	r_s	Q_{gw}
P	r_a	
DDF		
p_{runoff}		

For every month and the whole year, the summary statistics mean, standard deviation, minimum and maximum value of the model results were calculated and evaluated (see Table 3.11). The uncertainty interval width of each month and the whole year was calculated as the difference between minimum and maximum value. The average interval width *IW* of the uncertainty intervals is calculated as (Choi and Beven 2007 and Blasone and Vrugt 2008 in Lin et al. [56]):

$$IW = \frac{\sum_{i=1}^{n} [Q_{up}(i) - Q_{low}(i)]}{n}. \tag{3.42}$$

Although most of the input variables and parameters of the model are derived from physical properties, their uncertainty range $x_{min/max}$ leads to quite large uncertainty of the resulting factors. Especially the error of ET_p is very large as some extreme values of the input factors r_a and *n/N* can cause high ET_p values, but are actually limited to very small local areas. The water balance model is very sensitive to these extreme input values as the spatial distribution and their actual frequency is not taken into account (see Fig. 3.8). Uncertainties of climate parameters such as monthly mean air temperature and monthly precipitation sums of only several degrees Celsius and millimetres can lead to ET_p varying over 400 mm per month. The error of surface runoff is smaller as it tends towards zero in the winter and summer months due to temperatures below 0 °C and high ET_a, respectively.

Table 3.12 shows the uncertainty for the monthly values of surface runoff, summarising the monthly mean, standard deviation, minimum and maximum values. The mean standard deviation of monthly Q_{sf} and the mean monthly possible range (minimum to maximum value) are larger than the mean monthly mean value. For the annual sum, the standard deviation is only 16.10 mm and at least lower than the mean annual surface runoff of 25.26 mm, but still surpassed by the interval width (115.42 mm). Obviously, **the uncertainty and possible error of the runoff is large, as extreme runoff values are possible within the**

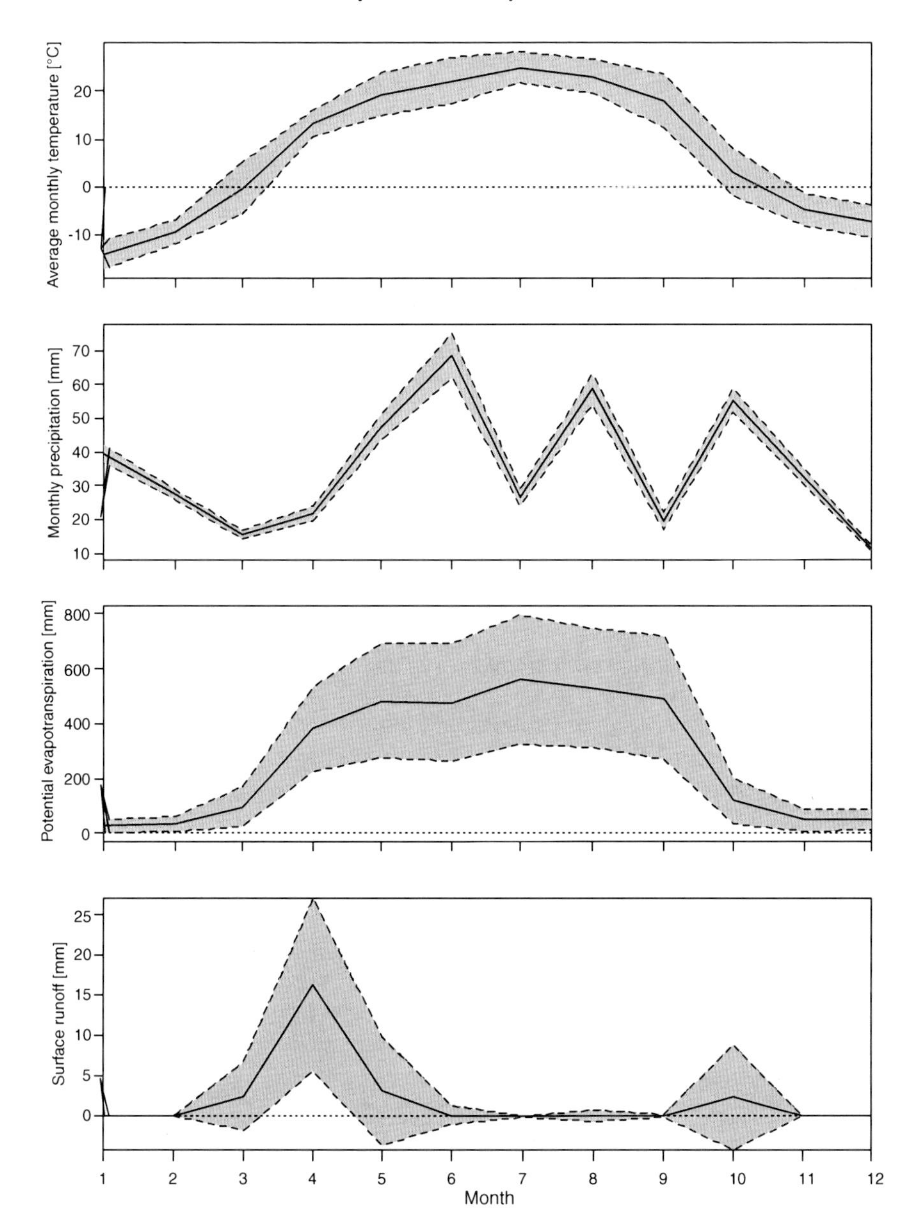

Fig. 3.8 Input variables monthly temperature and precipitation and results of the uncertainty analysis for ET_p and ET_a, *solid line* mean values, *dashed line* standard deviation, respectively, *grey polygon* mean ± standard deviation, *dotted liney* = 0. Negative values for evapotranspiration indicate condensation of water vapour

uncertainty of the input factors. Although they can be smoothed out over the extent of a large catchment, these finding show that the results' large uncertainty has to be considered when evaluating the modelled results.

3.3 Input Variables and Parameters

Ideally, the necessary input data and parameters for a process-based approach are directly measured to model the hydrological system and its processes as closely as possible. However, the procedure needed to be adopted to local circumstances for this research: On the one hand, index- and not only process-based modules were chosen to represent the hydrological processes as described in the previous section. On the other hand, most of the factors for the model could not be measured or calibrated in detail, but had to be acquired via alternative ways which are described in this section.

First, the elevation data which was used to define the catchment boundaries and hydrological units in the research area is analysed and evaluated. Then the input data is presented whose spatial variability could be taken into account: temperature and precipitation, solar radiation and other climate variables and characteristics of land use and cover. Most of the climate variables are only available for one point in the research area, as they are measured at a climatological station. However, the distributed model is sensitive to the spatial variability of its input factors and secondary information from other related attributes could supplement these point measurements [20, 57]. The choice of the variables whose spatial variability should be taken into account was based on the importance of the variables for the model results and the possibility to expand the point information into space. The effects of climate change on the hydrological system are mainly controlled by changes in precipitation and temperature, a relationship which is also confirmed for the water balance model by the sensitivity analysis in Sect. 3.2.1. Therefore, information about the spatial variation of land surface temperature and precipitation values were derived from the remote sensing products MODIS[2] 11A1 and TRMM[3] 3B43 (see Sect. 3.3.2). Additionally, the distribution of incoming solar radiation R_0 was calculated based on the elevation data as the estimation of ET_p showed to be sensitive to this input variable and the necessary data and procedure were readily available. The spatial distribution of other important climate variables such as station pressure and wind speed was not estimated as additional information could not be obtained and the degree of improvement was questionable.

The parameters required for the index-modules could also not be directly measured or calibrated. Existing thematic maps describing the land surface characteristics turned out to be not detailed enough for the size of the research

[2] Moderate-Resolution Imaging Spectroradiometer.

[3] Tropical Rainfall Measuring Mission.

Table 3.12 Summary statistics of surface runoff in mm for the uncertainty analysis with x ± σ

Month	1	2	3	4	5	6	7	8	9	10	11	12	Annual sum	Monthly mean
Mean	0.00	0.00	2.42	16.44	2.95	0.09	0.00	0.06	0.01	2.50	0.00	0.00	25.26	2.04
Std	0.00	0.00	4.35	10.74	6.45	1.20	0.08	0.89	0.23	6.77	0.00	0.00	16.10	2.56
Min	0.00	0.00	0.00	0.00	0.00	0.00	0.00	0.00	0.00	0.00	0.00	0.00	0.00	0.00
Max	0.00	0.00	67.16	51.40	74.00	35.07	4.23	25.05	10.04	81.37	0.00	0.00	115.42	29.03

area. Thus, a land use and cover classification was carried out for Urumqi Region at several dates throughout the year to capture the seasonal changes of the land surface characteristics and the required parameters derived from this information (see Sect. 3.3.3). For an overview of all data sets used as input for the water balance model, refer to Table A.1 in the appendix.

3.3.1 Elevation Data

Elevation data was necessary to define the catchment boundaries and streams in the research area, for which the water balance should be modelled. For the research area, the digital elevation models (DEM) SRTM-3 DEM (58) and ASTER Global DEM (GDEM) [59] are used. The DEMs differ in resolution and accuracy and were applied for different uses depending on their suitability, which is discussed in this section.

The ASTER GDEM is produced by the Ministry of Economy, Trade and Industry (METI) of Japan and the United States National Aeronautics and Space Administration (NASA) from optical stereo data acquired by the Advanced Spaceborne Thermal Emission and Reflection Radiometer (ASTER). The processing includes stereo-correlation to produce scene-based ASTER DEMs, cloud masking, stacking all DEMs, removing residual bad values and outliers, averaging selected data, correcting residual anomalies before partitioning the data [60]. So far, ASTER GDEM has been released by METI and NASA at a global resolution of 30 m × 30 m for public use and further evaluation [60]. The ASTER GDEM data were obtained through the online data pool at the NASA Land Processes Distributed Active Archive Center (LP DAAC), USGS/Earth Resources Observation and Science (EROS) Center, Sioux Falls, South Dakota[4] via the USGS Earth Explorer.[5]

The SRTM DEM provided by the U.S. Geological Survey (USGS) was produced by the Shuttle Radar Topography Mission (SRTM), a joint project of NASA, the National Geospatial Intelligence Agency (NGA) of the U.S. Department of Defense, and the German Aerospace Center (DLR). It operates with a synthetic aperture radar (SAR) system which was designed to function with dual radar antennas as a single-pass interferometer. The surface height is derived from the phase difference of returning signals detected by antennas at the same time from a slightly different vantage point [61, 62]. SRTM-3 with a spatial resolution of about 90 m was downloaded from the Global Land Cover Facility[6] (see Table 3.13) as the SRTM-1 with higher resolution (30 m × 30 m) was not available for the study region.

---

[4] http://lpdaac.usgs.gov/get_data, accessed 15.07.2009.

[5] http://earthexplorer.usgs.gov/, accessed 15.07.2009.

[6] http://glcf.umiacs.umd.edu/data/srtm/index.shtml, accessed 31.01.2008.

Table 3.13 Overview for the digital elevation models used for the hydrological modelling

	SRTM-3	ASTER GDEM
Spatial resolution	3 arc s ~ 90 m	30 m
Z accuracy	6.2 m (90 %)	20 m (95 %)
X–Y accuracy	8.8 m (90 %)	30 m (95 %)
Method	Synthetic aperture radar, dual radar antennas, interferometer	Stereo-correlation
Availability and scene size	60° N–56° S; 1° × 1°	83° N–83° S; 1° × 1°
Websites for data access	http://glcf.umiacs.umd.edu/data/srtm/index.shtml	http://reverb.echo.nasa.gov/reverb/; http://earthexplorer.usgs.gov/

Source Aster Validation Team [60], Rodriguez et al. [163]

Both ASTER GDEM and SRTM-3 are actually digital surface models (DSM) as they report canopy surfaces and the height of surface features. If possible, they should be filtered for sinks and outlier before hydrological modelling (cf. [63, 64]). Due to the size of the research area and the resolution of the calculation cell size which is larger than the expected error from vegetation heights, filtering was not conducted. Before choosing a DEM, the accuracy and its implications on the processes of the models had to be assessed as the DEMs differ in spatial resolution and consequently in their ability to represent topographic and hydrological features such as catchments and streams. The spacing of the original data used in constructing a DEM should be smaller than the hillslope length of the process to be modelled [65]. Neglecting these effects may lead to problems when delineating only slightly sloped catchment boundaries and streams as will be discussed in Sect. 3.3.1.2.

3.3.1.1 Accuracy and Anomalies

An accuracy assessment using ground truth points was not possible due to the insufficient coverage of the research area with measured points. Hence, results from other projects and publications had to be consulted to evaluate the usability of the DEMs. Jacobsen [66] stated that ASTER GDEM has advantages over SRTM-3 as the latter shows data gaps in mountaineous regions caused by radar lay over and viewing shadows. However, the error was not detected in the research area and was therefore not relevant for the choice of the DEM.

According to Jacobsen [66], the approximate relative horizontal accuracy of the ASTER GDEM (one point in relation to the neighbouring point) differs only by 0.4 m. He explains the higher relative accuracy for ASTER by the point spacing of 1″ in relation to 3″ for SRTM-3. With regard to the morphologic detail, the ASTER GDEM offers only slightly more detail as in the SRTM DSMs [66] and also the [60: 26] concludes from their analysis "that the spatial detail resolvable by the ASTER GDEM, at least of the data tested, is slightly better than 120 m".

The vertical accuracy of both DEMs varies with spatial location and topographic characteristics. The ASTER Validation Team [60] summarizes that SRTM-3 shows high vertical accuracies for built-up areas and bare soil and the lowest for forest. The accuracies of the ASTER GDEM on the other hand were highest for forest and lowest for open land. In general, ASTER GDEM shows a general negative bias, i.e. underestimates the surface elevation, while SRTM-3 overestimates them. The mean difference and RMSE was slightly smaller for SRTM-3 data (see also Table 3.14).

Wang et al. [67] found relatively large discrepancies in steep-sloped areas between the vertical accuracies of ASTER GDEM and SRTM-3. Fairly large differences between the ASTER GDEM and SRTM-3 of over 10 m were also observed in the mountaineous regions of the research area (see Fig. 3.9). Systematic aberrations are visible on the mountain slopes, pointing to the different methods used to derive the DEMs. The under- or overestimation of the terrain height seems to depend on terrain inclination and to increase with steepness. For the ASTER GDEM, the elevation accuracy varies with topography and terrain inclination, which could be the cause of the differences here [66]. In the plane and flat regions on mountain tops, grate and valleys, the differences are far less pronounced.

Also observable are rectangular structures which are inherent to the ASTER GDEM data. These artificial lines in vertical and horizontal direction are not easy to detect in the elevation data, but clearly appear in the first derivatives of the

Table 3.14 Evaluation of the height accuracy (z-error) of the ASTER and SRTM-3 DSMs compared to several reference models

ASTER (z-error)	Description	Source
±5.9 m	Comparison to DGM 50 (surveying and mapping agency of north-rhine Westphalia, Germany)	Bolten and Bubenzer [164]
±15.4 m (RMSE)	Comparison to 1: 50,000 scale DEM (DEM5) constructed from aerial photography	Wang et al. [67]
±10.87 m (RMSE)	Comparison to USGS national elevation dataset	Aster Validation Team [60]
±16.57 m (RMSE)	Comparison to 10 m resolution DEM by Japan's geographical survey institute for three ASTER tiles, mean estimated RMSE	
±9.35 m (RMSE)	Comparison to benchmark elevations from the U.S.	
SRTM-3 (z-error)		
±6.9 m	Comparison to DGM 50 (surveying and mapping agency of north-rhine Westphalia, Germany)	Bolten and Bubenzer [164]
±13.5 m	Comparison to 1: 50,000 scale DEM (DEM5) constructed from aerial photography	Wang et al. [67]
±7.10 m (RMSE)	Comparison to USGS national elevation dataset	Aster Validation Team [60]
±11.84 m (RMSE)	Comparison to 10 m resolution DEM by Japan's geographical survey institute for three ASTER tiles, mean estimated RMSE	
±4.12 m (RMSE)	Comparison to benchmark elevations from the U.S.	

Source Aster Validation Team [60], Bolten and Bubenzer [164], Wang et al. [67]

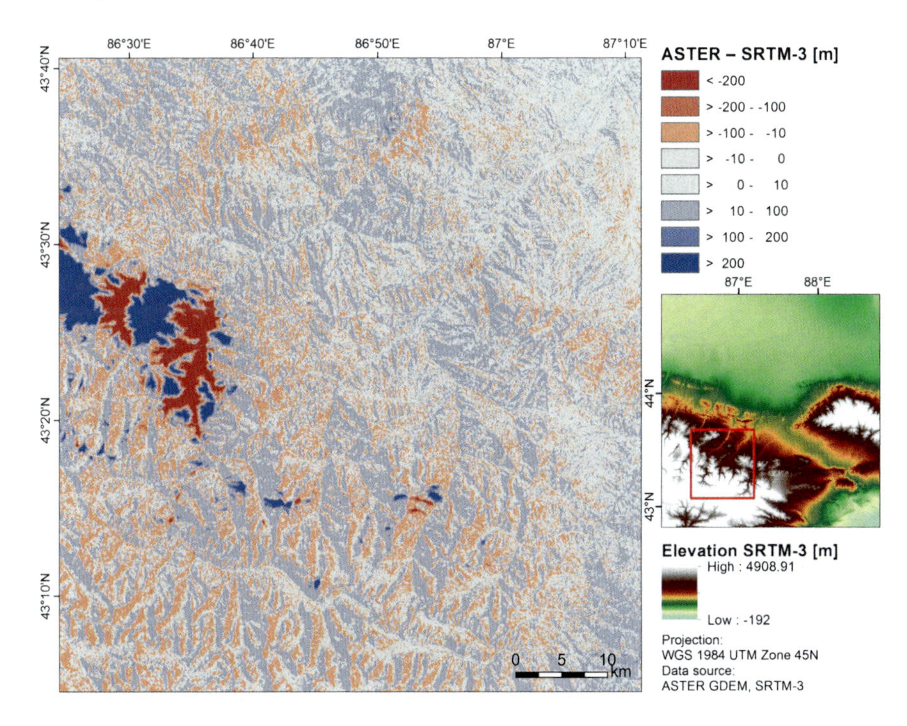

Fig. 3.9 Difference between ASTER GDEM and STRM-3 in mountain area. The high deviation values in the western part of the map but outside the research area are due to clouds in the SRTM-3 data, while the ASTER GDEM shows no such problems there (*Data source* [58, 59])

ASTER GDEM such as slope, especially in the mountainous regions (see Fig. 3.10). The ASTER Validation Team [60] does mention that the GDEM contains residual anomalies and artefacts, degrading its overall accuracy and effective utilisation. Apart from possible cloud residuals, which are not found for the research area, the artefacts possibly stem from the "algorithm used to generate the final GDEM from the variable number of individual ASTER DEMs available to contribute to the final elevation value for any given pixel" [60: 27]. These artefacts are then related to borders between scenes that are visible in the GDEM (cf. also [63]) and to the fact that "the magnitude of the associated elevation error can be relatively large" and the Version 1 of the ASTER GDEM should be viewed as "experimental" or "research grade" [60: 27]. The artefacts observed in this study were usually one to two pixels wide and ran over 10–30 pixels length. They were probably residuals of rectangular structures in the DEM, difficult to detect before the slope calculation. To eliminate these structures from the DEM, a 7×7 kernel mean filter was run over the GDEM twice to completely remove the artifacts and to achieve a smooth surface in the resulting slope data (Fig. 3.10). The results show clearly that the line structures can be eliminated but at the cost of reducing the effective resolution, the absolute range of elevation and slope and

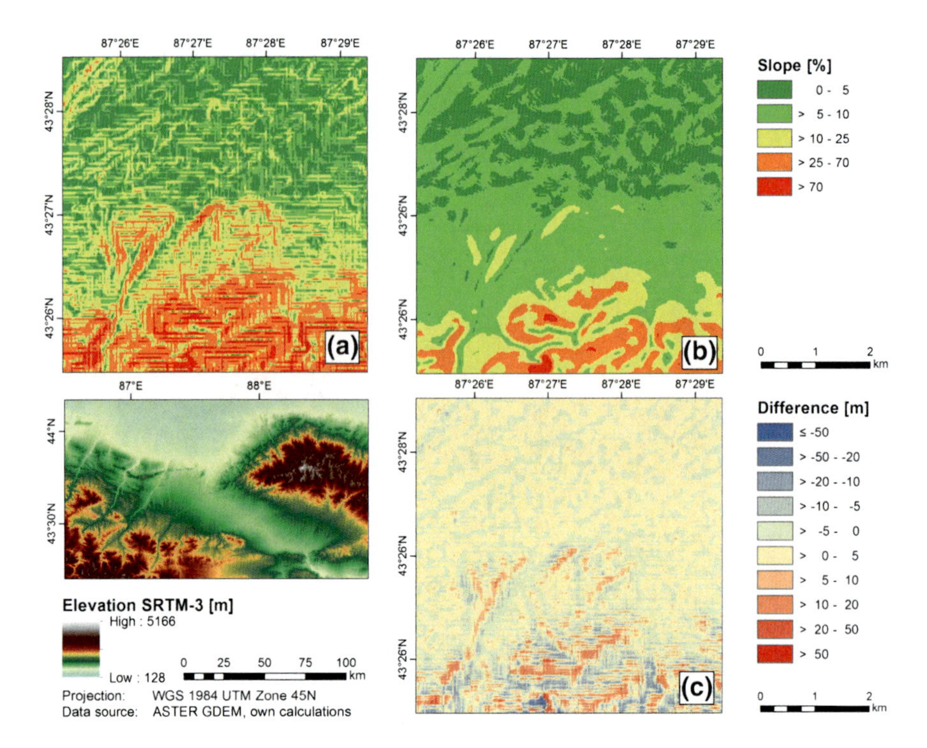

Fig. 3.10 Linear artefacts in the ASTER GDEM. **a** Example of slope values [%] calculated from ASTER GDEM in the catchment area of Urumqi River where the artefacts in the ASTER GDEM can be seen. **b** Slope derived from ASTER GDEM after calculating a 7 × 7 neighbourhood mean filter twice using the tool neighbourhood from ERDAS IMAGINE with sum function. **c** Difference between the original and the processed ASTER GDEM in m (*Data source* 59, own calculations and design)

detail of the DEM which can be seen in the difference between the original and the processed ASTER GDEM (see Fig. 3.10c).

3.3.1.2 Terrain Parameters

As has been shown in the previous section, not only the accuracy of the different DEMs is important, but also the effects of the DEM on derived terrain parameters, e.g. the first order derivatives slope and aspect [68, 69]. The interpolation algorithm used to generate the terrain derivatives is said to have less effect than the grid cell sizes and DEM sources on the variation of terrain variables [69]. Kienzle [69] compared the results for slope analysis using the Horn (1981) (*ArcGIS* standard algorithm) and the Zevenbergen and Thorne (1987) methods and found no significant differences in slope values.

Therefore, the next paragraphs focus on the effects of grid cell size on slope, other relevant terrain parameters and results of the hydrologic modelling. It is difficult to directly analyse the effects of input data on the results and several publications dealing with the sensitivity of hydrological models to DEM resolution will be presented in the following section.

Slope

Several authors show that slope steepness decreases with increasing cell size, as larger grid cell sizes do not represent steep slopes successfully [69] and cell sizes above 50 m lead to underestimation of slope values (cf. also [65, 68, 69]). The correct representation of steep terrains requires a cell size of 7.5 m and of flat terrains a cell size of 20 m [69]. This requisition is met by none of the available DEMs and would require data resampling. According to de Vente et al. [70], a SRTM-3 DEM with a resolution of 90 m provides more accurate estimates of slope gradient and upslope drainage area than the ASTER GDEM with a resolution of 30 m in 14 investigated Spanish catchments. However, the results for the effect of DEM resolution on predicted runoff in hydrological models, for example SWAT, are inconsistent. Some found out that an increase in runoff comes along with coarser DEM resolutions [71, 72], some found out that the predicted runoffs were not sensitive to DEM resolutions for monthly or annual runoff output [56, 73]. Di Luzio et al. [68] found that when modelling with SWAT it does not significantly influence most of the soil parameters relevant for runoff. The correct soil data base seemed to be of more importance for the correct estimation of surface runoff (see also [74]). Also the degradation of modelling efficiency of the model TOPMODEL is more sensitive to an increase in the time step than to an increase in grid size [75]. The SRTM-3 seemed also to be slightly superior to the ASTER GDEM with regard to accuracy and anomalies (see Sect. 3.3.1.1). Additionally, in the case of Urumqi Region, the slope calculated based on ASTER GDEM would either have artefacts or underestimate the slope value. Hence, the SRTM-3 was chosen as the basis for slope calculation.

Aspect, Stream Networks and Catchment Boundaries

Additional to the calculation of slope, the correct calculation of aspect is necessary for the subsequent derivation of hydrological variables such as flow direction and accumulation, stream networks and catchment boundaries or environmental variables such as solar radiation [69]. Di Luzio et al. [68] state that the correct delineation of watersheds and subwatersheds and a detailed soil map is necessary for precise and diversified simulations and management strategies, especially in smaller watersheds. In their study, the DEM with a resolution of 90 m affected the delineation results (watershed boundary and some geomorphological parameters) significantly and resulted in an erroneous calculation of the watershed area. Hengl and Reuter [63] give an average error of 60–100 m for the location of streams derived from ASTER GDEM and SRTM-3. This was the same case for the calculation of streams and watershed boundaries from the SRTM-3 DEM in the research area (see Fig. 3.11). The watershed boundaries and drainage lines were calculated with [76] to get a general overview over the catchments of the research area. SRTM derivatives showed incorrect stream lines when compared to maps from the Autonomous Bureau of Surveying And Mapping [77] and high resolution satellite imagery (*GoogleEarth*™) and quite different catchment areas than the ones derived from ASTER GDEM.

The differences were especially pronounced in the Dabancheng Corridor (see Fig. 3.11) between the mountain ridges, where the larger lakes and reservoirs are located, and for the alluvial plain to the North. In this flat terrain as described in the landscape classification (Sect. 2.1), very small changes in aspect are decisive for the calculation of flow direction and accumulation, as well as the watershed boundaries and drainage lines. According to Bubenzer and Bolten [78], the quality of DEMs may suffer in very smooth terrain or under problematic conditions, e.g. diffuse cloud cover, which is both the case for the above mentioned areas. The coarse resolution of the SRTM-3 DEM does not represent the underlying elevation accurately enough. Based on the comparison of catchment areas and drainage lines with actual rivers and streams mapped by the Autonomous Region Bureau of

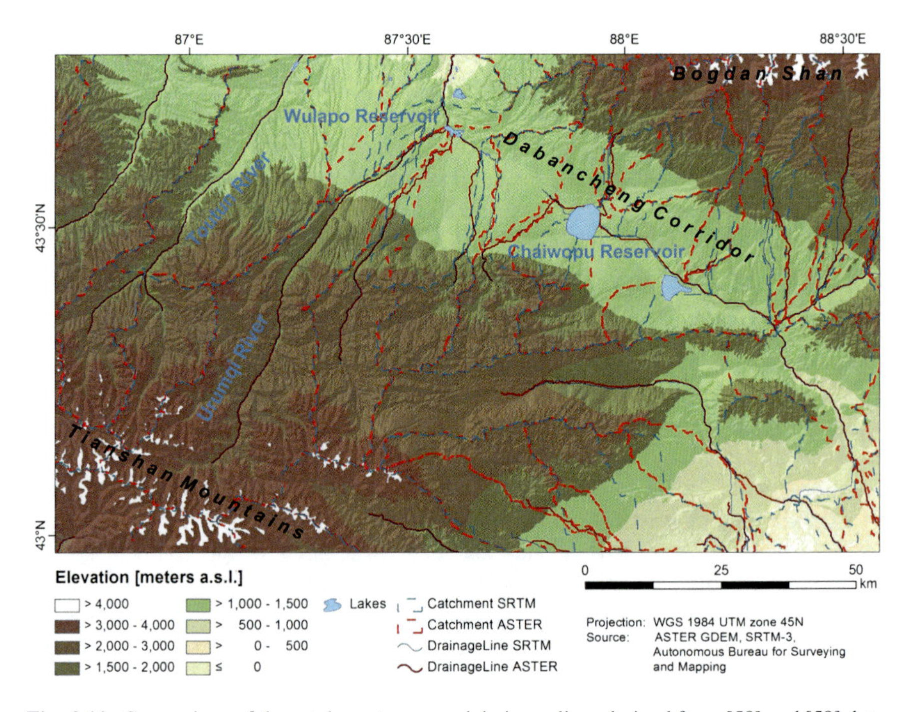

Fig. 3.11 Comparison of the catchment areas and drainage lines derived from [58] and [59] data. The *solid turquoise line* is representing the catchment area of Wulapo Reservoir

Surveying and Mapping the catchment areas based on ASTER data were chosen for further research and calculations. By combining elevation and slope from SRTM-3 and watershed and streams from ASTER GDEM, accuracy and practicability are both accounted for.

3.3.2 Climate Variables

The climate data used were downloaded from the U.S. National Environmental Satellite, Data, and Information Service (NESDIS) using the data set "Surface Data, Global Summary of the Day".[7] Near Urumqi Region, station data are available for Wulumuqi, Fukang, Shihezi and Diwopu (reservoir) station 79–82.

The station Wulumuqi at 87.65° N 43.80° E and 947 m a.s.l. is the only one located within the research area and offers data available from 1956 until today with a data gap 1965–1972. Mean, minimum, maximum, and dew point temperature, wind speed, station pressure, visibility, precipitation and snow depth could be downloaded on a daily time step.

Normally, data used for analysing climate change and simulating a hydrological system should be consistent, and homogeneous where variations are caused only by variations in weather and climate (cf. [83, 84]). Therefore, the climate data used as input for the hydrological modelling was tested for relative consistency and homogeneity with a screening procedure according to Dahmen and Hall [84]. The procedure included a rough screening, plotting monthly and annual totals to note any trends and discontinuities, double-mass analysis and analysis of proportionality factors compared to other stations [84].

After the rough screening of the original daily data sets for Wulumuqi station, only the years 1973–2010 were chosen for further testing as prior data showed large data gaps. To test the homogeneity and consistency of the climate data, the data from Wulumuqi station was compared with observations from two other stations less than 100 km away, Diwopu and Fukang station (Fig. 3.12). The climate data for these stations was only available for the time period 1973–1977 and 1979–1988 for Fukang station and 1990–2009 for Diwopu station. Data from other, similar stations was not available and stations with consistent time series were located only in considerable distance to Wulumuqi station (see also [85]). The visual comparison of mean annual temperature did not show any obvious steps or changes in the time series. The comparison of the annual precipitation sums and wind speed showed inconsistencies, but could not be further investigated because the data base was too small. Station pressure could not be compared as the other station did not measure this variable.

[7] The data is available for over 9,000 stations, including many airport and additional city locations worldwide, and provides daily summaries for mean, min, max, and dewpoint temperature, wind speed, pressure, visibility, precipitation and snowdepth (http://hurricane.ncdc.noaa.gov/cdo/info.html#GSOD, accessed 08.09.2011).

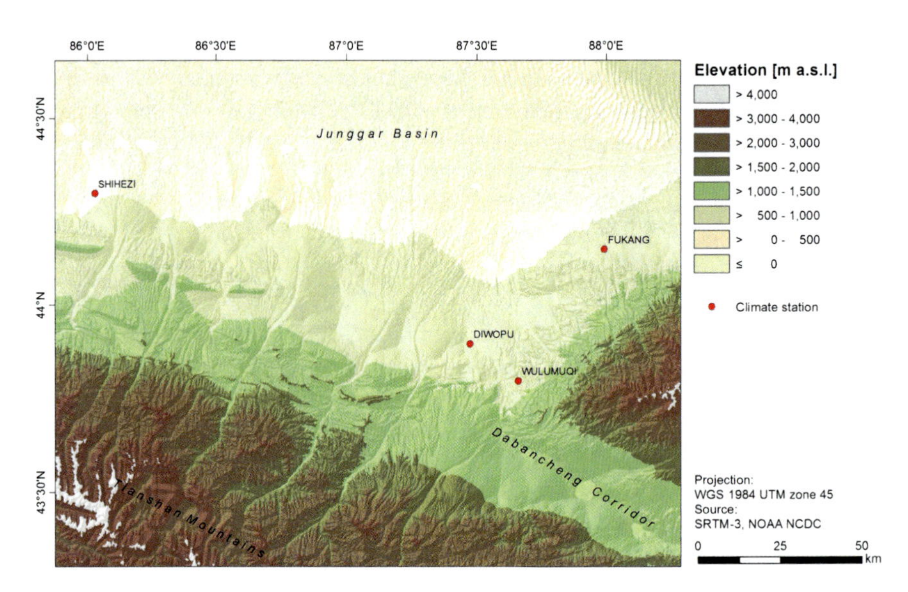

Fig. 3.12 Location of the climate stations Diwopu, Fukang, Shihezi and Wulumuqi around Urumqi City available from NOAA national climatic data centre (*Data source* 58, [72–82])

The **relative consistency and homogeneity or relative trends were tested for the time series** of the above named three stations with double-mass analysis and analysis of proportionality factors on the annual mean air temperature and precipitation sums. The data from Fukang station showed a slightly increasing proportionality factor for mean annual temperature in the compared time period, while the proportionality factor for Diwopu station was decreasing. These relative temperature changes were not assessed as data inconsistencies, but as the effects of climate change and its regional effects on temperature. As already mentioned in Sect. 2.4.1, the effect of climate change and global warming on the temperatures measured at climate stations in Xinjiang differ depending on the locality and situation of the station. For Wulumuqi station, a siginificant trend could be found for the last 36 years, but the degree of increase might be different at other stations. No gross errors were detected; therefore the data was aggregated to monthly and annual totals and plotted (see annual mean air temperature and precipitation sums in Fig. 3.13).

The spatial climate data is one of the key drivers for the water balance model, but only data from one station is available for the research area. Spatial climate patterns are affected by terrain and water bodies through elevation effects, terrain-induced climate transitions, cold air drainage, inversions and coastal effects [86]. From large scale temperature and precipitation distribution maps it can be observed that the climate parameters strongly change in the transition zone from the basins to the mountain regions. The Tianshan Mountains and the Dabancheng Corridor are significant terrain features that induce precipitation and temperature patterns as described in Chap. 2, which cannot be modelled just accounting for

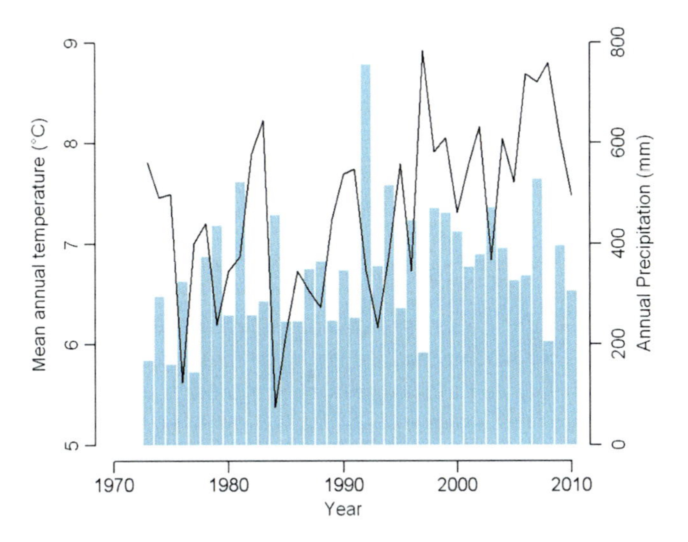

Fig. 3.13 Mean annual temperature and annual precipitation sums 1973–2010 at Wulumuqi station (*Data source* [82])

Table 3.15 Mean annual climate statistics at Wulumuqi station 1973–2010

	Mean temperature (°C)	Dew point temperature (°C)	Station pressure (kPa)	Wind speed (m/s)	Maximum temperature (°C)	Minimum temperature (°C)	Precipitation (mm)
Mean	7.30	–2.53	91.33	2.39	12.05	2.43	341.24
Std	0.92	0.81	0.13	0.40	1.87	1.00	130.50

Data source [82]

elevation effects. The station spacing does also not offer the possibility to interpolate between stations.

As temperature and precipitation are the main climate variables driving the hydrological system as explained in the beginning of this section [57], their spatial variability has been taken into account by extrapolating their variation with remote sensing products as described in the next two sections. Also, the spatial variation of solar radiation was estimated based on the elevation data acquired for the modelling. Other climate variables (wind speed and station pressure as required by the Penman-Monteith equation for the calculation of evapotranspiration) were used in the hydrological model with their climate station measurements and were not extrapolated (Table 3.15).

3.3.2.1 Temperature

Temperature is an important variable for calculating evapotranspiration and for the water balance model as analysed in the sensitivity analysis (see Sect. 3.2.1). However, it is strongly controlled by topographic attributes, longitude, latitude and distance from the coast. Temperature varies both in space and time, and is generally directly affected by elevation [19]. In particular in Urumqi Region, the temperature decrease with elevation should be significant as the altitude ranges from under 400 to over 4,000 m a.s.l. Towards the centre of the basin, increasing temperatures are expected due to a heat island in the middle of the basin. As the distribution of available climate stations and measured data sets are insufficient to appropriately interpolate and to represent the temperature differences, the distribution of temperature values were extrapolated on the basis of remote sensing data acquired by the Moderate-resolution Imaging Spectroradiometer (MODIS).

The data product used is MOD11A1, a MODIS/Terra Version 005 Land Surface Temperature and Emissivity (LST/E) product providing per pixel temperature tile-based and gridded in the Sinusoidal projection, and produced daily at 1 km spatial resolution. MODIS uses the generalised split-window LST algorithm and validates the product through field campaigns and radiance-based validation studies [87]. The data sets were selected and downloaded from U.S. Geological Survey Earth Resources Observation and Science Center.[8] For every month, a representative and preferably complete MOD11A1 data set was chosen. During winter season, the coverage was low and only few data sets available. Hence, the months October till December were represented by one raster data set and the months January till March by another (Table A.2).

From the available science data set layers, daytime land surface temperature was chosen. Land surface temperature is not equivalent to air temperature and shows a delay in temperature changes as the thermal conductivity of the land surface is much lower than in the air. Additionally, the temperature distribution might also vary from day to night which is not represented by the daytime data set. However, daytime temperatures have the largest influence on ET_p and were chosen as the representative distribution when the MOD11A1 raster data sets were added into *ArcGIS* 10 [76]. As most of the raster data sets showed data gaps over the research area, they were filled using the tool *Focal statistics* which uses the values within a specific neighbourhood around it to calculate the value for each input cell location. The data sets were clipped to the extent of the research area and the gaps were filled based only on the closest neighbours within a 3×3 rectangle employing a standalone *Python* script.

The unit of the MOD11A1 data set is Kelvin with a scale factor of 0.02 to directly calculate daytime LST [87]. The calculated values were selectively tested against climate station data, but showed large differences as the MOD11A1 represents daytime land surface temperature and the climate station data set provided

[8] http://glovis.usgs.gov/, accessed 4.10.2009.

daily mean, minimum or maximum air temperature. Therefore, the MOD11A1 data set was used as a spatial gradient to extrapolate the measured station data according to the following equation:

$$\frac{T_K}{T_{st,K}} = \frac{LST}{LST_{st}} \quad (3.43)$$

$$T_K = T_{st,K} * \frac{LST}{LST_{st}}, \quad (3.44)$$

with the extrapolated temperature T_K [K], the temperature measured at climate station $T_{st,K}$ [K], land surface temperature from MOD11A1 raster LST [K], and land surface temperature from MOD11A1 raster at climate station LST_{st} [K]. The extrapolated temperature T_K [K] can be converted to T [°C] using the equation:

$$T_K = T + 273.15, \quad (3.45)$$

which also applies to the conversion of the temperature measured at climate station $T_{st,K}$ [K] and T_{st} [°C]:

$$T_{st,K} = T_{st} + 273.15. \quad (3.46)$$

Equation (3.46) is then solved for extrapolated temperature T measured in °C:

$$T = \left((T_{st} + 273.15) * \frac{LST}{LST_{st}} \right) - 273.15 \quad (3.47)$$

and used within the hydrological model to extrapolate the distributed temperature for the research area.

3.3.2.2 Precipitation

Beven and Hornberger (1982) stated that the correct assessment of the rainfall volume (in a highly spatial variable pattern) is more important than the pattern itself for simulating streamflow hydrographs (cited in Ajami et al. [88]). For the input precipitation, the data from Wulumuqi station was compared to the climate data published in the Urumqi Statistical Yearbooks [89–94] and an isohyets map [77]. It can be assumed to be a fairly reliable reference for Urumqi City. However, it was not representative for the mountain areas where almost twice the amount of precipitation can be expected. In mountainous regions, the amount of precipitation can vary widely between elevation bands and depends on the main direction of the incoming weather, advective processes as well as the exposition of the surface. Other studies have used the DEM to interpolate meteorological data to obtain the spatial distribution of parameters (cf. PRISM used by Zhang et al. [95]). However, this method needs to be calibrated using the relation between at least two stations located at different altitudes. Unfortunately, no representative stations were available for this research. There are interpolated station data available (e.g.

Table 3.16 Technical specifications for the used remote sensing data sets MODIS and TRMM

	MODIS	TRMM (TMI)
Orbit	705 km	403 km
Repetition rate	1–2 days	92.5 min
Swath width	2,330 km (cross track) × 10 km (along track at nadir)	878 km
Temporal resolution	Daily, 8-day, 16-day, monthly, quarterly, yearly	3-h, daily, monthly
Spatial resolution	250 m (band 1–2), 500 m (bands 3–7), 1,000 m (bands 8–36)	0.25° × 0.25°, 0.5° × 0.5°, 5° × 5°
Product	MOD11A1	TRMM 3B43
Specification	Daytime land surface temperature (K)	Precipitation rate (mm h^{-1})
Temporal resolution	Daily	Monthly accumulated
Spatial resolution	Ca. 1,000 m × 1,000 m	0.25° × 0.25°
Scene size	~1,100 km × 1,100 km	See spatial coverage
Spatial coverage	Global	50° S–50° N (prior to February 2000 40° S–40° N)

Source U.S. Geological Survey [82, 165], NASA [166]

WorldClim[9]), but for Xinjiang the station density is very low and the interpolated product would have high inaccuracies.

To retrieve information about the different precipitation volumes over the area and the precipitation distribution, TRMM (Tropical Rainfall Measuring Mission) Microwave Imager (TMI) data was chosen to complement the station data [96]. TRMM is a joint U.S.-Japan satellite mission to monitor tropical and subtropical precipitation from 50° South to 50° North. The TRMM TMI is a passive microwave sensor quantifying water vapour, cloud water, and rainfall intensity in the atmosphere. Its product 3B43 provides the best-estimate precipitation rate [mm h^{-1}] and root-mean-square (RMS) precipitation error estimates at the surface combining the 3-hourly merged high-quality/IR estimates with the monthly accumulated Climate Assessment and Monitoring System (CAMS) or Global Precipitation Climatology Centre (GPCC) rain gauge analysis (3A-45[10]). The data was accessed and downloaded via the Goddard Earth Sciences Data and Information Center Distributed Active Archive Centre (GES DISC DAAC[11]) (Table 3.16).

The temporal resolution is one calendar month, matching the modelling time step. The spatial resolution of the gridded data is 0.25° × 0.25°, which is considerably larger than the resolution of the other data used. Therefore, the grid was interpolated to a higher resolution in order to match the resolution of the Landsat

[9] http://www.worldclim.org/, accessed 24.05.2010.

[10] http://mirador.gsfc.nasa.gov/collections/TRMM_3B43__006.shtml, accessed 08.09.2011.

[11] http://mirador.gsfc.nasa.gov/cgi-bin/mirador/presentNavigation.pl?project=TRMM&tree=project, accessed 24.05.2010.

TM images also used for the model and avoid jumps in the spatial distribution of the model results. A semivariogram model and kriging interpolation was chosen as interpolation method for precipitation due to its statistical quality of predictions and its advantages in terms of local geometry and smoothness over local neighbourhood interpolation approaches such as inverse distance weighted interpolation, natural neighbor interpolation and interpolation based on a triangulated irregular network.

Kriging is the first choice for interpolation of geospatial data as it considers the spatial relationships between data points. Splining would also provide a smooth and exact surface and be easily applied (cf. [97, 98]). However, the standard splining procedure implemented in *ArcGIS* could also provide negative values for the interpolation and is therefore not applicable for precipitation data. Kriging is based on a concept of random functions, which represent together with their covariance the surface to be interpolated. It is a stochastic method, using statistical criteria (in this case a semivariogram, "a measure of spatial correlation between pairs of points describing the variance over a distance") to determine weight factors [97: 5]. It has been used for example to interpolate groundwater levels [99], geological data [98] or climate data [97]. Reviews of interpolation techniques also showed that kriging produced the best quantitative and accurate results [35, 97, 100].

Before the data processing with *ArcGIS* 10, several kriging methods and variations using the *Geostatistical Analyst* were tested. They included different methods for trend removal and transformation type. However, none of the combinations proved to be overall superior to the automatic optimisation of the semivariogram when comparing the standardised mean nearest to zero, the smallest root-mean-square prediction error, the average prediction standard error nearest the root-mean-square prediction error, and the standardised root-mean-square prediction error nearest to one in the *Cross Validation Comparison* [101]. Hence, the automatic optimisation was chosen to enable an automatic processing of the data including conversion in point files, clipping to the research area and optimising semivariogramm kriging using an *ArcPython* script. The nugget and sill of model were automatically optimised to the average values of the semivariogram calculated from the available input points. For the distances between pairs of points where the values/semivariogram did not exist, the optimisation process fitted the curve as calculated for the existing values. The files are finally converted to an ESRI Grid with the resolution of $0.0003° \times 0.0003°$, which corresponds to about 33.3 m $\times$ 24.1 m. Using the exact resolution of the Landsat TM images was not possible due to the transformation of the raster files from geographic to projected coordinate system.

The evaluation of TRMM rainfall estimates compared to gauge measurement in Iran showed that the TRMM product 3B42 (basis for the calculation of the 3B43 product as explained above) underestimated mean annual precipitation by 0.15–0.39 mm day^{-1} for different regions. The mean annual rainfall spatial correlation coefficients are 0.57–0.77, respectively [102]. Approximately the same differences were observed for the precipitation values for Urumqi Region. When compared with the Wulumuqi station data, the TRMM rainfall estimates underestimated the monthly precipitation by the factor 0.69–0.71 on average. Therefore, the TRMM raster data sets were calibrated using the monthly precipitation values at Wulumuqi station similar to the extrapolation of temperature values.

In this case, we assume that the relative difference between the precipitation and TRMM values at the station and distributed precipitation and TRMM values is proportional:

$$\frac{\Delta P}{P_{st}} = \frac{\Delta TRMM}{TRMM_{st}} \tag{3.48}$$

with ΔP, the difference between precipitation measured at climate station P_{st} [mm] and distributed precipitation P [mm]:

$$\Delta P = P - P_{st}, \tag{3.49}$$

and $\Delta TRMM$, the difference between $TRMM_{st}$, the $TRMM$ value of the climate station pixel [mm h^{-1}] and $TRMM$, the distributed $TRMM$ values [mm h^{-1}]:

$$\Delta TRMM = TRMM - TRMM_{st}, \tag{3.50}$$

As the $TRMM$ raster is non-negative, the absolute value of a negative difference $\Delta TRMM$ can never be larger than $TRMM_{st}$. Thus, the absolute value of a negative ΔP can never be larger than P_{st} and P is always larger or equal to 0 as already required for the interpolation. Equations (3.49) and (50) are inserted in [49] and the equation is solved for P:

$$\frac{P - P_{st}}{P_{st}} = \frac{TRMM - TRMM_{st}}{TRMM_{st}} \tag{3.51}$$

$$P - P_{st} = \left(\frac{TRMM - TRMM_{st}}{TRMM_{st}}\right) * P_{st} \tag{3.52}$$

$$P = \left(\left(\frac{TRMM - TRMM_{st}}{TRMM_{st}}\right) * P_{st}\right) + P_{st}. \tag{3.53}$$

According to Eq. (3.53), monthly raster data sets are calculated based on the TRMM data sets and the according station values for measured precipitation and TRMM pixel value in *ArcGIS* using a *Python* script.

3.3.2.3 Solar Radiation and Other Climate Variables

Solar radiation plays an important role in the processes of the hydrological cycle, especially for evapotranspiration through the transport of sensible heat. At a global scale, the amount of incoming solar radiation at the top of the atmosphere is determined by latitudinal gradient effects, the orbital parameters of Earth and its sphericity. At the smaller scale, apart from atmospheric influences, the topography (slope angle and orientation, shadows cast by topographic agents) influences the distribution of incoming solar radiation at the land surface (cf. [103]). Aguilar et al. [103] used a digital elevation model (DEM) with a horizontal resolution of 30 m × 30 m and 1 m of vertical precision to evaluate the topographic effects on solar radiation fields on reference evapotranspiration estimates. They compared the estimation results with field measurements and with reference potential evapotranspiration calculated with values obtained through classical interpolation techniques. The standard deviation was higher with topographic solar radiation values, but the interpolated solar radiation fields led to a general overestimation of evapotranspiration, especially "in summer, when the importance of ET_0 for the water balance is greater" [103: 2491]. As the catchment area of the Urumqi River is located in the Tianshan Mountains with strong vertical gradients and steep mountain slopes, topographic effects on incoming solar radiation are to be expected.

Station data for solar radiation was not available at the climatological station and the distributed solar radiation R was calculated using the *ArcGIS* tool *Area solar radiation*. The tool uses a DEM, the time of the year and latitude for the calculation of incoming solar radiation R [Wh m^{-2}] and day duration N [h]. The SRTM-3 DEM was used as input for the calculation to avoid artefacts in the spatially distributed radiation data (see Sect. 3.3.1). The maximum possible duration of sunshine per day N [h] was also calculated based on the SRTM-3 DEM while the actual duration of sunshine n was taken from the Urumqi Statistical Yearbooks 2000–2010 [104].

The overall incoming solar radiation and day duration was calculated for every month, divided by the number of days to obtain the average daily radiation R_0 and converted into the units demanded by the evapotranspiration equation [MJ m^{-2} day^{-1}]:

$$R_0 = \frac{R}{d_m} * 0.0036, \tag{3.54}$$

with the solar radiation R_0 top-of-the-atmosphere [MJ m^{-2} day^{-1}], R as the mean daily incoming solar radiation per month [Wh m^{-2}] and d_m as the number of days per month. The factor 0.0036 is necessary to convert from Wh to MJ. Because of the latitude and steep slopes, areas in the high mountains are calculated to receive little to none direct sunshine during the winter months.

3.3.3 Land Use and Land Cover

Most processes of the hydrological cycle such as infiltration, erosion, and evapotranspiration depend on land use and surface characteristics, including land use and vegetation [105]. Hence in this section, the land use and cover classification conducted to retrieve information about the land surface characteristics and the most important relationships will be discussed. The albedo of the land surface influences the surface heat balance and thereby the potential evapotranspiration. Furthermore, the resistance parameters of vegetation type and surface cover set constraints for the actual evapotranspiration besides the available water. In addition, albedo, snow and land cover type control the degree and melting rate during snow melt processes. Vegetation, soil type, permeability and roughness of the surface influence rates and limits of infiltration and runoff. For example, a decrease of vegetation cover causes reduced infiltration and increased runoff, but also the amount of soil and sediment that can be moved and eroded depends on the land cover (Hunsaker and Levine 1995 in [105]). The relationship between vegetation and the amount and timing of runoff is especially sensitive in semi-arid regions. For a temperature of 10 °C, a curve of precipitation versus runoff and sediment yield reaches its peak at 300 mm, because at lower precipitation values the runoff total is equally smaller and at higher precipitation values the vegetation cover offers better protection from erosion (Langbein and Schumm 1958 in Burns et al. [105]).

Due to the model concept, several parameters steering the hydrological processes are derived from the characteristics of land use and land cover (LULC). Preferably, the information about LULC should also represent the seasonal changes of vegetation and other land covers (e.g. seasonally changing characteristics of snow, water bodies or soil). A reliable local dataset was not available for the research area at the required scale and there are also few remote sensing products which offer data and information about the land and vegetation cover that can be used directly by hydrological models at sufficient resolution. Global Land Cover Characterisation (GLCC) or NDVI Composites for example are only available at a resolution of 1 km as they are derived from AVHRR data. Therefore, land use classifications of Landsat data with a resolution of 30 m × 30 m were conducted to deduce characteristics from land use classes and substitute necessary data for the modelling. However, all classifications and biophysics parameters retrieved from remote sensing are subject to uncertainties in the parameters and assumptions of the used model and references including conceptual differences [106]. Nonetheless, the classification of remote sensing images offered the necessary information about land use and cover at the required resolution for the hydrological model. The selection, preprocessing, and classification process as well as the classification accuracy and retrieved parameters are described in the next paragraphs.

3.3.3.1 Available Remote Sensing Data and Its Characteristics

Due to extent of the research area, the processing capacities and the data availability, images from the Landsat 7 mission [107] were chosen as basis for the LULC classification (see Table 3.17). They were available over a long time period, with frequent images and low costs, adequate cell size and scene distribution. Only two scenes are necessary to cover the total research area which reduces the

Table 3.17 Technical specifications for the Landsat 7 mission

Name	Landsat 7	
Sensor	Enhanced thematic mapper plus (ETM+, before scan line corrector (SLC) failure May 31, 2003)	
Altitude and orbit	705 km, polar orbit, sun-synchronous	
Repeat coverage	16 days	
Scene size	183 km × 170 km	
Swath	183 km	
Band number	*Wavelength* (μm)	*Resolution* (m)
1	0.45–0.515	30
2	0.525–0.605	30
3	0.63–0.69	30
4	0.75–0.90	30
5	1.55–1.75	30
6	10.4–12.5	60
7	2.09–2.35	30
8	0.52–0.9	15

Source NASA [167]

need of combining or mosaicing the respective scenes. The research area of Urumqi Region is covered by two scenes (path 143 row 29 and 30), which include the headwaters and catchment areas of the two most important rivers, Urumqi River and Toutun River. The eastern parts of the administrative area are cut off and are not included in the hydrological model, as they drain through the Dabancheng Corridor into the Turpan Depression. The pixel size of 30 m × 30 m is equivalent to the resolution of the digital elevation model (ASTER GDEM) used for the calculation of the catchment areas and are sufficient for a research area of over 7,000 km^2. A resolution merge could be calculated with the high resolution band included in the Landsat ETM+ images to improve the resolution of the satellite images. However, the resolution merge proved to enlarge the size of the satellite images and the subsequently calculated layer-stack to a degree that caused unreasonably long computing time. Hence, the resolution of 30 m × 30 m was kept for further calculations. To achieve a good representation of the seasonal changes of the land cover, several images taken at different dates from autumn 1999 to winter 2000/2001 have been selected for the classification and downloaded: winter with snow cover (December 1999/January 2000, January 2001), at the beginning, in the middle and at the end of the vegetation season (October 1999, March, June and September 2000) (see Table A.2 in the appendix).

This way, growth cycles and irrigation practices could be taken into account. This was also necessary in order to cover the hydrological year from November to October and the following winter months to allow the computation of the hydrological processes of the complete calendaric year January to December. Based on data availability and quality (mainly cloud cover and sensor errors) the dates presented in Table A.2 were chosen for classification of preliminary land use and cover classification.

3.3.3.2 Preprocessing

Before the classification, the satellite images had to be preprocessed. Sensor accuracies at the fringes were subset with the program [108] to avoid corruption of the histogram values in these areas and to enable matching of the two scenes. Then a layer stack was build of Landsat bands 1–7 before addressing possible aberrations of the images. These include radiometric, topographic and atmospheric corrections, which have been researched and discussed in a vast number of papers. Here, only standard references, considered and necessary procedures will be mentioned.

As a first step, the raw sensor data is subject to a geometric correction to remove errors of perspective due to the Earth's curvature and sensor motion [61]. The necessary data is stored in the sensor metadata and commonly processed at the sensor's data processing center. The geometric accuracy of the retrieved scenes was compared with each other, a limited number of ground control points with an accuracy of 5–20 m taken during field trips in Urumqi Region and characteristic land marks in *GoogleEarth™*. Among each other, the Landsat images were perfectly congruent. The error compared to the reference points and landmarks was less than one pixel of the remote sensing data as far as the reference points and marks could be identified and therefore considered acceptable.

The intensity of electromagnetic radiation from the earth' surface is recorded by the sensor as digital numbers (DNs) for each spectral band and calibrated on board of the satellite. DNs can be converted to absolute units of at-sensor spectral radiance when comparing scenes taken by different sensors or for standardised comparison of objects [109]. This was not necessary as images from the same sensor were used, and not intended or possible as the spectral characteristics of the land use classes changed over the year and were derived from training areas separately. In a region with mountains or large topographic differences such as the Tianshan and adjacent basins, topographic effects appear on the satellite images, for example shaded areas of mountain ridges due to the solar incident angle or deep valleys that are covered by their slopes due to the sensor angle. Some of these effects can be reduced by software based *topographic corrections* using a DEM which is also provided by *ERDAS IMAGINE* or other programs such as *ATCOR3*. In the scenes covering Urumqi Region, atmospheric effects could also be observed, hazing the recorded radiation and reflection intensities. These occur when incoming and outgoing radiation is absorbed or scattered by components of the atmosphere [110]. The influence of the atmospheric scattering is inversely proportional to the wave length, and therefore influencing each band of the images to a different extend. Possible correction algorithms include dark pixel/object subtraction, linear regression, empirical line calibration, relative calibration/histogram matching and atmospheric/physical modelling (cf. [110–116]). The dark pixel subtraction and invariant objects method were not possible as the chosen images for the classification did not provide reliable and sufficient dark or invariant objects. Knowledge of ground truth reflectance for a reflectance conversion or empirical line calibration was also not available; hence the images could not be calibrated along these characteristics (cf. [117]). The overlap region for a relative calibration via histogram matching proved to be too small and not representative for the remaining scene as the differences in reflectance and atmospheric disturbance over the mountainous areas and the semi-arid basin seemed to be too large, also preventing the use of the linear regression method.

Therefore, a physical atmospheric model was chosen for the corrections. *ATCOR3* has been used by several authors (a.o. [118, 119]) for vegetation and land use studies and was chosen due to its availability (add-on module for *ERDAS IMAGINE*) and user friendliness. Atmospheric modelling is computationally intensive and requires atmospheric parameters as input such as atmospheric profile, aerosol type, solar zenith angle, and sensor viewing angle. The necessary inputs where taken from the metadata or estimated and calibrated in the program itself (cf. description by Neubert and Meinel [120]). Simultaneously, the topographic correction based on the SRTM-3 DEM was computed by ATCOR3. After the corrections, a mosaic was calculated with the resulting radiances for the different classification dates using the mosaic tool and histogram matching for the overlapping areas from *ERDAS IMAGINE*. In some of the images, cloud areas remained after the atmospheric correction which had to be removed. The high levels of reflectance of cloud tops across the wavelength range 0.4–3.0 μm and in the thermal-infrared can lead to spectral confusion with snow areas (cf. Belward [121]). They were visually identified, manually digitised and masked with an area of interest (AOI) and then substracted as areas of noninterest (AON) using the *ERDAS IMAGINE* Subset tool.

3.3.3.3 Classification

In this section, the steps and methods for the classification of the Landsat TM images are described: band combination and indices used, integration of slope and elevation, supervised classification and accuracy assessment. The aim of the classification was to define the LULC classes in Urumqi Region as detailed as possible with regard to their different effects on the hydrological processes (albedo, resistance height, stomatal resistance, degree-day-factor, and proportion of runoff), based on the ability to differentiate them. The derived classes were agriculture, dry agriculture, partly vegetated areas, grassland, soil, coniferous forest, alpine grass, rock, sealed area, ice and snow, old snow, irrigated areas, water and dark surfaces (unclassified urban areas and open-cast mining). It proved to be not possible to accurately differentiate further LULC classes (e.g. urban areas with different uses or different crop types), therefore they were summarised as one land surface class and average parameter values were assigned accordingly.

For the classification of the Landsat images, the Landsat bands 1–7, the normalised difference vegetation index (NDVI), the soil water content index (SWCI) (cf. [122]), the principal component analysis (PCA) 1 and 2, the Tasseled Cap (TC) bands 1 (brightness, index for soil) and 2 (greenness, index for vegetation) (cf. [123]) as well as slope and elevation (from SRTM-3 ca. 90 m $\times$ 90 m) were

Table 3.18 Classification bands joined in the layer stack for the selection of training sites and the classification

Name	Source	Description
L1	Landsat ETM+	Visible band, blue
L2	Landsat ETM+	Visible band, green
L3	Landsat ETM+	Visible band, red
L4	Landsat ETM+	Visible band, near-infrared
L5	Landsat ETM+	Visible band, mid-infrared
L6	Landsat ETM+	Thermal infrared
L7	Landsat ETM+	Visible band, mid-infrared
NDVI	Index calculated with Landsat ETM+	Vegetation index
SWCI	Index calculated with Landsat ETM+	Soil water content index
PC 1	Transformation calculated with Landsat ETM+	First principal component
PC 2	Transformation calculated with Landsat ETM+	Second principal component
TC 1	Transformation calculated with Landsat ETM+	Tasseled cap transformation, brightness component
TC 2	Transformation calculated with Landsat ETM+	Tasseled cap transformation, greenness component
Elevation	SRTM-3	Elevation
Slope	Derivate from SRTM-3	Slope

used. The additional bands apart from the usual Landsat bands were necessary to differentiate between vegetation and soil types as well as to assess the different types of land cover in the mountains and lowlands which had very similar spectral characteristics. The indices PCA and TC were calculated with *ERDAS IMAGINE* tools to enhance the spectral characteristics and to increase the separability of the classes. The NDVI is one of the most commonly used vegetation descriptor (cf. [124–126]) and defined as $(R_{nir}-R_{red})/(R_{nir}+R_{red})$, where R_{nir} and R_{red} are red and near-infrared reflectances obtained from remote sensing (cf. [62, 122, 127]). For the Landsat ETM+ images, they are represented by band 3 and 4, respectively (see Tables 3.17 and 3.18). Additionally, the SWCI describing the soil water content was calculated as $(R_{nir}-R_{SWIR})/(R_{nir}+R_{SWIR})$ with short-wavelength infrared reflectances R_{SWIR} (Landsat ETM+ band 7) [122]. The PCA was conducted to compress information in multispectral data sets and combine the most important spectral characteristics of the bands [62]. As part of the bands to be evaluated, the brightness component and the greenness component were used to enhance the contrast between bright-degraded soil and vegetated areas (similarly implemented by Karnieli et al. [128]).

The slope values from SRTM-3 have already been calculated for the landscape classification (see Chap. 2). Together with the elevation data, they are used to differentiate vegetated agricultural fields from densely vegetated slopes and alpine grasslands. The reasons for calculating slope from SRTM-3 have been discussed before, but here the SRTM-3 DEM was also used for elevation information. The

Table 3.19 Classification classes, land use, land cover and other characteristics

Class	Description and special characteristics
Agriculture	Agricultural cultivation, covered with crops, fully vegetated (NDVI >0.5)
Agriculture dry	Agricultural cultivation, covered with crops, but more sparsely vegetation than 'agriculture' (NDVI 0.35–0.5)
Sparse vegetation	Sparsely vegetated (NDVI 0.2–0.35)
Grassland	Alpine grasslands in the mountains (NDVI >0.35)
Soil	Uncovered soil and barren land (NDVI <0.2)
Coniferous forest	Mostly coniferous forest areas in the mountains, darker than other vegetated areas
Rock	Unvegetated, impermeable areas (NDVI <0.1)
Sealed	Sealed urban areas and transportation (NDVI <0.1)
Ice	Ice (visible only in summer)
Snow	Snow covered areas
Snow old	Snow covered areas, darker and lower albedo values
Irrigation	Water with vegetation characteristics (NDVI >0)
Water	Water surface (NDVI close to –1)
Dark	Open-cast mines and other open, dark areas

resolution of SRTM-3 is adequate for the size of the investigated catchment area and other possible regions for predictions, and keeps the amount of used data at a practicable size. The supervised classification was conducted using the *Signature Editor* in the program *ERDAS IMAGINE*. The spectral characteristics (signatures) of 10–15 training sites for every class were assessed and grouped to define the class limits in each band. As suggested by Congalton and Green [129], the sample units were cluster of pixel groups created with the *Region growth tool* or polygons.

Training sites with known land use and cover class were taken mainly from the previous land use mapping and classification conducted by Fricke [130] or could be identified based on their visual call (*GoogleEarth*^{TM} and the Landsat bands) and the information of the additional bands (see Table 3.18 and 3.19) to maximize the number of 'ground truth' points [129]. The classification was mutually exclusive and totally exhaustive, i.e. every part of the research area was included into the classification and assigned to one class (cf. [131]). The decision boundaries could not be defined including all possible pixel values due to the limited number and variety of training sites. To ensure the classification of all pixels, the *maximum likelihood* decision rule was chosen as *parametric rule*, which assumes a normal distribution of the training data statistics and assigns the classes based on probability [132]. After each test run, the classification of the training sites was reviewed and the test sites were adapted accordingly until a satisfactory result was reached. This procedure proved to produce the best classification results.

No post-classification sorting or other methods were applied as additional information such as slope, elevation or band ratios were already included in the classification process (Fig. 3.14). For the *accuracy assessment*, 20 random test points per class were used as reference data and the classification tested against them. The *accuracy assessment* was done with *equalised randompoints* to assess

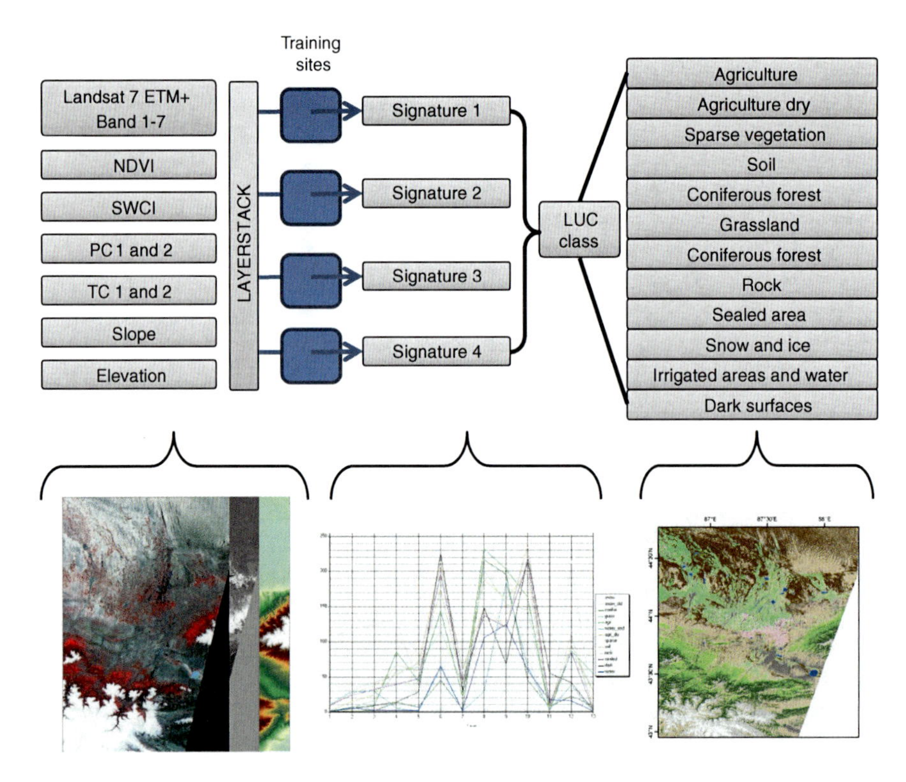

Fig. 3.14 Flow chart of the classification process with examples of the input raster data sets (landsat band 4, 3 and 2, NDVI and elevation), a signature plot derived from the chosen training sites and classification result below (from *left* to *right*)

the accuracy of all classes. Random points and stratified random points tended to neglect and undersample classes with few classified pixels but possibly very important areas in the images such as dark, water, sealed, and coniferous forest. The point sampling scheme, stratified random, can only be applied after the map is completed and provides satisfactory results with regard to statistical validity [129]. The map classes of the test point could then be defined fairly well from their visual call similar to the identification of the trainings sites, so ground measurements were not required. The reference data is assumed to be more correct than the map, but there is no guarantee that they are perfect or the 'truth' and should be used with caution [129] (Tables 3.19, 3.20).

The *overall classification accuracy* and *kappa statistics* are used to evaluate the classification accuracy of the derived maps. The *overall classification accuracy* is calculated by dividing the number of correctly classified pixel by the total number of pixels in the error matrix [129, 131, 133]:

$$\text{Overall accuracy} = \frac{\sum_{i=1}^{k} n_{ii}}{n} \tag{3.55}$$

with the number of classes k, the correctly assigned pixels in each class n_{ii} and the total number of samples n. It ranged from 92.27 (March 2003) to 95.56 % (January 2001) (see Table 3.21). To account for pixels randomly assigned to the correct class and to test the statistical significance of the error matrix, the *kappa statistics* $\widehat{K}$ are computed [132]:

$$\widehat{K} = \frac{n \sum_{i=1}^{k} n_{ii} - \sum_{i=1}^{k} \left(n_{+i} \cdot n_{+j}\right)}{n^2 - \sum_{i=1}^{k} \left(n_{+i} \cdot n_{+j}\right)} \tag{3.56}$$

with the total number of pixel that were classified into that category n_{+i} and the total number of reference pixels of a category n_{+j}. The *kappa coefficient* measures chance-corrected nominal scale agreement between two raters, i.e. whether a classification process reduces the error compared with the error of a completely random classification [134]. The use of *kappa analysis* assumes a multinomial sampling model, which is strictly speaking not satisfied by the equalised random sampling as the test points of small classes are clustered around the few areas they occupy in the classified map [131]. However, the results included only a few small classes and the normalised distribution was achieved with only minor aberration. An *overall accuracy* of 85 % has been accepted by Bastiaanssen and Bandara [135] for a heterogeneous watershed with abundant small-scale landscape features and described as the acceptable overall accuracy level by Congalton and Green [129]. Foody [136] criticised that the 85 % target has not been not proposed as being universally applicable and encourages developing confidence levels specifically for each application. The range of both the *overall classification accuracy* and *kappa statistics* are comparable with the results of classifications using similar satellite imagery and covering Northwest China (cf. [32]). Therefore, the accuracy levels were regarded as sufficient (cf. Table 3.21).

To evaluate the differentiation between the individual classes, the *producer's* and *user's accuracy* were calculated for every class (cf. [129, 131]). The *producer's accuracy* or *error of omission* is calculated by dividing the total number of correctly classified pixels in a category n_{jj} by the total number of reference pixels of that category n_{+j}:

$$\mathit{Producer's\ accuracy}\ j = \frac{n_{jj}}{n_{+j}} \tag{3.57}$$

Accordingly, the user's *accuracy or error of commission* is calculated dividing the total number of correct pixels that were classified into one category n_{ii} by the total number of pixel that were classified into that category n_{+i}:

$$\mathit{User's\ accuracy}\ i = \frac{n_{ii}}{n_{+i}} \tag{3.58}$$

The classes sealed, ice, snow, irrigation, water, and dark showed a good *producer's accuracy*, i.e. almost all pixels belonging to one of these classes were also classified as such. A poor *producer's accuracy* had the classes rock, soil, and

Table 3.20 Accuracy assessment of the classes for all classifications

Classification		1999/10	2000/01	2000/03	2000/06	2000/09	2001/01	Overall
Agriculture	Producer's accuracy	100.00			100.00	95.24		98.41
	User's accuracy	95.00			90.00	100.00		95.00
Agriculture dry	Producer's accuracy	100.00		90.91	100.00	100.00		97.73
	User's accuracy	100.00		100.00	100.00	100.00		100.00
Sparse vegetation	Producer's accuracy	90.91	95.24	95.27	100.00	100.00	90.48	95.32
	User's accuracy	100.00	100.00	100.00	90.00	100.00	100.00	98.33
Soil	Producer's accuracy	76.92	100.00	80.00	76.92	100.00	86.96	86.80
	User's accuracy	100.00	75.00	100.00	100.00	100.00	100.00	95.83
Coniferous forest	Producer's accuracy	100.00	100.00	90.00	95.24	100.00	100.00	97.54
	User's accuracy	85.00	95.00	90.00	100.00	100.00	85.00	92.50
Grassland	Producer's accuracy	86.96			90.91	100.00		92.62
	User's accuracy	100.00			100.00	100.00		100.00
Rock	Producer's accuracy	90.00	94.44	76.92	76.92	68.97	95.00	83.71
	User's accuracy	90.00	85.00	100.00	100.00	100.00	95.00	95.00
Sealed area	Producer's accuracy	100.00	100.00	100.00	100.00	100.00	94.44	99.07
	User's accuracy	55.00	100.00	65.00	45.00	25.00	85.00	62.50
Ice	Producer's accuracy				100.00			100.00
	User's accuracy				95.00			95.00
Snow	Producer's accuracy	100.00	100.00	100.00	100.00	100.00	100.00	100.00
	User's accuracy	95.00	100.00	100.00	100.00	100.00	100.00	99.17
Snow old	Producer's accuracy	100.00	71.43	100.00	95.41	74.07	95.25	89.36
	User's accuracy	100.00	100.00	95.00	100.00	100.00	100.00	99.17
Irrigation	Producer's accuracy					100.00		100.00
	User's accuracy					90.00		90.00
Water	Producer's accuracy	100.00	100.00	94.74	100.00	100.00	100.00	99.12
	User's accuracy	100.00	95.00	90.00	100.00	100.00	100.00	97.50
Dark areas	Producer's accuracy	100.00		100.00	100.00	100.00		100.00
	User's accuracy	100.00		75.00	100.00	100.00		93.75

Not all classes existed in all classified scenes, for the complete results of the accuracy assessment see Table A.4, A.5, A.6, A.7, A.8, and A.9 in the appendix

Table 3.21 Overall classification accuracies for all classified images

Classification	Overall accuracy (%)	Kappa statistics
1999/10	93.85	0.9333
1999/12/ 2000/01	94.44	0.9375
2000/03	92.27	0.9150
2000/06	94.29	0.9385
2000/09	93.93	0.9346
2001/01	95.56	0.9500

sealed, which means they were often classified as other classes. The classes agriculture dry, grassland, snow, and snow old show a good *user's accuracy*, as pixels assigned to these classes also proved to match their reference points. Coniferous forest, sealed and irrigation had a relatively poor *user's accuracy* and included pixels actually not belonging to their class (see Table 3.20). Pixels belonging to this class were often confused with the water or agriculture, rock, and water areas, respectively. The errors in the classification arose mainly from the not completely achieved differentiation between classes which had very similar spectral characteristics such as rock and sealed, soil, sparse vegetation, dry agriculture and grass, irrigation and water. The pixels size of the satellite images is also larger than the size of individual features (trees, shrubs, plants) or patches composed of only one feature type [137]. Several land use classes are located within one 'mixed pixel', which is not able to represent the spectral characteristics of only one class, making the separation more difficult [61]. Atmospheric effects could be still inherent to the image even after the atmospheric correction and could also mask subtle differences between similar classes. Another possible source of errors and inaccuracies is the lack of training sites with real ground truth information [129]. The influence of the last two error sources could not be determined within this study.

As the number of training sites was limited and concentrating on areas with known surface characteristics, thus not covering the whole research area and the whole variety of e.g. vegetation types. Also due to the lack of test points, the *accuracy assessment* was mainly done comparing classification results with expert knowledge of the area and optical evaluation of the satellite image, the additional bands and indexes and comparable sources (i.e. *GoogleEarth*^{TM}). As the reference points have to be assumed to be correct for the *accuracy assessment* [131], errors due to the training sites and reference data are not evaluated by the *accuracy assessment*.

3.3.3.4 Post-Processing and Results

As some of the classified images exhibited missing areas from the cloud removal, they had to be post-processed before using them for the landuse evaluation and the hydrological modelling. The missing areas within the research area were filled

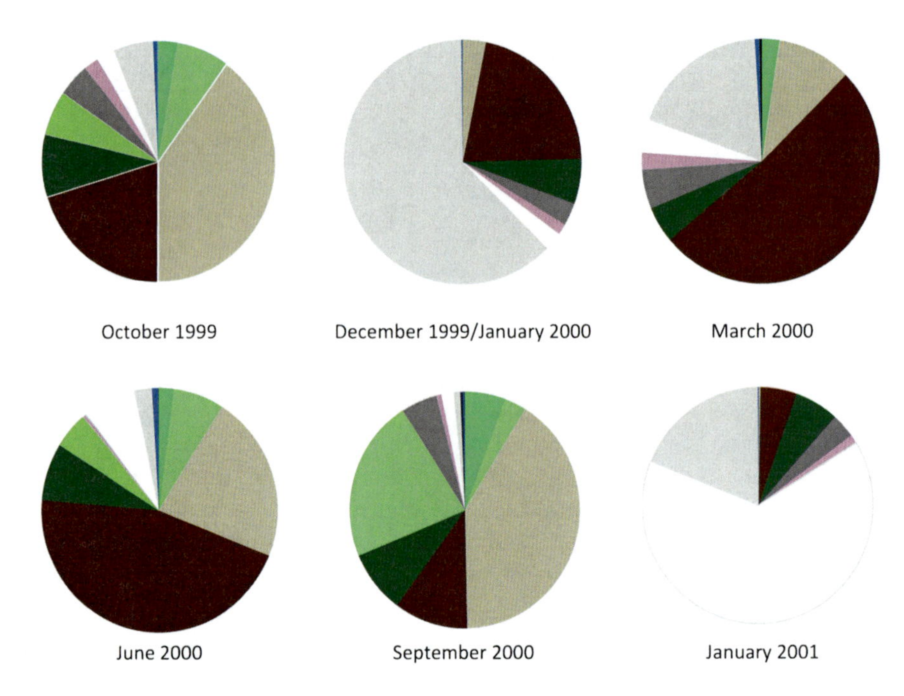

Fig. 3.15 Proportion of LULC classes in the classified images (see Fig. 3.16)

with the *Focal Statistics* tool in *ArcGIS* according to the neighbouring land use classes. The results of the classification are information about the land surface characteristics in Urumqi Region at six different dates throughout the years 1999/2000. The seasonal changes of the land cover are clearly visible, in December and January an extensive snow layer covers Urumqi Region and the vegetation cover (sparse vegetation, grass, agriculture and dry agriculture) increases from March to June and October. In June and September, the sealed areas in the city centre could not be identified as the urban green areas cause classification of mixed pixels as vegetated areas or they were confused with the classes soil or rock. The coverage of the class coniferous forest remains almost the same throughout the year because evergreen conifers predominate and the extent changes only marginally (Figs. 3.15 and 3.16).

The classification results for the different dates confirm again the necessity to take seasonal changes into account as the dominant land cover classes and their relevant hydrological characteristic change throughout the year. The most important classes are sparse vegetation, soil and snow cover. The parameters necessary for the hydrological modelling were assigned to the land use classes as presented in Sects. 3.1.1, 3.1.2 and 3.1.3.

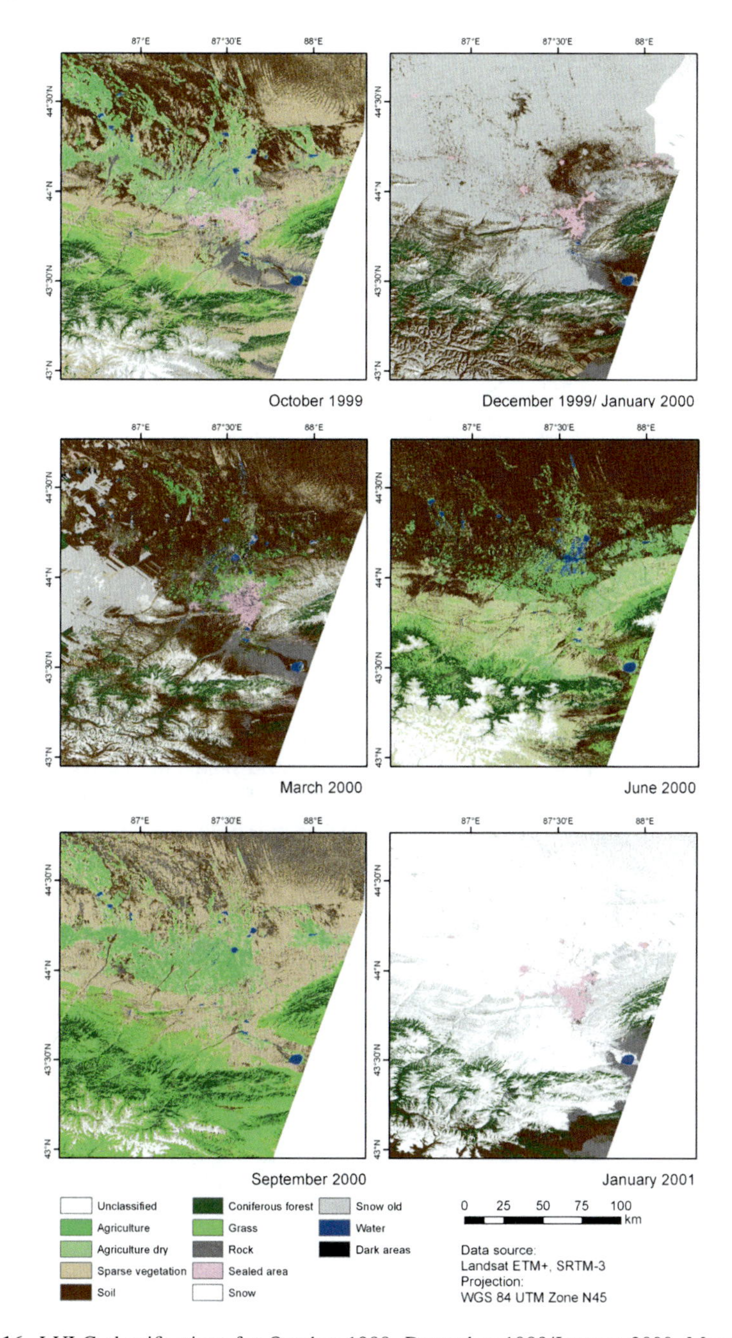

Fig. 3.16 LULC classifications for October 1999, December 1999/January 2000, March 2000, June 2000, September 2000 and January 2001 (*Data source* Landsat 7 ETM+ [107] path 143 and row 29/30, SRTM-3 [58])

3.4 Model Evaluation

Predictions based on model results only make sense when the model has been calibrated and validated, i.e. the predictive capability has been tested. In order to assess the quality of the hydrological model and the accuracy of the model results, the monthly simulated values of evapotranspiration and runoff as well as annual values for runoff, evapotranspiration, precipitation, snow melt water and groundwater recharge were evaluated. Observed data were available from different sources and on different scales ranging from point measurements, over the (sub-)catchment scale to the whole administrative region of Urumqi, depending on the scale of the available reference data. The data for calibration and validation was taken from different sources as a continuous measurement series was not available. As measures of accuracy, the absolute and relative volume error of the monthly or annual results, the coefficient of determination and the Nash-Sutcliffe coefficient for model efficiency were used based on the available reference information.

When evaluating the results of the hydrological model, it is not enough to describe and assess the calculated output totals and the result's spatial distribution. Abbaspour et al. introduced three errors that have to be considered when evaluation model performance: (i) the error in the driving variables, measured or estimated, whose effects have been assessed in the sensitivity analysis in Sect. 3.2.1. Then (ii), the error of the measured or observed output which cannot be assessed for this specific case, but can only be estimated, and (iii) the predictive uncertainty (see also Sect. 3.2.2) and error in the conceptual model, which can be evaluated by comparing the simulated runoff values with the measured ones (cf. [138]).

Measurements of evapotranspiration were not available, hence the derived potential and actual evapotranspiration (ET_p and ET_a) were compared to ET from other models. The first comparison is with the MODIS product MOD16, which is a post-launch product estimating potential and actual evapotranspiration (PET and ET). Secondly, the modelling results from another hydrological model, SWAT, are compared to the results of the water balance model, including ET_p, P, ET_a, Q_n, Q_{sf} and Q_{gw} for one subcatchment of the Urumqi River, Wulapo Reservoir (see Figs. 3.2, 3.3 and 3.4), where inflow measurements were available for the calibration of the hydrological models. For the subsequent validation and evaluation of the water balance model, the results of the average water balance 2001–2010 for Urumqi Region were compared to the information from the Water Reports during the same time period (Water Affairs Bureau Urumqi [139–142]).

The quality of the reference data is assessed in the specific subsections separately, but there are some possible sources of errors and reasons for aberrations that can affect the comparison (cf. [8]). First, observation errors may have a negative impact on the modelling results and obstruct a comparison or they may be based on a different concept to assess the water resources and therefore lead to different results. Secondly, the units and areas that the comparison is based on can differ due to erroneous delineation of surface and groundwater catchments, unknown drainage to other catchment areas. These error source have to be kept in mind when evaluating the model quality.

3.4.1 Measures of Accuracy

Several measures of accuracy can be applied to compare the total flows and water resources such as root of the mean difference between observed and simulated values (*RMSE*, not presented here) and **absolute volume error** (VE_{abs}) (cf. [5, 57, 143]):

$$VE_{abs} = \sum_{i=1}^{n} (Q_m(i) - Q_o(i)) \tag{3.59}$$

with the absolute volume error VE_{abs}, the number of modelled monthly or annual values n, modelled runoff Q_m and observed runoff Q_o. For the evaluation of the hydrological model and comparison with other catchment areas or methods, the relative accuracy is of more interest. This was measured using the **relative volume error** VE_{rel}, which is similar to the volume ratio [144] or deviation of runoff values [145]. It gives information about the quantity of differences, but not about the quality of the modelled runoff dynamics. It is calculated as follows [54]:

$$VE_{rel} = 100 * \frac{\sum_{i=1}^{n} [Q_m(i) - Q_o(i)]}{\sum_{i=1}^{n} Q_o(i)}. \tag{3.60}$$

The compensation of huge negative and positive value when computing VE_{rel} for $n > 1$ is desired: even when n is equal to 12 and the monthly results calculated for 1 year are evaluated, the final annual VE_{rel} was considered most important.

For the evaluation of the modelled monthly runoff values and in order to assess the model efficiency, the determination coefficient of linear regression analysis R^2 between observed and modelled values and the **Nash-Sutcliffe coefficient of model efficiency** (*NSE*) were used (as applied by Bäse [5], Bieri and Schleiss [144], Harlin and Kung [54], Liersch [144]). The determination coefficient is the square value of the Pearson's correlation coefficient and measures how well the relationship between modelled and measured values can be described by a linear model. It reacts strongly to extreme values compared to mean values and measures the accuracy with regard to peak and not to mean values [5]. *NSE* assesses the predictive capability of the modelled values compared to the observed mean. It is calculated as:

$$NSE = 1 - \frac{\sum_{i=1}^{n} [Q_m(i) - Q_o(i)]^2}{\sum_{i=1}^{n} \left[Q_o(i) - \overset{n}{\bar{Q}_o}\right]^2} \tag{3.61}$$

with $\overline{Q_o}$ as the mean of observed values (cf. [5, 54, 143–145]). The *NSE* reacts more sensitive to deviations of observed and simulated means and variances than the determination coefficient and is therefore a better measure for accuracy (Legates and McCabe 1999 in Bäse [5]. It ranges from 1 to negative infinity, with 1 representing optimal congruency between observed and modelled values, 0 means the observed mean has the same predictive capability as simulated values and everything below 0 means the model has a lower predictive capability than the observed means.

For additional information, the **systematic volume error** (VE_{sys}) was calculated similarly to VE_{rel} using the double sum gradient of observed and modelled values to assess whether the model under- or overestimates observed values [5]. The double sum gradient is calculated as the linear regression of the added values of observed and modelled values.

3.4.2 Comparison of ET Values with MOD16 ET

In this section, the results for ET_p and ET_a were compared with another remote sensing product estimating evapotranspiration, the MOD16 Global Terrestrial Evapotranspiration Data Set. Both estimation results were compared in Urumqi Region for the year 2000 which was also used for calibration (see Sect. 3.4.3). MOD16 ET has been computed by Mu et al. [146] using the improved ET algorithm which is based on the Penman-Monteith equation [27]. MOD16 is a post-launch product and is calculated using land cover type, *LAI* and albedo products from MODIS and NASA's MERRA GMAO (GEOS-5) daily meteorological reanalysis data [147]. The data set with the sum of monthly potential evapotranspiration (PET) and actual evapotranspiration (ET) was downloaded from the project's website in *.hdf format and imported into *ArcGIS*. Within the research area, there are several patches of water bodies, barren or sparsely vegetated land, urban or built-up land. For these land use classes, MOD16 did not provide any values for evapotranspiration. NODATA was assigned to these pixels to exclude them from the subsequent analysis. The MOD16 product PET and ET was assumed to be the 'observed' evapotranspiration, and the difference to calculated monthly ET_p and ET_a was computed and statistically evaluated. As the MOD16 data product also uses the Penman-Monteith equation, differences due to the model structure were expected to be small. However, the absolute values differed considerably (see Fig. 3.17). Especially the difference between calculated ET_p and MOD16 PET ranged from −400 mm to over 200 mm and was on average − 121.76 mm per month for the whole year 2000 (average VE_{abs}) and had a relative volume error VE_{rel} of −72.73 %, mainly due to an extreme range of deviations in the summer months. PET was approximately twice as high as calculated ET_p, which could be due to the meteorological input data for PET which can be based only for a limited number of observation station in or near the research area.

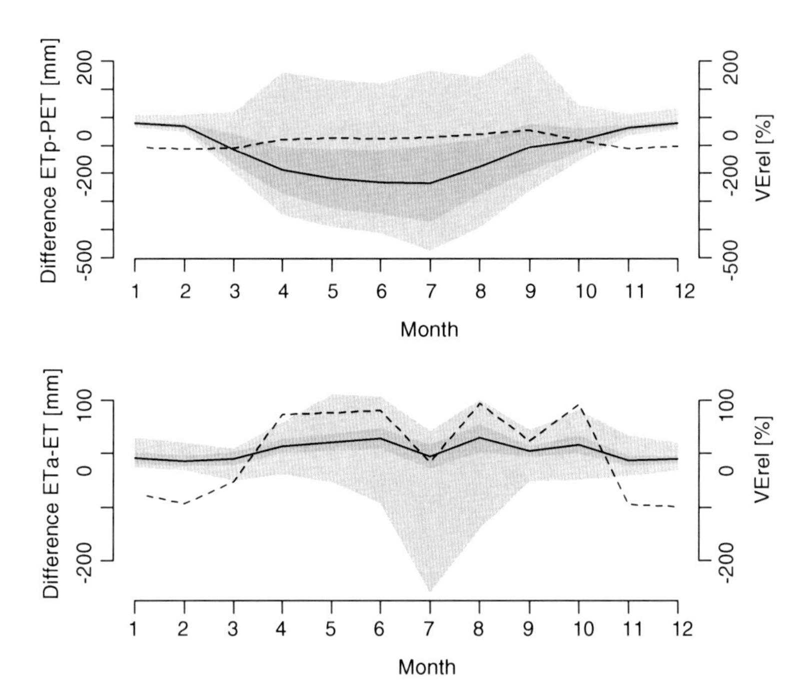

Fig. 3.17 Differences between calculated potential and actual evapotranspiration (ET_p/ET_a) and MOD16 PET/ET with mean difference (*bold line*), standard deviation (*dark grey*), minimum to maximum values (*light grey*) in mm, and the average relative monthly volume error (*dashed line*). The values are calculated from the spatially distributed monthly values in Urumqi Region (*Data source* MOD16A2 [146], own calculations)

Again, it is unsure whether both models are taken the same meteorological station as input and whether PET takes the strong climatic gradient in the research area into account. The differences for actual evapotranspiration ($ET_a - ET$) were considerably smaller with a VE_{abs} of only +4.32 mm on average per month and a VE_{rel} of 20.19 %. From April to Ocotber, the VE_{abs} was 15.35 mm on average per month while it was -10.25 mm from November to March. The differences stem from the strong dependency of ET_a on the available water resources that were estimated to have a higher value by the water balance model in summer and could have been underestimated by MOD16 ET due to differing input data.

The spatial patterns of both potential and actual evapotranspiration (not shown here) were captured in approximately the same way by both data sets. For the densely vegetated areas as identified by the land use classification (mainly forest and agriculture), MOD16 estimates were relatively low. These were also the areas with the largest differences for both potential and actual evapotranspiration, causing the maximal differences between the two datasets.

3.4.3 Comparison with Reference Measurements

To calibrate the simulated runoff, the modelled runoff values for 2000 were compared with observed monthly inflow data at Wulapo Reservoir in the same year, provided by project partners from the the College of Hydraulic and Civil Engineering, Xinjiang Agricultural University. This way, the model performance was evaluated with regard to the simulation of runoff dynamics and the absolute annual runoff quantity. The challenge of the calibration was to remain true to the physical and processed-based nature of the modules while calibrating the model results to the specific catchment area (cf. [3]). In the end, the degree-day-factors were adapted in several test runs, while the other parameters were only slightly changed.

First, total runoff values are compared (see Table 3.22): The simulated runoff overestimates the observed inflow data and has a total volume error (VE_{abs}) of $+5.29 \cdot 10^6$ m^3 and a relative volume error (VE_{rel}) of +3.19 % of the total observed runoff volume. The comparison of monthly values gives a different picture: The mean difference between observed and simulated values ($RMSE$) is $16.24 \cdot 10^6$ m^3, which is higher than the mean runoff volume. The first modelled runoff peak arrives too early and the second one is too low, but they are almost balanced out at the end of the year (Fig. 3.18).

The modelled runoff is directly fed by snow melt water and precipitation, while in the observed runoff a base flow from ground water and/or interflow can be detected and there are obviously buffers to the direct runoff. It seems the **water balance model is overestimating snow melt runoff in spring** as the observed values do not have a peak in spring and the runoff from ice and glacier melt later in summer is not accounted for. On the other hand, baseflow is underestimated, as the model **does not include intermediate storage in soil and underground** and there is almost no runoff modelled from December to February.

First conclusion is that the model is good for modelling annual values, but not suitable for modelling monthly runoff or runoff dynamics. This is confirmed by the correlation coefficient between modelled and observed flows, which is low ($r_m = 0.26$) for monthly values and high ($r_c = 0.95$) for cumulative monthly values. The systematic volume error for monthly values ($VE_{sys} = 0.80$) is lower than 1 as the model is underestimating the runoff values. This is obviously the case except for April and May (see Fig. 3.18a). For the cumulative values however, the model tends to overestimate the observed flows ($VE_{sys} = 1.14$) due to the runoff peak in April and May which is carried on for the rest of the year. As a final

Table 3.22 Observed and modelled surface runoff input flow at Wulapo Reservoir for the year 2000

Month	1	2	3	4	5	6	7	8	9	10	11	12	Total
Observed	6.87	5.17	7.88	6.13	7.08	20.92	33.86	35.31	17.58	11.69	6.84	6.65	165.98
Modelled	0.00	0.00	3.47	36.95	47.47	8.92	22.33	28.75	8.79	14.20	0.40	0.00	171.27

Units 10^6 m^3

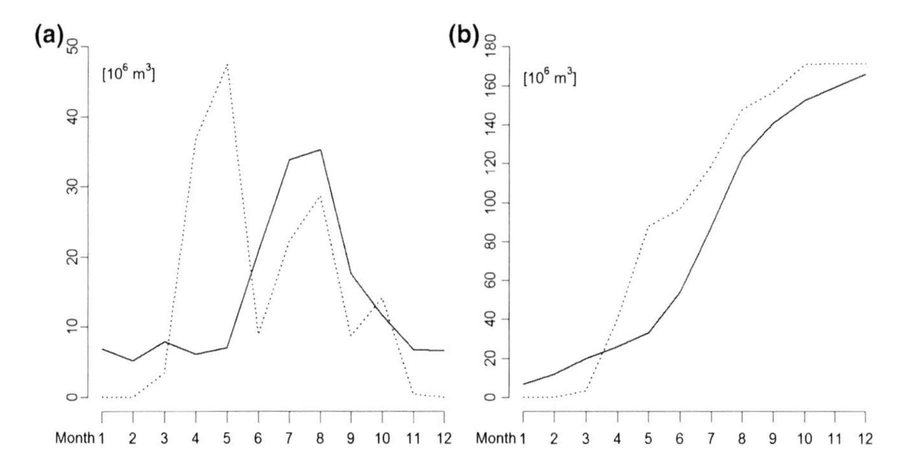

Fig. 3.18 Comparison of observed and modelled runoff values at Wulapo Reservoir for monthly and cumulative values of the year 2000 (*solid line* observed, *dotted line* modelled; *Source* own calculations, observed data from Xinjiang Agricultural University, Urumqi)

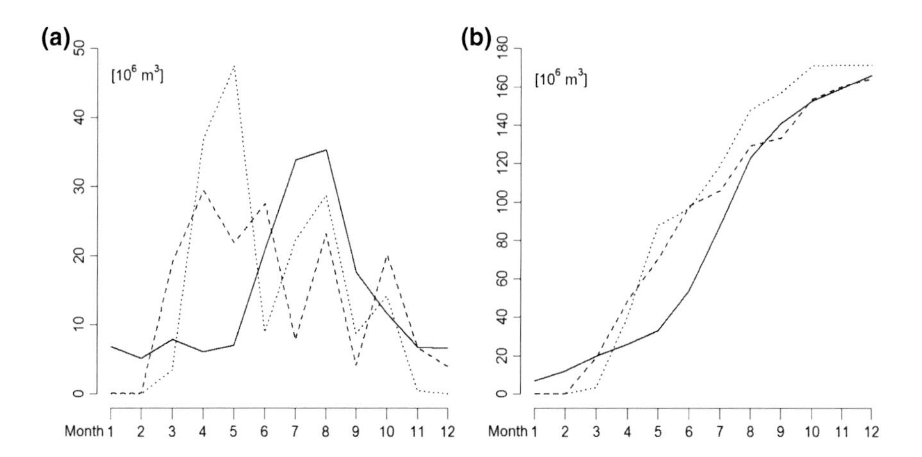

Fig. 3.19 Comparison of observed and modelled surface runoff for Wulapo Reservoir with observed inflow at Wulapo Reservoir (*solid black line*), water yield (inflow in the reservoir) simulated by ArcSWAT (*dashed line*) and surface water flow simulated by the water balance model (*dotted line*) for monthly (**a**) and cumulative (**b**) values in the year 2000 (*Source* own calculations, observed data from Xinjiang Agricultural University, Urumqi)

measure of the model quality, the *NSE* has been calculated. For a strongly controlled catchment area, values over 0.8 are desired (2010). For the hydrological model and its monthly results however, the coefficient is very low at $NSE = -1.44$ because the peak runoff value and the runoff dynamics are not modelled correctly. This actually means that the annual mean value would be a better predictor for monthly runoff than the modelled values. For the accumulated monthly runoff values however, the Nash–Sutcliffe model efficiency index is much higher at $NSE = 0.92$. Therefore, **the accumulated sums can be modelled quite well, which was also the minimal aim of the modelling**.

3.4.4 Comparison with SWAT Results

To assess the model performance in relation to another well known and widely used hydrological model, simulated precipitation, evapotranspiration and runoff were compared to the SWAT results for the year 2000. Second aim was to evaluate the accuracy of the absolute model results of actual evapotranspiration, distributed precipitation and groundwater recharge for which measurement data were not available. The Soil and Water Assessment Tool (SWAT) is a semi-distributed eco-hydrological model developed to assess the effect of management and climate on water supplies, sediments, and agricultural chemical yields [19]. It is a semi-physically based and lumped-parameter model with good accuracy for ungauged rural basins [105, 148]. Complex watersheds at the basin scale "with varying soils, land use and management conditions over long periods of time" [112: 1] can be divided into a large number of sub-watersheds, allowing a high level of spatial detail [138]. On the temporal scale, it is a continuous time model and can be used to study long-term impacts but is not designed to simulate single-event flooding [19]. *ArcSWAT* [149] was used via an interface in *ArcView* [76] to assess the components of the water balance also simulated by the water balance model. Based on the characteristics such as land use, vegetation type, soil type, aspect and slope, hydrological response units (HRUs) are defined. Each HRU "is considered homogeneous with respect to its hydrological response" [150: 2278]. The concept of partitioning a watershed into units is very common in hydrological models as it reduces the complexity and increases the computational efficiency. They are not identified spatially within a SWAT simulation [148], which was the main reason why SWAT was not used as the model of choice in this research.

The SWAT model is also based on the water balance equation and accounts for precipitation, evapotranspiration, surface runoff, soil water, percolation and groundwater return flow [19]. However, some of the sub-modules differ. For example, the runoff volume is controlled by the Soil Conservation Sercive (SCS) runoff equation which is an empirical model that has been widely used to model the rainfall-runoff relationships of watersheds and to estimate the runoff under varying land use and soil types [19]. The SCS curve number takes the soil's permeability, land use and antecedent soil water content into account. For snow and glacier melt processes, a temperature index approach is used (see [42]).

Further and comprehensive information on the SWAT model, its advantages, disadvantages and problems has been published in numerous research and review articles (e.g. [148]). For example, it has also been used to model the impacts of climate change on the availability of water resources (e.g. [151]). The detailed information about input data and the calibration process is described in the

Table 3.23 Results from *ArcSWAT* simulation run with climate data from the year 2000 for the catchment area of Wulapo Reservoir

	SWAT simulation (10^6 m^3)		Water balance model (10^6 m^3)		VE_{rel} (%)
Precipitation	PREC	1,034	P	1,204	16.1
Potential evapotranspiration	PET	2,354	ET_p	1,861	−20.9
Actual evapotranspiration	=ET+SUBL	761	ET_a	794	4.4
Net available water	=PREC−ET−SUBL	273	Q_n	330	21.1
Surface flow	SURQ	102			
Lateral flow	LATQ	59			
	=SURQ+LATQ	162	Q_{sf}	171	2.2
Groundwater flow	GWQ	7			
Deep aquifer recharge	DEEP AQ RECHARGE	143			
Groundwater recharge	TOTAL AQ RECHARGE	150	Q_{gw}	159	6.0

SWAT simulation results do not necessarily have to add up as changes in several intermediate storages are not presented in detail here. Average values for a 10 year run are shown (*Source* own calculations)

appendix. Some of the input data was already provided for the water balance model, but especially soil data had to be derived from global data sets. The SWAT simulation was again calibrated with monthly inflow data for Wulapo Reservoir, provided by the College of Hydraulic and Civil Engineering, Xinjiang Agricultural University.

The results of the annual water balance from the SWAT simulation are compared with the modelled water balance (see Table 3.23). The water balance model predicts a higher precipitation than *ArcSWAT* because SWAT uses only the precipitation value from Wulumuqi station and a precipitation lapse rate, while the hydrological model uses precipitation values with a more detailed spatial distribution as described in Sect. 3.3. This leads to higher precipitation values in the mountainous regions, which extend over a large part of the catchment area. The potential evapotranspiration in SWAT on the other hand is higher, probably due to the different methods for the distribution of temperature. In this SWAT simulation, six different elevation bands were defined according to temperature and precipitation lapse rates derived from the same input data used for the water balance model. Nevertheless, the actual evapotranspiration is lower in the SWAT simulation, as less water is available for evapotranspiration than in the water balance model. Probably due to the higher precipitation, the net available water flow Q_n in the water balance model is higher than the available water modelled by *ArcSWAT*. For the comparison of surface and groundwater flows, the results for surface and lateral flow and groundwater flow and deep aquifer recharge were summed up and compared to Q_{sf} and Q_{gw} (see Table 3.23).

The water balance model only divides into surface and groundwater flow and does not account for intermediate storage and flows in soil or groundwater bodies. In general, the flow results of SWAT were lower (see comparison of total available water in SWAT and Q_n), but the percentage of surface (51.9 and 51.8 % for

Table 3.24 Evaluation measures for the used hydrological models: Nash-Sutcliffe coefficient for model efficiency for monthly and cumulated values, relative volume error VE_{rel}, monthly and annual absolute colume error VE_{abs}, RMSE, Pearson's product correlation coefficient and systematic volume error VE_{sys}

Simulation	NSE (−)		VE_{rel}	VE_{abs} (10^6 m^3)		$RMSE$	Pearson's correlation coefficient (−)		VE_{sys} (−)	
	Monthly	Cum	(%)	Monthly	Annual	(10^6 m^3)	Monthly	Cum	Monthly	Cum
Water balance model	−1.44	0.92	3.19	0.44	5.29	16.24	0.26	0.95	0.80	1.14
SWAT	−0.61	0.88	−1.09	0.15	1.81	13.20	0.20	0.96	0.70	1.04

SWAT and water balance model) and groundwater flow was almost the same (48.1 and 48.2 %). The SWAT results were also evaluated against the measured inflow at Wulapo reservoir and same measures applied as to the water balance model in Sect. 3.4.3. The results in Table 3.24 showed that the **SWAT model could be calibrated better and had slightly lower relative and absolute volume errors** (VE_{rel} and VE_{abs}) and RMSE. However, the SWAT model proved also not to be useful to simulate the dynamic monthly runoff values for Wulapo Reservoir as **the storage function of interflow and groundwater flow are not well represented.** The *NSE* was also negative and for the cumulated values even lower than for the results of the water balance model. Abbaspour et al. [138: 413–414] concluded from their simulation of a watershed in Switzerland, "with good data quality and availability and relatively small model uncertainty—it is feasible to use SWAT as a flow and transport simulator". Here, the input data was apparently not sufficient for a reasonable calibration and representation of runoff peaks. This could be associated with the limited data availability and frequent use of estimation of input parameters for the watershed or with groundwater watersheds deviating from surface watersheds and groundwater flows that could not be accounted for. According to the experience of previous regional studies by the College of Hydraulic and Civil Engineering and this SWAT simulation, the chosen model concepts may also not be suitable to accurately represent the complex relationships between surface and groundwater resources in Urumqi Region (see Sect. 2.3).

3.4.5 Comparison with Water Resource Data in Urumqi Region

As one year is insufficient for the calibration and validation of a hydrological model, additional information on water resources from the Water Reports [139–142] published by the Water Affairs Bureau Urumqi were used. There were also figures available from the Xinjiang Statistical Yearbooks (Statistics Bureau of

Table 3.25 Water resources in Urumqi 2002–2008 for an area of 10,891 km^2 with total water resources, surface water with and without infiltrating water, groundwater and groundwater recharge through infiltrating water

Years	Xinjiang Statistical Yearbooks					Water Reports			
	Total	Surface water			Ground water	Total	Surface water		Ground water
			Without infiltrating water	Infiltrating water				Without infiltrating water	
2002							1,218		
2003							1,176		
2004	1,059	1,024	547	477	512	1,066	1,031	554	512
2005	1,128	1,092	531	561	597	1,051	1,016	454	597
2006							876		
2007	1,277	1,235	589	646	688	1,185	1,142	497	688
2008	964	934	539	395	425				

Units 10^6 m^3; *Source* Statistics Bureau of Xinjiang Uygur Autonomous Region [152–156], Water Affairs Bureau Urumqi [124–126, 146]

Xinjiang Uygur Autonomous Region [152–156]), which are presented together with the Water Report data in Table 3.25. However, this data set was not used because it is based on the whole administrative region of Urumqi, which was not fully included in the hydrological model. Limited data availability and computing requirements restricted the research and possible simulation area to the catchments of Urumqi and Toutun River as well as to some smaller rivers in the northeast of Urumqi. As the Water Reports did not state exact information about the methods and catchment areas that were included in the publication, the comparability of the statistics and simulated results is difficult to assess.

Even the statistical data itself is not consistent when comparing data from Statistics Bureau of Xinjiang Uygur Autonomous Region and Water Affairs Bureau Urumqi, possibly due to different data sources and definitions of administrative areas and watersheds. **On the basis of the statistical data, only annual values could be compared** and evaluation measures based on monthly values (cf. Sect. 3.4.3) could not be calculated.

According to the statistics, surface and groundwater resources largely overlap in Urumqi Region and the total amount of water resources is only 3.2–5.4 % larger than the surface water resources (see Table 3.25). This active exchange is probably due to high infiltration rates, interflow and percolation out of shallow groundwater bodies in the foothills and alluvial fans which comprise a large part of the catchment area (see landscape classification in Sect. 2.1 and description of water resources in Sect. 2.3). An exact delineation of the water resources is therefore not possible and only the amount of total water resources could be compared to the simulated net runoff Q_n.

The selection of the spatial extent to be compared was also difficult. Because the administrative boundaries do not necessarily coincide with watershed delineations, the district level data was not used. Also the location and extend of

Table 3.26 Comparison annual of precipitation P and total water runoff Q_n according to measured average of several years (Water Reports 2003–2007) and simulated for 2001-2010

		Area	P		Q_n	
		(km²)	(mm)	(10^6 m³)	(mm)	(10^6 m³)
Urumqi river	Simulated	3,126	425	1,329	140	439
	Measured	3,906	299	1,169	109	427
	VE_{rel}		42.0 %	13.7 %	28.3 %	2.8 %
Toutun river	Simulated	1,069	503	537	167	178
	Measured	1,050	343	360	144	151
	VE_{rel}		47.0 %	49.3 %	15.8 %	18.1 %
Urumqi region	Simulated	7,552	410	3,093	131	989
	Measured	10,890	256	2,793	93	1,017
	VE_{rel}		60.0 %	10.7 %	40.2 %	–2.8 %

Source Water Affairs Bureau Urumqi [124–126, 146], own calculations

catchments were unsure as figures for watershed areas changed from source to source and year to year. They were not withstanding internal cross-checks, probably because they were calculated with different methods, not clearly differentiating between surface and groundwater or estimated and extrapolated from the same average values. The accuracy and the error of the data are also not defined. Hence, only a rough comparison of the statistical data and simulated results and calculation of the relative volume error (VE_{rel}) was possible.

By using mean climate data from 2001–2010 as input the modelled total and surface water flow results were compared with data on the total and surface water resources from Water Affairs Bureau Urumqi. In Table 3.26 the runoff on catchment level is evaluated for the years from 2003 to 2007 where data is available. Either the mean values for 2003–2007 were available or the average of several years during that period, not specified in more detail. When calculating the water balance with the climate variable data of a single year, the runoff results are strongly influenced by the precipitation amount of that year (see sensitivity analysis in Sect. 3.2.1), while in reality runoff is buffered by intermediate storage in soil and underground and not taken into account by the model. The averaged climate variables as input are supposed to average out the direct reaction of the modelled flow results to the variability of the precipitation values. The catchments of Toutun and Urumqi River were chosen for the comparison as their watersheds lie within the research area and can be deduced fairly well from the catchment size stated by Water Affairs Bureau Urumqi. The surface runoff documented by the Water Reports represents surface runoff that flows into surface water bodies and partly infiltrates into the underground and becomes groundwater. Therefore, the net runoff Q_n is expected to be comparable with the total runoff in the catchment area, but should be overestimating the Water Report's data on surface water resources. This hypothesis is confirmed by the simulation results for the catchment areas of Urumqi and Toutun River (see Table 3.26).

When comparing the average flow values based on the input data from 2001–2010 with the averaged measured flow, the VE_{rel} is 2.8 % for Urumqi River catchment and 1.1 % for the Toutun River catchment. For the catchment of Urumqi Region, the model is underestimating the runoff values by 2.8 %, probably due to the different sizes of the catchments the calculations are based on. However, the main problem is that the water balance model simulates a higher amount of precipitation in the catchment areas (see also Table 3.26). The precipitation values from Wulumuqi station on which the simulation is based are around 100 mm higher than the average precipitation values published in the Water Reports. Whether and how the latter are measured, derived or extrapolated is not known and the data quality could not be assessed. As reliable information about the climate data was not available, this deviation could not be evaluated and the problem could not be solved. When simulating and evaluating the water balance of Urumqi Region, the focus has to lie even more on the comparison of scenarios and the relative changes.

In this chapter, a relatively simple water balance model including modules for evapotranspiration, snow accumulation and snow melt, and runoff formation has been presented. Special focus was on the most important input parameters and the calibration and validation of the model. For this purpose, the performance of the water balance model compared to measurements and statistical figures was assessed, but also compared to other models and calculations. The evaluation shows that the water balance model has little predictive capacity for the monthly balance (see evaluation measures in Table 3.24) and tends to overestimate the precipitation and subsequent flows in the watersheds, athough robust measurements confirming this hypothesis were not available. However, the annual water balance results are comparable to measurements, statistical figures and SWAT simulation results. Both SWAT and the water balance model had problems modelling the exchange processes between surface runoff, intermediate storage and groundwater and displaye similarvalues for the accuracy measures.

References

1. Bremicker, M., Luce, A., Haag, I., & Sieber, A. (2005). *Das Wasserhaushaltsmodell Larsim: Modellgrundlagen* (80 p) Retrieved 15 May, 2011, from http://www.hvz.baden-wuerttemberg.de/pdf/LARSIM_DE_2005-06-24.pdf.
2. Uhlenbrook, S. (1999). *Untersuchung und Modellierung der Abflussbildung in einem mesoskaligen Einzugsgebiet* (201 p). Institut für Hydrologie der Universität Freiburg i. Br., Freiburg (Freiburger Schriften zur Hydrologie, 10).
3. Beven, K. J. (2001). How far can we go in distributed hydrological modelling? *Hydrology and Earth System Sciences, 5*(1), 1–12.
4. Aerts, J. C., Kriek, M., & Schepel, M. (1999). Spatial tools for river basins and environment and analysis of management options (STREAM): Set up and requirements. *Physics and Chemistry of the Earth Part B Hydrology Oceans and Atmosphere, 24*(6), 591–595.

5. Bäse, F. (2005). *Beurteilung der Parametersensitivität und der Vorhersagesicherheit am Beispiel des hydrologischen Modells J2000* (102 p). Unpublished diploma thesis, Friedrich-Schiller Universität Jena.
6. Bouwer, L. M., Aerts, J. C., Droogers, P., & Dolman, A. J. (2006). Detecting the long-term impacts from climate variability and increasing water consumption on runoff in the Krishna river basin (India). *Hydrology and Earth System Sciences, 10*, 703–713.
7. Glugla, G. & Fürtig, G. (1997). *Dokumentation zur Anwendung des Rechenprogramms ABIMO* (37 p), Bundesanstalt für Gewässerkunde, Berlin.
8. Meßer, J. (1997). *Auswirkungen der Urbanisierung auf die Grundwasserneubildung im Ruhrgebiet unter besonderer Berücksichtigung der Castroper Hochfläche und des Stadtgebietes Herne, Essen* (58 p). Deutsche Montan Technologie GmbH (DMT-Berichte aus Forschung und Entwicklung).
9. Meßer, J. (2008). *Ein vereinfachtes Verfahren zur Berechnung der flächendifferenzierten Grundwasserneubildung in Mitteleuropa* (60 p). Lippe: Lippe Wassertechnik.
10. Haase, D. (2009). Effects of urbanisation on the water balance: A long-term trajectory. *Environmental Impact Assessment Review, 29*, 211–219.
11. Schmidt, G., Gretzschel, O., Volk, M., & Uhl, M. (2003). Konzept zur skalenspezifischen Modellierung des Wasser- und Stoffhaushaltes im Projekt FLUMAGIS: A concept for the scale-specific simulation of water-bound material fluxes in the project FLUMAGIS. In K. Hennrich, M. Rode, A. Bronstert (Eds.), *6. Workshop zur großskaligen Modellierung in der Hydrologie—Flussgebietsmanagement* (pp. 7–20). Kassel: University Press.
12. Glugla, G., Jankiewicz, P., Rachimow, C., Lojek, K., Richter, K., Fürtig, G., & Krahe, P. (2003). *Wasserhaushaltsverfahren zur Berechnung vieljähriger Mittelwerte der tatsächlichen Verdunstung und des Gesamtabflusses* (102 p). BfG, Koblenz (BfG-Bericht, 1342).
13. Kite, G. (2000). Using a basin-scale hydrological model to estimate crop transpiration and soil evaporation. *Journal of Hydrology, 229*, 59–69.
14. Simonneaux, V., Duchemin, B., Helson, D., Er-Raki, S., Olioso, A., & Chehbouni, A. (2008). The use of high-resolution image time series for crop classification and evapotranspiration estimate over an irrigated area in Central Morocco. *International Journal of Remote Sensing, 29*(1), 95–116.
15. Zhao, C., Nan, Z., & Cheng, G. (2005). Methods for estimating irrigation needs of spring wheat in the middle Heihe Basin, China. *Agricultural Water Management, 75*, 54–70.
16. Armbruster, V. (2002). *Grundwasserneubildung in Baden-Württemberg* (140 p). Dissertation. Institut für Hydrologie der Universität Freiburg i. Br., (Freiburger Schriften zur Hydrologie, 17).
17. Merz, R., Blöschl, G., & Parajka, J. (2006). Spatio-temporal variability of event runoff coefficients. *Journal of Hydrology, 331*(3–4), 591–604.
18. Huintjes, E., Li, H., Sauter, T., Li, Z., & Schneider, C. (2010). Degree-day modelling of the surface mass balance of Urumqi Glacier No. 1, Tian Shan, China. *The Cryosphere Discussions, 4*, 207–232.
19. Neitsch, S. L., Arnold, J. G., Kiniry, J. R., & Williams, J. R. (2005). *Soil and water assessment tool theoretical documentation: Version 2005* (476 p). Temple: Grassland, Soil and Water Research Laboratory.
20. Buttafuoco, G., Caloiero, T., & Coscarelli, R. (2010). Spatial uncertainty assessment in modelling reference evapotranspiration at regional scale. *Hydrology and Earth System Sciences, 14*(11), 2319–2327.
21. Aguilar, C., & Polo, M. J. (2011). Calculation of reference evapotranspiration surfaces in distributed hydrological modelling at different temporal scales. *Hydrology and Earth System Sciences Discussions, 8*, 4813–4850.
22. Contreras, S., Jobbágy, E. G., Villagra, P. E., Nosetto, M. D., & Puigdefábregas, J. (2011). Remote sensing estimates of supplementary water consumption by arid ecosystems of central Argentina. *Journal of Hydrology, 397*(1–2), 10–22.

23. Allen, R. G. (1998). *Crop evapotranspiration: Guidelines for computing crop water requirements* (300 p). Rome (FAO irrigation and drainage paper, 56): Food and Agriculture Organisation of the United Nations.
24. DVWK (1996). *Ermittlung der Verdunstung von Land- und Wasserflächen: DVWK-Merkblätter zur Wasserwirtschaft* (135 p). Wirtschafts- und Verl.-Ges. Gas und Wasser, Bonn (Gas und Wasser, 238).
25. Iqbal, M. (1983). *An introduction to solar radiation* (390 p). Toronto: Academic Press.
26. Schulla, J. (1997). *Hydrologische Modellierung von Flussgebieten zur Abschätzung der Folgen von Klimaänderungen* (163 p). Dissertation, Dresden (Zürcher Geographische Schriften, 69).
27. Allen, R. G. (1986). A Penman for all seasons. *Journal of Irrigation and Drainage Engineering, 112*(4), 348–368.
28. Monteith, J. L. (1965). Evaporation and environment. In G. E. Fogg (Ed.), *Symposium of the society for experimental biology: The state and movement of water in living organisms* (Vol. 19, pp. 205–234). New York: Academic Press.
29. Penman, H. L. (1948). Natural evaporation from open water, bare soil and grass. *Proceedings of the Royal Society of London. Series A Mathematical and Physical Science,* 193(1032), 120–145.
30. Hurtalová, T., Matejka, F., Roznovsky, J., & Janous, D. (2003). Aerodynamic resistance of spruce forest stand in relation to roughness length and airflow. *Contributions to Geophysics and Geodesy, 33*(3), 147–160.
31. Lindroth, A. (1993). Aerodynamic and canopy resistance of short-rotation forest in relation to leaf area index and climate. *Boundary-Layer Meteorology, 66*, 265–279.
32. Tiyip, T., Taff, G. N., Kung, H.-T., & Zhang, F. (2010). Remote sensing assessment of salinisation impacts in the Tarim Basin: The Delta Oasis of the Ugan and Kuqa Rivers. In G. Schneier-Madanes & M.-F. Courel (Eds.), *Water and sustainability in arid regions. Bridging the gap between physical and social sciences* (pp. 15–32). Dordrecht: Springer Science+Business Media B.V.
33. Thom, A. S., & Oliver, H. R. (1977). On Penman's equation for estimating regional evapotranspiration. *Quarterly Journal of the Royal Meteorological Society, 103*, 345–357.
34. Vietinghoff, H. (2000). *Die Verdunstung freier Wasserflächen: Grundlagen, Einflußfaktoren und Methoden der Ermittlung* (113 p). UFO, Atelier für Gestaltung und Verlag, Allensbach (Ufo Naturwissenschaft, 201).
35. Wimmer, F., Schlaffer, S., Aus der Beek, T., & Menzel, L. (2009). Distributed modelling of climate change impacts on snow sublimation in Northern Mongolia. *Advances in Geosciences, 21*, 117–124.
36. Knauf, D. (1976). *Die Abflußbildung in schneebedeckten Einzugsgebieten des Mittelgebirges* (155 p). Institut für Hydraulik und Hydrologie Darmstadt, Darmstadt (Technische Berichte, 17).
37. Rango, A., & Martinec, J. (1995). Revisiting the degree-day method for snowmelt computations. *Water Resources Bulletin, 31*(4), 657–669.
38. Bagchi, A. K. (1983). Areal value of degree-day factor. *Hydrological Sciences Journal/ Journal Sciences Hydrologiques, 28*(4), 499–511.
39. Kuusisto, E. (1980). On the values and variability of degree-day melting factor in Finland. *Nordic Hydrology, 11*, 235–242.
40. Martinec, J., Rango, A., & Roberts, R. (2008). *Snowmelt runoff model (SRM) user's manual: Edited by Enrique Gómez-Landesa & Max P. Bleiweiss* (178 p). Las Cruces: New Mexico State University.
41. Shin, H. J., Park, M. J., Ha, R., Yi, J. E., Kim, G. S., & Kim, S.-J. (2011). *Evaluation of snow melt contribution to streamflow in a heavy snowfall watershed of South Korea using SWAT model.* Toledo, Spain: SWAT 2011.
42. Neitsch, S. L., Arnold, J. G., Kiniry, J. R., Srinivasan, R., & Williams, J. R. (2009). *Soil and water assessment tool: Input/output file documentation Version 2009* (604 p). Temple: Grassland, Soil and Water Research Laboratory.

43. Merz, R., & Blöschl, G. (2004). Regionalisation of catchment model parameters. *Journal of Hydrology, 287*, 95–123.
44. Kondo, J., & Yamazaki, T. (1990). A prediction model for snowmelt, snow surface temperature and freezing depth using a heat balance method. *Journal of Applied Meteorology, 29*, 375–384.
45. Hock, R. (2003). Temperature index melt modelling in mountain areas. *Journal of Hydrology, 282*, 102–115.
46. Beven, K. J. (1991). Spatially distributed modelling: Conceptual approach to runoff prediction. In D. S. Bowles (Ed.), *Recent advances in the modelling of hydrologic systems. Proceedings of the NATO Advanced Study Institute on Recent Advances in the Modelling of Hydrologic Systems, Sintra, Portugal 10–23 July, 1988* (pp. 373–387). Dordrecht: Kluwer.
47. International Soil Reference and Information Centre (ISRIC) (2011). Soil and terrain database for China. Version 1.0—scale 1:1 million. ISRIC Wageningen. Retrieved from 5 Oct, 2011. http://www.isric.org/data/soil-and-terrain-database-china.
48. Batjes, N. (2002). *Soil parameter estimates for the soil types of the world for use in global and regional modelling* (Version 2.1., 46 p). Wageningen: ISRIC Report, 2002/02c, International Food Policy Research Institute and International Soil Reference and Information Centre. Retrieved 25 Mar, 2011, from http://www.isric.eu/isric/webdocs/docs/ISRIC_Report_2002_02c.pdf.
49. Meinrath, G., & Schneider, P. (2007). *Quality assurance for chemistry and environmental science: Metrology from pH measurement to nuclear waste disposal* (326 p). Berlin, Heidelberg: Springer.
50. Hamby, D. M. (1994). A review of techniques for parameter sensitivity analysis of environmental models. *Environmental Monitoring Assessment, 32*, 135–154.
51. Campolongo, F., Kleijnen, J., & Andres, T. (2004). Screening Methods. In A. Saltelli (Ed.), *Sensitivity analysis* (pp. 65–80). Chichester: Wiley.
52. Campolongo, F., Saltelli, A., Sorensen, T., & Tarantola, S. (2004). Hitchhiker's guide to sensitivity analysis. In A. Saltelli (Ed.), *Sensitivity Analysis* (pp. 15–47). Chichester: Wiley.
53. Lenhart, T., Eckhardt, K., Fohrer, N., & Frede, H. G. (2002). Comparison of two different approaches of sensitivity analysis. *Physics and Chemistry of the Earth, Parts A/B/C,27*(9–10), 645–654.
54. Harlin, J., & Kung, C.-S. (1992). Parameter uncertainty and simulation of design floods in Sweden. *Journal of Hydrology, 137*, 209–230.
55. Frühwirth, R., & Regler, M. (1983). *Monte-Carlo-Methoden: Eine Einführung* (171 p). Mannheim: Bibliogr. Inst.
56. Lin, S., Jing, C., Chaplot, V., Yu, X., Zhang, Z., Moore, N., et al. (2010). Effect of DEM resolution on SWAT outputs of runoff sediment and nutrients. *Hydrology and Earth System Sciences Discussions, 7*, 4411–4435.
57. Liu, T., Willems, P., Pan, X. L., Bao, A. M., Chen, X., Veroustraete, F., et al. (2011). Climate change impact on water resource extremes in a headwater region of the Tarim basin in China. *Hydrology and Earth System Sciences, 15*, 3511–3527.
58. U.S. Geological Survey (2005). Shuttle Radar Topography Mission, version 2, 3-arc second resolution. Global Land Cover Facility, University of Maryland, College Park, Maryland Retrieved 31 Jan, 2008, from http://glcf.umiacs.umd.edu/data/srtm/index.shtml.
59. U.S. Geological Survey & Japan ASTER Programme (2009). ASTER GDEM Version 1. NASA Land Processes Distributed Active Archive Center, Sioux Falls, Retrieved 15 Jul, 2009, from http://earthexplorer.usgs.gov/.
60. Aster Validation Team (2009). *ASTER global dem validation summary report: ASTER GDEM validation team: METI/ERSDAC, NASA/LPDAAC, USGS/EROS in cooperation with NGA and other collaborators* (28 p). Retrieved 19 Oct, 2011, from https://lpdaac.usgs.gov/lpdaac/content/download/4009/20069/version/3/file/ASTER+GDEM+Validation+Summary+Report.pdf.
61. Albertz, J. (2001). *Einführung in die Fernerkundung: Grundlagen der Interpretation von Luft- und Satellitenbildern* (249 p). Darmstadt: Wiss. Buchges.

62. Sabins, F. F. (2007). *Remote sensing: Principles and interpretation* (494 p). Ill: Waveland Press, Long Grove.
63. Hengl, T., & Reuter, H. (2011). How accurate and usable is GDEM? A statistical assessment of GDEM using LiDAR data. *Geomorphometry,* 45–48.
64. Hirt, C., Filmer, M. S., & Featherstone, W. E. (2010). Comparison and validation of the recent freely available ASTER-GDEM ver1, SRTM ver4.1 and GEODATA DEM-9S ver3 digital elevation models over Australia: Australian Journal of Earth Sciences. *Australian Journal of Earth Sciences, 57*(3), 337–347.
65. Zhang, W., & Montgomery, D. R. (1994). Digital elevation model grid size, landscape representation, and hydrologic simulations. *Water Resources Research, 30*(4), 1019–1028.
66. Jacobsen, K. (2010). Comparison of ASTER GDEMs with SRTM Height Models. In R. Reuter (Ed.), *30th EARSeL symposium remote sensing for science, education, and natural and cultural heritage* (pp. 521–526). Paris, France: UNESCO.
67. Wang, G. Q., Zhang, J. Y., Jin, J. L., Pagano, T. C., Calow, R., Bao, Z. X., et al. (2012). Assessing water resources in China using PRECIS projections and a VIC model. *Hydrology and Earth System Sciences, 16*, 231–232.
68. Di Luzio, M., Arnold, J. G., & Srinivasan, R. (2005). Effect of GIS data quality on small watershed stream flow and sediment simulations. *Hydrological Processes,19*(3), 629–650.
69. Kienzle, S. (2004). The Effect of DEM Raster Resolution on First Order, Second Order and Compound Terrain Derivatives. *Transactions in GIS*, *8*(1), 83–111.
70. de Vente, J., Poesen, J., Govers, G., & Boix-Fayos, C. (2009). The implications of data selection for regional erosion and sediment yield modelling. *Earth Surface Processes and Landforms*, *34*(15), 1994–2007.
71. Bosch, D. D., Sheridan, J. M., Batten, H. L., & Arnold, J. G. (2004). Evaluation of the SWAT model on a coastal plain agricultural watershed. *Transactions of the American Society of Agricultural Engineers*, *47*(5), 1493–1506.
72. Dixon, B., & Earls, J. (2009). Resample or not?! Effects of resolution of DEMs in watershed modelling. *Hydrological Processes*, *23*(12), 1714–1724.
73. Tulu, M. D. (February, 2005). *SRTM DEM suitability in runoff studies* (87 p). Retrieved 19 Oct, 2011, from http://www.itc.nl/library/papers_2005/msc/wrem/mesay.pdf.
74. Terribile, F., Coppola, A., Langella, G., Martina, M., & Basile, A. (2011). Potential and limitations of using soil mapping information to understand landscape hydrology. *Hydrology and Earth System Sciences Discussions*, *8*, 4927–4977.
75. Bruneau, P., Gascuel-Odoux, C., Robin, P., Merot, P., & Beven, K. J. (1995). Sensitivity to space and time resolution of a hydrological model using digital elevation data. *Hydrological Processes, 9*, 69–81.
76. ESRI (2010): ArcGIS, Version 10.0 [Computer program], Redlands, CA.
77. Autonomous Region Bureau of Surveying and Mapping (2004). *Xinjiang-Weiyu'er-Zizhiqu-dituji: Xinjiang Uygur Autonomous Region Atlas* (307 p). Beijing: Zhongguo ditu chubanshe.
78. Bubenzer, O., & Bolten, A. (2008). The use of new elevation data (SRTM/ASTER) for the detection and morphometric quantification of Pleistocene megadunes (draa) in the eastern Sahara and the southern Namib. *Geomorphology, 102*, 221–231.
79. National Oceanic and Atmospheric Administration of the U.S. Department of Commerce National Climatic Data Centre (NOAA NCDC) (2011). Global Summary of the Day Diwopu Station, Station Number 514635, 29.02.1985–31.12.2010. Retrieved 16 May, 2011, from http://www7.ncdc.noaa.gov/CDO/cdo.
80. National Oceanic and Atmospheric Administration of the U.S. Department of Commerce National Climatic Data Centre (NOAA NCDC) (2011). Global Summary of the Day Fukang Station, Station Number 513650, 28.03.1962–13.05.1997. Retrieved 23 Feb, 2010, from http://www7.ncdc.noaa.gov/CDO/cdo.
81. National Oceanic and Atmospheric Administration of the U.S. Department of Commerce National Climatic Data Centre (NOAA NCDC) (2011). Global Summary of the Day Shihezi

Station, Station Number 513560, 21.08.1956–19.08.1996. Retrieved 23 Feb, 2010, from http://www7.ncdc.noaa.gov/CDO/cdo.

82. National Oceanic and Atmospheric Administration of the U.S. Department of Commerce National Climatic Data Centre (NOAA NCDC) (2011). Global Summary of the Day Wulumuqi Station, Station Number 514630, 22.08.1956–31.12.2010. Retrieved 16 May, 2011, from http://www7.ncdc.noaa.gov/CDO/cdo.
83. Peterson, T. C., Easterling, D. R., Karls, T. R., Groisman, P., Nicholls, N., Plummer, N., et al. (1998). Homogeneity adjustments of in situ atmospheric climate data: A review. *International Journal of Climatology, 18*, 1493–1517.
84. Dahmen, E. R., & Hall, M. J. (1990). *Screening of hydrological data: Tests for stationarity and relative consistency* (58 p). ILRI: Wageningen.
85. Barth, N. C. (2011). *Auswertung der Temperatur- und Niederschlagsdaten von 15 Klimastationen im Umkreis von Urumqi (AR Xinjiang, China)* (65 p). Heidelberg: Unpublished bachelor thesis, Ruprecht-Karls-Universität.
86. Daly, C. (2006). Guidelines for assessing the suitability of spatial climate data sets. *International Journal of Climatology, 26*, 707–721.
87. U.S. Geological Survey (2012a). *Land surface temperature and emissivity daily L3 global 1 km Grid SIN*. Retrieved 19 Mar, 2012, from https://lpdaac.usgs.gov/products/modis_products_table/mod11a1.
88. Ajami, N. K., Gupta, H., Wagener, T., & Sorooshian, S. (2004). Calibration of a semi-distributed hydrologic model for streamflow estimation along a river system. *Journal of Hydrology, 298*, 112–135.
89. Statistics Bureau of Urumqi (2000). *Urumqi Statistical Yearbook 2000* (325 p). China Statistics Press: Beijing
90. Statistics Bureau of Urumqi (2005). *Urumqi Statistical Yearbook 2005* (366 p). China Statistics Press: Beijing
91. Statistics Bureau of Urumqi (2006). *Urumqi Statistical Yearbook 2006* (439 p). China Statistics Press: Beijing
92. Statistics Bureau of Urumqi (2008). *Urumqi Statistical Yearbook 2008* (455 p). China Statistics Press: Beijing
93. Statistics Bureau of Urumqi (2009). *Urumqi Statistical Yearbook 2009* (506 p). Beijing: China Statistics Press
94. Statistics Bureau of Urumqi (2010). *Urumqi Statistical Yearbook 2010* (456 p). Beijing: China Statistics Press
95. Zhang, W., Ogawa, K., Ye, B., & Yamaguchi, Y. (2000). A monthly stream flow model for estimating the potential changes of river runoff on the projected global warming. *Hydrological Processes, 14*(10), 1851–1868.
96. NASA (2011). TRMM Rainfall Product TRMM_3B43, Monthly Average Rainfall, 0.25°×0.25° resolution. Retrieved 19 April, 2012, from http://mirador.gsfc.nasa.gov/collections/TRMM_3B43__006.shtml.
97. Hartkamp, A. D., Beurs, K., de Stein, A., & White, J. W. (1999). *Interpolation techniques for climate variables* (26 p). CIMMYT, Mexico, D.F. (NRG-GIS Series, 99-01).
98. Mitas, L., & Mitasova, H. (1999). Spatial interpolation. In P. A. Longley, M. F. Goodchild, D. J. Maguire & D. W. Rhind (Eds.), *Geographical information systems. Principles, techniques, management and applications* (pp. 481–492). Hoboken, NJ: Wiley.
99. Yan, J., Chen, X., Luo, G., & Guo, Q. (2006). Temporal and spatial variability response of groundwater level to land use/land cover change in oases of arid areas. *Chinese Science Bulletin, 51*(1), 51–59
100. Hunukumbura, P. B., Tachikawa, Y., & Shiiba, M. (2011). Distributed hydrological model transferability across basins with different hydro-climatic characteristics. *Hydrological Processes*. doi: 10.1002/hyp.8294.
101. ESRI. (2003). *AcrGIS 9 Using ArcGIS Geostatistical Analyst* (300 p). Redlands, CA: ESRI.

102. Javanmard, S., Yatagai, A., Nodzu, M. I., BodaghJamali, J., & Kawamoto, H. (2010). Comparing high-resolution gridded precipitation data with satellite rainfall estimates of TRMM_3B42 over Iran. *Advances in Geosciences, 25*, 119–125.
103. Aguilar, C., Herrero, J., & Polo, M. J. (2010). Topographic effects on solar radiation distribution in mountainous watersheds and their influence on reference evapotranspiration estimates at watershed scale. *Hydrology and Earth System Sciences, 14*(12), 2479–2494.
104. Statistics Bureau of Urumqi (2010). *Urumqi Statistical Yearbook 2010* (456 p). Beijing: China Statistics Press.
105. Burns, I. S., Scott, S. N., Levick, L. R., Semmens, D. J., Miller, S. N., Hernandez, M., et al. (2007). *Automated geospatial watershed assessment (AGWA) documentation* (Version 2.0, 145 p). Tuscon, Arizona.
106. van Dijk, A. I., & Renzullo, L. J. (2011). Water resource monitoring systems and the role of satellite observations. *Hydrology and Earth System Sciences, 15*, 39–55.
107. NASA Landsat Program (2010). *Landsat ETM+ scenes L1T*, USGS, Sioux Falls. Retrieved from http://glovis.usgs.gov/
108. ERDAS (2010): *ERDAS Imagine* (Version 10) [Computer program], Norcross.
109. Chander, G., Markham, B.L., & Helder, D. L. (2009). Summary of current radiometric calibration coefficients for Landsat MSS, TM, ETM+, and EO-1 ALI sensors. *Remote Sensing of the Environment, 113*, 893–903.
110. Furby, S. L., & Campbell, N. A. (2001). Calibrating images from different dates to 'like-value' digital counts. *Remote Sensing of Environment, 77*, 186–196.
111. Beisl, U., Telaar, J., & Schönermark, M. v. (2008). Atmospheric correction, reflectance calibration and BRDF correction for ADS40 image data. *The International Archives of the Photogrammetry, Remote Sensing and Spatial Information Sciences, 37*(B7), 7–12.
112. Callahan, K. E. (2003). *Validation of a radiometric normalisation procedure for satellite derived imagery within a change detection framework* (61 p). Master thesis: Utah State University.
113. El Hajj, M., Bégué, A., Lafrance, B., HAgolle, O., Dediau, G., & Rumeau, M. (2008). Relative radiometric normalisation and atmospheric correction of a SPOT 5 time series. *Sensors, 8*, 2774–2791.
114. Hong, G., & Zhang, Y. (2008). A comparative study on radiometric normalisation using high resolution satellite images. *International Journal of Remote Sensing, 29*(2), 425–438.
115. Schott, J. R., Salvaggio, C., & Volchok, W. J. (1988). Radiometric scene normalisation using Pseudoinvariant features. *Remote Sensing of Environment, 26*, 1–16.
116. Yuan, D., & Elvidge, C. D. (1996). Comparison of relative radiometric normalisation techniques. *ISPRS Journal of Photogrammetry and Remote Sensing, 51*, 117–126.
117. Brütt, A. (2005). *Erstellung eines Moduls zur atmosphärischen Korrektur für optische Satellitendaten* (67 p). Bachelorarbeit: Georg-August-Universität Göttingen.
118. Schöttker, B. (2002). *Erfassung der Landbedeckung und Ableitung von Vegetationsveränder ungen anhand multitemporaler LANDSAT-Daten in Westafrika (Benin)* (152 p). Diploma thesis, Rheinische Friedrich-Wilhelms-Universität Bonn. Retrieved 19 Dec, 2008, from http://www.giub.uni-bonn.de/rsrg/www/docs/Schoettker_Diplomarbeit_2002.pdf.
119. Vogel, M. (2005). *Erfassung von Vegetationsveränderungen in Namibia mit Hilfe von Fernerkundungs-Change-Detection-Verfahren und unter Berücksichtigung rezenter Niederschlagsereignisse* (366 p). Dissertation, Julius-Maximilians-Universität Würzburg.
120. Neubert, M., & Meinel, G. (2005). Atmosphärische und topographische Korrektur von IKONOS-Daten mit ATCOR. In C. Heipke, K. Jacobsen & M. Gerke (Eds.), *ISPRS Hannover Workshop 2005*. High-Resolution Earth Imaging for Geospatial Information (17–20 May, 2005). *Hannover International Archives of Photogrammetry and Remote Sensing, 36*, 1–6.
121. Belward, A. S. (1991). Spectral characteristics of vegetation, soil and water in the visible, near-infrared and middle-infrared wavelength. In A. S. Belward & C. R. Valenzuela (Eds.), *Remote sensing and geographical information systems for resource management in developing countries* (pp. 31–53). Dordrecht: Kluwer.

122. Su, Z. (2000). Remote sensing of land use and vegetation for mesoscale hydrological studies. *International Journal of Remote Sensing, 21*(2), 213–233.
123. Kauth, R. J., & Thomas, G. S. (1976). The tasseled cap: a graphic description of the spectral-temporal development of agricultural crops as seen by LANDSAT. *Symposium on Machine Processing of Remotely Sensed Data,1976*, 4B-41–4B-51 (June 29–July 1).
124. Franke, J. (2003). *Analyse der Aussagefähigkeit verschiedener satellitengestützter Vegetationsindizes bezüglich der räumlichen Vegetationsverteilung* (121 p). Unpublished dimploma thesis, Rheinische Friedirch-Wilhelms-Universität Bonn. Retrieved 19 Dec, 2008, from http://www.giub.uni-bonn.de/rsrg/www/docs/Franke_Diplomarbeit_2003.pdf.
125. Runnstrom, M. C. (2003). Rangeland development of the Mu Us Sandy Land in semi-arid China: An analysis using landsat and NOAA remote sensing data. *Land Degradation and Development, 14*(2), 189–202.
126. Sobrino, J. A., Jiménez-Muñoz, J. C., & Paolini, L. (2004). Land surface temperature retrieval from LANDSAT TM 5. *Remote Sensing of Environment,90*(4), 434–440.
127. Huete, A. R. (1988). A soil-adjusted vegetation index (SAVI). *Remote Sensing of Environment, 25*, 295–309.
128. Karnieli, A., Gilad, U., Ponzet, M., Svoray, T., Mirzadinov, R., & Fedorina, O. (2008). Assessing land-cover change and degradation in the Central Asian deserts using satellite image processing and geostatistical methods. *Journal of Arid Environments, 72*, 2093–2105.
129. Congalton, R. G., & Green, K. (2009). *Assessing the accuracy of remotely sensed data: Principles and practices* (183 p). Boca Raton, FL: CRC Press.
130. Fricke, K. (2008). *The development of Midong New District, Urumqi, PR China: Ecological and historical context and environmental consequences* (165 p). Heidelberg University: Unpublished diploma thesis.
131. Congalton, R. G. (1991). A review of assessing the accuracy of classifications of remotely sensed data. *Remote Sensing of Environment, 37*, 35–46.
132. Jensen, J. R. (2005). *Introductory digital image processing: A remote sensing perspective* (526 p). Upper Saddle River, NJ: Prentice Hall.
133. Star, J. L. (Ed.) (2010). *Integration of geographic information systems and remote sensing* (225 p). Cambridge: Cambridge University Press.
134. Banerjee, M., Capozzoli, M., McSweeney, L., & Sinha, D. (1999). Beyond kappa. A review of interrater agreement measures. *The Canadian Journal of Statistics, 27*(1), 3–23.
135. Bastiaanssen, W. G., & Bandara, K. M. (2001). Evaporative depletion assessments for irrigated watersheds in Sri Lanka. *Irrigation Science, 21*, 1–15.
136. Foody, G. (2008). Harshness in image classification accuracy assessment. *International Journal of Remote Sensing, 29*(11), 3137–3158.
137. Shoshany, M. (2000). Satellite remote sensing of natural Mediterranean vegetation: A review within an ecological context. *Progress in Physical Geography, 24*(2), 153–178.
138. Abbaspour, K. C., Yang, J., Maximov, I., Siber, R., Bogner, K., Mieleitner, J., et al. (2007). Modelling hydrology and water quality in the pre-alpine/alpine Thur watershed using SWAT. *Journal of Hydrology, 333*(2–4), 413–430.
139. Water Affairs Bureau Urumqi (2003). *Water Report 2003*. Urumqi: Water Affairs Bureau Urumqi City, digital.
140. Water Affairs Bureau Urumqi (2004). *Water Report 2004* (27 p). Urumqi: Water Affairs Bureau Urumqi City.
141. Water Affairs Bureau Urumqi (2005). *Water Report 2005* (27 p). Urumqi: Water Affairs Bureau Urumqi City.
142. Water Affairs Bureau Urumqi (2007). *Water Report 2007* (24 p). Urumqi: Water Affairs Bureau Urumqi City.
143. Liersch, S. (2005). *Auswirkungen von Landnutzungsänderungen und umweltgerechten Bewirtschaftungsmethoden auf den Wasser- und Stoffhaushalt des Einzugsgebietes der Ems in Nordrhein-Westfalen: Modellierung einer Landnutzungssituation, die den Umweltstandards der EG-Wasserrahmenrichtlinie entspricht* (120 p). Diploma thesis, Universität Potsdam.

144. Bieri, M., & Schleiss, A. J. (2010). Hydrologisch-hydraulische Modellierung von alpinen Einzugsgebieten mit komplexen Kraftwerksanlagen. In K. Weber, E. Fenrich, T. Gebler, M. Kramer & M. Noack (Eds.), *12. Treffen junger WissenschaftlerInnen an Wasserbauinstituten* (pp. 99–105). Stuttgart: Institut für Wasserbau der Universität Stuttgart.
145. Martinec, J., & Rango, A. (1989). Merits of statistical criteria for the performance of hydrological models. *Water Resources Bulletin, 25*(2), 421–432.
146. Mu, Q., Zhao, M., & Running, S. W. (2011). MOD Global Terrestrial Evapotranspiration Dataset. Retrieved 27 Apr, 2012, from http://www.ntsg.umt.edu/project/mod16.
147. Mu, Q., Zhao, M., & Running, S. W. (2011). Improvements to a MODIS global terrestrial evapotranspiration algorithm. *Remote Sensing of Environment, 115*, 1781–1800.
148. Gassmann, P. W., Reyes, M. R., Green, C. H., & Arnold, J. G. (2007). The soil and water assessment tool: Historical development, applications, and future research directions. *Transactions of the American Society of Agricultural and Biological Engineers,50*(4), 1211–1250.
149. Grassland, Soil and Water Research Laboratory (2011). ArcSWAT (Soil and Water Assessment Tool), Version 2009. Retrieved from http://www.swat.tamu.edu/software/arcswat.
150. Legesse, D., Abiye, T. A., Vallet-Coulomb, C., & Abate, H. (2010). Streamflow sensitivity to climate and land cover changes: Meki River, Ethiopia. *Hydrology and Earth System Sciences,14*, 2277–2287.
151. Zhang, X., Chen, Y., Xia, J., & Yang, Q. (2011a). Impacts of climate change on the availability of water resources and water resources planning. In L. Ren, W. Wang & F. Yuan (Eds.), *Proceedings of IWRM2010 Hydrological Cycle and Water Resources Sustainability in Changing Environments* (Wallingford pp. 324–329, November 2010). Nanjing, China: IAHS Press.
152. Statistics Bureau of Xinjiang Uygur Autonomous Region (2005). *Xinjiang Statistical Yearbook 2005*. China Statistics Press, Beijing, CD-ROM.
153. Statistics Bureau of Xinjiang Uygur Autonomous Region (2006). *Xinjiang Statistical Yearbook 2006*. China Statistics Press, Beijing, CD-ROM.
154. Statistics Bureau of Xinjiang Uygur Autonomous Region (2007). *Xinjiang Statistical Yearbook 2007*. China Statistics Press, Beijing, CD-ROM.
155. Statistics Bureau of Xinjiang Uygur Autonomous Region (2009). *Xinjiang Statistical Yearbook 2009*. China Statistics Press, Beijing, CD-ROM.
156. Statistics Bureau of Xinjiang Uygur Autonomous Region (2010). *Xinjiang Statistical Yearbook 2010*. China Statistics Press, Beijing, CD-ROM.
157. Roth, K. (2011). *Soil physics lecture notes: V2.0*. Heidelberg: Institute of Environmental Physics, Heidelberg University.
158. Foken, T. (1990). *Turbulenter Energieaustausch zwischen Atmosphäre und Unterlage: Methoden, meßtechnische Realisierung sowie ihre Grenzen und Anwendungsmöglichkeiten* (287 p). Offenbach am Main: DWD.
159. Kelliher, F. M., Leuning, R., & Schulze, E. D. (1993). Evaporation and canopy characteristics of coniferous forests and grasslands. *Oecologia, 95*, 153–163.
160. Nakai, T., Sumida, A., Matsumoto, K., Daikuko, K., Iida, S., Park, H., et al. (2008). Aerodynamic scaling for estimating the mean height of dense canopy. *Boundary-Layer Meteorology, 128*, 423–443.
161. Rahman, K. (2011). Runoff simulation in a glacier dominated watershed using semi distributed hydrological model. *International SWAT Conference 2011, Toledo, Spain*. Retrieved 5 Oct, 2011, from http://swat.tamu.edu/media/40645/rahman.pdf.
162. Johnsson, H., & Lundin, L.-C. (1991). Surface runoff and soil water percolation as affected by snow and soil frost. *Journal of Hydrology, 122*(1–4), 141–159.
163. Rodriguez, E., Morris, C. S., Belz, J. E., Chapin, E. C., Martin, J. M., Daffer, W., & Hensley, S. (2005). *An assessment of the SRTM topographic products* (143 p). Pasadena, California: Technical Report JPL D-31639, Jet Propulsion Laboratory.

164. Bolten, A., & Bubenzer, O. (2006). New elevation data (SRTM/ASTER) for geomorphological and geoarchaeological research in arid regions. *Zeitschrift für Geomorphologie Suppl., 142*, 265–279.
165. U.S. Geological Survey (2012b). *MODIS overview*. Retrieved 19 Mar, 2012, from https://lpdaac.usgs.gov/products/modis_overview.
166. NASA (2011). *Product TRMM_3B43*. Retrieved 19 Apr, 2012, from http://mirador.gsfc.nasa.gov/collections/TRMM_3B43__006.shtml.
167. NASA (2012). *Landsat 7*. Retrieved 19 Apr, 2012, from http://landsat.gsfc.nasa.gov/about/L7_td.html.

Chapter 4
Scenarios

To simulate and evaluate the effects of climate and land use change on the water balance in Urumqi Region, several scenarios have been used. Two climate scenarios representing the climate situation in the past have been calculated and three possible climate scenarios for the year 2050. The planned land use changes are presented in Sect. 4.3.

4.1 Climate Scenarios

The long-term climatic situation was represented by the average monthly values for the period 1975–2010 (long-term climate scenario, SC_{LT}) and the short-term situation was simulated using the average monthly values from the last decade 2001–2010 (short-term climate scenario, SC_{ST}). Average temperature and precipitation values are presented in Fig. 4.1, but also the averages of minimum, maximum and dew point temperature, wind speed, and station pressure were calculated and used in the simulation as described before (see Chap. 3 and Sect. 3.3.2). When comparing the long- and the short-term average, increasing temperatures and precipitation values can be observed.

Monthly precipitation sums increased from November to May and especially in July, while decreasing in June, September and October. It is assumed that the spatial distribution of temperature and precipitation does not change and the extrapolation method as described in Sect. 3.3.2 and the same TRMM and MODIS data sets are applied.

4.2 Climate Change Scenarios

The effects of climate change on hydrological processes are mainly due to changes in precipitation and evaporation [2]. The most immediate effects are the increasing ability of air to absorb water as temperature rises. By rule of thumb the potential

K. Fricke, *Analysis and Modelling of Water Supply and Demand Under Climate Change, Land Use Transformation and Socio-Economic Development*, Springer Theses, DOI: 10.1007/978-3-319-01610-8_4,

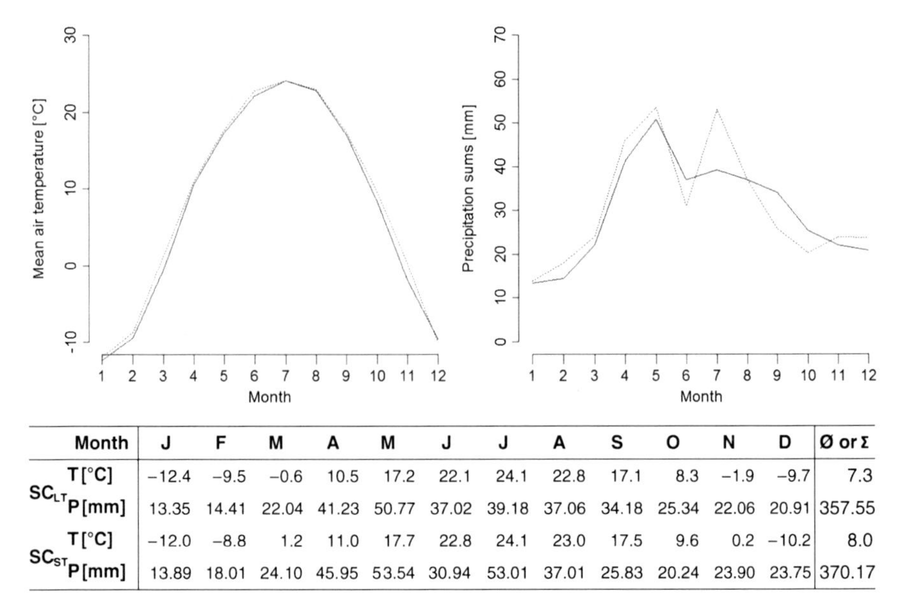

Month		J	F	M	A	M	J	J	A	S	O	N	D	Ø or Σ
SC_{LT}	T[°C]	−12.4	−9.5	−0.6	10.5	17.2	22.1	24.1	22.8	17.1	8.3	−1.9	−9.7	7.3
	P[mm]	13.35	14.41	22.04	41.23	50.77	37.02	39.18	37.06	34.18	25.34	22.06	20.91	357.55
SC_{ST}	T[°C]	−12.0	−8.8	1.2	11.0	17.7	22.8	24.1	23.0	17.5	9.6	0.2	−10.2	8.0
	P[mm]	13.89	18.01	24.10	45.95	53.54	30.94	53.01	37.01	25.83	20.24	23.90	23.75	370.17

Fig. 4.1 Mean monthly air temperature and monthly precipitation sums of the periods 1975–2010 and 2001–2010. *Solid line* 1975–2010, *dashed line* 2001–2010 (*Data source* NOAA NCDC [15])

evapotranspiration increases by 4 % for every °C increase in temperature. Additional changes happen in net radiation, humidity, wind speed and cloudiness and vegetation characteristics with the climate change and in turn alter the rate of potential evapotranspiration [2]. The runoff generation is also influenced by soil and land cover properties that may be affected by global warming. Higher temperatures and rainfall can lead to a loss of soil organic matter and decrease the ability of the soil to hold moisture depending on the soil type. Increased desiccation in summer may also lead to greater infiltration and increased rainfall may foster the development of gleyed layers that limit downward infiltration [2]. However, these effects are difficult to assess and to model, especially without detailed information about the soil properties and field research and are therefore not included in the modelling. Only the effects of anthropogenic land use changes are discussed in Sect. 5.3.

Although many **climatic variables** influence evaporation (see Sect. 3.1.1), it would not be efficient to use all of them in the climate change scenarios. They **were selected according to their importance for the hydrological model and the exhibition of trends in the past** as determined by the sensitivity analysis and the *Mann-Kendall test* for trend detection, respectively. The *Mann-Kendall test* is a test for non-parametric data sets, i.e. normal distribution is not necessary, which accommodates the potentially not completely normally distributed datasets. The annual and monthly mean or sums as well as the grouped monthly values 1975–2010 were analysed in the program R [16] using the Mann-Kendall

Table 4.1 *Mann-Kendall test* statistics for annual means 1975–2010 from Wulumuqi station with tau-statistics denoting the agreement between time and value of the variable

Climate variable	Tau	*p*-value (2-sided)	CF[a] (%)	S
Temperature	0.408	0.0005	99.90	256
Maximum temperature	0.186	0.1141	77.19	117
Minimum temperature	0.519	0.0001	100.00	327
Dew point temperature	0.314	0.0073	98.54	198
Precipitation	0.171	0.1450	71.00	108
Station pressure	−0.490	0.0002	99.96	−198
Wind speed	−0.114	0.3203	35.94	−80

[a] *CF* confidence factor = (1 − (2 * *p*-value)) %
If the symmetric difference between the two data sets and rankings is smallest, the coefficient has the value 1 ([1]). The probability for $\alpha = 0.05$ is represented by *p* value, confidence factor CF and S value which indicates the direction of the trend

package.[1] Almost all values (annual, monthly, monthly grouped) showed negative or positive trends.

However, only mean annual mean, dew point and minimum temperature and station pressure as well as some monthly grouped values showed very certain (95 % confidence level) or probably (90 % confidence level) positive trends (Table 4.1, monthly results not shown). These findings are consistent with the significant increase of mean annual air temperature as evaluated by Barth [3] and described in Sect. 2.4.1.

The approach to model the hypothetical changes of monthly mean temperatures and precipitation sums was applied by [19] who consider this approach more complicated than just increasing mean air temperature, but more consistent with the observed trends. Feedback mechanisms between the climate variables within the climate system were not considered. A very popular method for trend analysis and projection in the future is the linear regression model which was considered for the creation of the climate change scenarios. The temperature changes are extracted from the linear trend over the time series for a given period. Non-linear regression and extrapolation models are either not suitable for projections in the far future or need too many parameters that have to be calibrated [8]. The considered approach is not undisputed as the behaviour of the climate system is actually non-linear. The modelled linear trend is not stable and sensitive to beginning and end points: "Adding or subtracting a few points can result in marked differences in the estimated trend" [20: 336].

To justify the use of linear regression models for the purpose of prediction the assumptions of linearity, independence of errors, homoscedasticity and normality have to be fulfilled by the data. The linearity was tested by visual evaluation when plotting the monthly and annual mean values. The independence of the errors and absence of serial/auto-correlation was tested with the *Durbin-Watson test* for mean

[1] http://cran.r-project.org/web/packages/Kendall/index.html, accessed 19.12.2011.

Table 4.2 Independence of errors and absence of auto-correlation, homoscedasticity and normality tested for annual and monthly mean values of the climate variables mean, maximum, minimum and dewpoint temperature, precipitation, station pressure and wind speed at Wulumuqi station

Climate variable (years used for test statistics)	Independence	Monthly homoscedasticity	Normality	
			Annual means	Non-normal months
Mean temperature (1975–2010)	+	+	+	2
Maximum temperature (1975–2010)	+	+	+	3
Minimum temperature (1975–2010)	(+)	+	+	1
Dewpoint temperature (1975–2010)	+	+	+	1
Precipitation (1973–2010)	(+)	–	+	9
Station pressure (1976–1981,1988–2010)	–	+	–	1
Wind speed (1973–2010)	–	–	–	8
	+ no statistical evidence for auto-correlation (+) test inconclusive or one month with positive auto-correlation – positively autocorrelated	+ homo-scedacticity – hetero-scedacticity	+ normal distribution – non-normal distribution	

All statistics were tested for 5 % significance level (*Source* raw data from NOAA NCDC [15], own calculations)

annual values and all monthly values separately. Homoscedasticity or homogeneity of variance was evaluated using *Levene's test*, and normality of the data distribution was tested with the *Shapiro-Wilk test*. All test statistics were evaluated at the 5 % significance level (see Table 4.2) and taken into account when chosing the climate variables to be projected.

Using the results of the sensitivity analysis, the number of important variables for potential evapotranspiration was brought down to canopy height, maximum temperature, wind speed, station pressure, minimum temperature, dew point temperature and solar radiation. For the resulting effective runoff, again canopy height, precipitation, snow water equivalent and mean air temperature were the most important variables. Canopy height is mainly managed by anthropogenic land use and therefore not discussed in this chapter. Solar radiation was calculated depending on topography, but not including solar and atmospheric factors. These are also not discussed here as the involving solar radiation system and changes and feedback processes due to atmospheric changes proved to be too complex. For the effects of climate change on wind speed and station pressure the information and research findings available for Urumqi Region and Xinjiang were quite limited. Additionally, wind speed data did not exhibit a significant trend at the 5 %-level

(Table 4.1) and showed a non-normal distribution of annual and monthly values (see Table 4.2), which made it unsuitable for a linear regression analysis. Station pressure exhibited a significant negative trend (Table 4.1), but the data series had large gaps which could not be filled with neighbouring stations' data or interpolated and had also a non-normal distribution for annual and monthly means (Table 4.2). For the climate scenarios, the values of these two climate variables were taken from the long-term average of the station measurements.

For temperatures and precipitation, the data availability at Wulumuqi station is good. Mean air temperature is the climate variable whose changes can be estimated most exactly [19], and together with precipitation it also influences the snow water equivalent. The *Mann-Kendall test* showed significant increasing trends for mean, minimum and dew point temperatures and all temperatures fulfil the preconditions for the linear regression analysis. Hence, the trend for temperature change in Urumqi was assessed using linear regression analysis. This approach could not be applied to the precipitation data as it was not normally distributed and a significant trend could not be observed with the *Mann-Kendall test* statistics. Therefore, the precipitation values were extrapolated based on the regional climate projections for the A1B emission scenario of the IPCC:

> The A1 storyline and scenario family describes a future world of very rapid economic growth, global population that peaks in mid-century and declines thereafter, and the rapid introduction of new and more efficient technologies. [...]The A1 scenario family develops into three groups that describe alternative directions of technological change in the energy system. The three A1 groups are distinguished by their technological emphasis: fossil intensive (A1FI), non-fossil energy sources (A1T), or a balance across all sources (A1B) [9, 10].

4.2.1 Temperature Projection

In this subsection, the preprocessing of the temperature data and the method for projecting future monthly temperature values are described. The preprocessing included padding and filtering of the data sets. By calculating monthly average temperatures, an average filter has already been applied to the daily values. As the average function takes every single value into account, strong deviations from the mean also influence the final filtered value. In the course of climate change, the frequency of extreme values can increase. Therefore we want to consider these extreme values in our calculation which is possible with the average filter.

However, the chosen period for the regression analysis and calculation of the trend line and values or even the last few data points strongly influence the resulting trend line. Extreme values at the margins should only be considered if their occurrence was persisting over more than a few years. Therefore, the **data was smoothed to remove inner-decadal trends and outlier and their effects on the trend lines** before calculating the trend values and prognosis [4, 11, 12, 20]. The smoothing filter has 13 weights 1/576 (1–6–19–42–71–96–106–96–71–42–19–6–1) and removes trends with a cycle of less than 4 years [20] (see Figs. 4.2, 4.3). It was adapted from Trenberth et al. [20] to remove fluctuations on less than decadal time scales from annual values

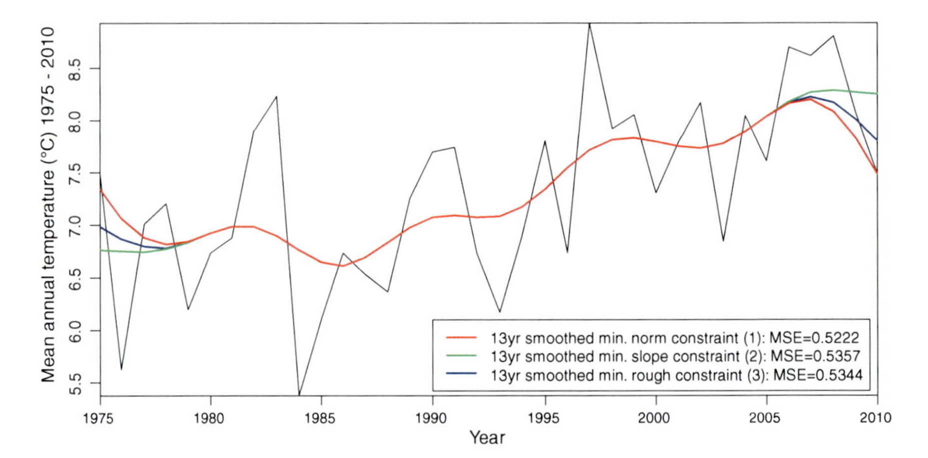

Fig. 4.2 Annual mean temperature series of Wulumuqi station shown along with 13-year smoothes of series based on alternative boundary constraints (1)–(3). Annual mean temperature series in black, associated mean squared error (MSE) scores favour use of the 'minimum norm' constraint (*Data source* NOAA NCDC [15], own calculations)

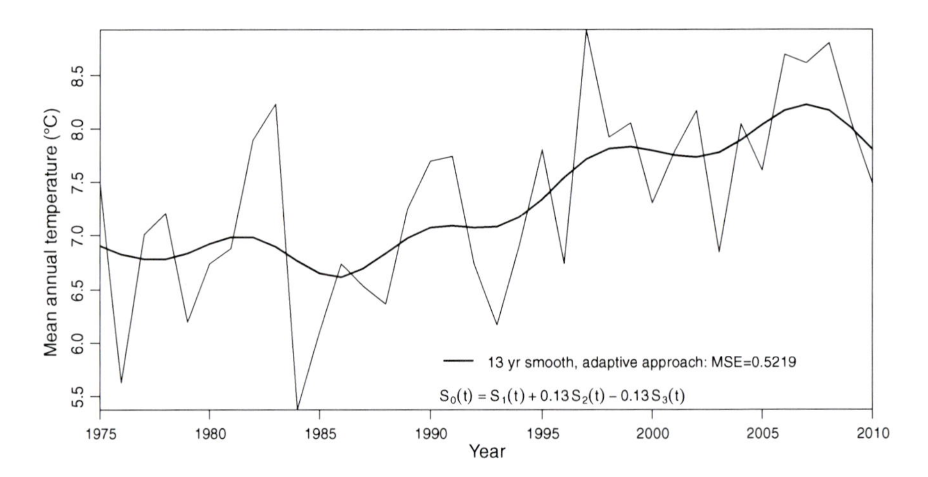

Fig. 4.3 Annual mean temperature series of Wulumuqi station shown with 13-year smooth using the adaptive approachwhich uses a combination of the three boundary constraints (thin line: temperature series, bold line: smooth; cf. [12] (*Data source* NOAA NCDC [15], own calculations)

or for comparison of multiple curves for the IPCC Report 2007. Fluctuations with longer cycles are possible indicators of climate change and not removed.

Actually, we can only smooth a time series from the start plus half the filter width to the end minus half the filter width scientifically correct. Unfortunately, during this time period (1981–2003) especially high temperature increases took place. The calculated linear trend model predicts temperature values for the future which are exceeding the minimum and maximum values of the regional climate

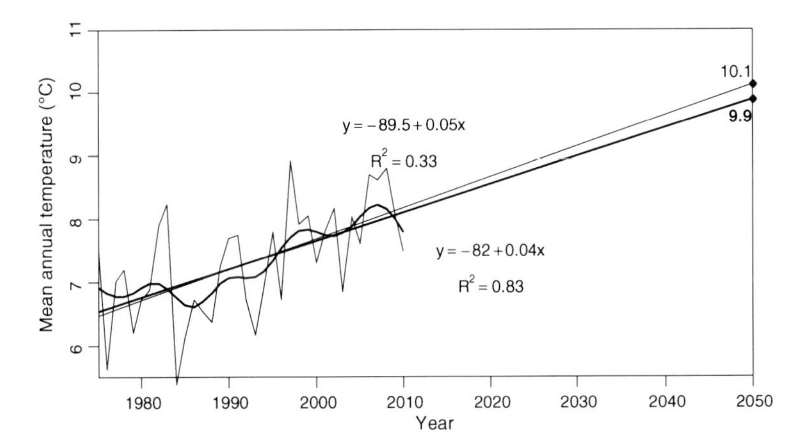

Fig. 4.4 Annual mean temperature series of Wulumuqi station and 13-year smooth using the adaptive approach by Mann [12] with their respective trend lines(linear regression), function, coefficient of determination R^2 and values for the year 2050 *data source* NOAA NCDC [15], own calculations

projections of Christensen et al. [5] for the region Tibet and Xinjiang. This time period obviously does not represent the changes in temperature.

To be able to smooth the data sets from starting to end point of the available data, they had to be padded at their respective ends with the chosen filter width to remove any influence of boundaries of padded series on the interior [11]. Mann [12] developed an approach to the smoothing of potentially non-stationary time series which provides an optimal choice among three alternatives, the three lowest order boundary constraints minimum norm, minimum slope and minimum roughness that can be applied to a smooth (see also Fig. 4.2). The adaptive approach weights the three boundary constraints to reduce the mean square error and to optimize the fit between smoothed and raw time series [12]. It was used to pad all used time series (see Fig. 4.3). The resulting smoothed data series is mathematically optimised, but corruption of the time series caused by the padding and smoothing cannot be evaluated as we have no information about the time before and after the time series.

When comparing the linear trend line and trend values for 2050 for the raw and the smoothed time series, the trend line for the smoothed series seemed to give more conservative (lower) prognosis for temperature variables (see as an example the annual mean temperature series in Fig. 4.4).

The procedure was repeated for mean annual maximum, minimum and dew point temperature and annual precipitation values to determine the trend line, trend value 1975–2010 and predicted values for 2050 (see Table 4.3). As the monthly grouped time series showed different trend directions which was also confirmed by the analysis of [3], trend line and values were also calculated for monthly grouped values of the mentioned climate variables (Tables A.17–A.20 in the appendix). The accuracy of the derived trend values for 2,050 is difficult to assess. To be mathematically correct, two additional regression lines taking into account the error of the input data could

Table 4.3 Trend line and trend values for smoothed time series of mean annual air, dew point, maximum and minimum temperature

	y-intercept	Slope	R^2	Mean 1975–2010	Trend values			
					1975	2010	Δ 1975–2010	2050
Temp	−81.96	0.04	0.83	7.3	6.5	8.1	1.6	9.9
Dewp	−62.57	0.03	0.53	−2.54	−3.1	−2.0	1.1	−0.8
Max	−35.36	0.02	0.54	12.4	11.9	12.8	0.9	13.7
Min	−116.43	0.06	0.79	2.4	1.4	3.5	2.1	5.9

have been calculated and used to determine the error of the regression analysis for the projected values. However, already the assumption that past and future temperature change can be described by a linear regression has been doubted [20] and the extrapolated values can merely be seen as estimations and possible scenarios. Thus further error evaluation was abandoned in favour of comparing the projected values with the regional results of global climate models (GCMs). When comparing the calculated trend values of annual mean surface temperature and the change relative to 1980–1999 to the multi-model ensemble means presented by the IPCC Report [13], it becomes obvious that the modelled global mean warming for several scenarios is much lower than the trend values. However, already the observed warming trend for 1975–2010 exceeded the linear trends in global land-surface air temperatures in the Northern Hemisphere, but matched the maximum range and regional modelled warming trends (see Sect. 2.4.1).

The calculated trend values were compared with the regional averages of temperature projections from a set of 21 global models in the multi model dataset (MMD) for the A1B scenario [5]. These global climate models are believed to be reliable as (i) the model fundamentals are based on established physical laws along with observations, (ii) they are able to simulate important aspects of the current climate and (iii) they are able to reproduce features of past climates and climate changes. Nonetheless, the GCMs also show significant errors for example on the smaller, regional scale and in the representation of clouds [17].

The calculated trend values are between the mean and the maximum model values (see Table 4.4). Only for the summer months, the calculated trend values were much lower than the modelling results. It is also mentioned that fine spatial scales in climate can also be generated by topography, which is probably the case in Urumqi. In this case, information is insufficient "on how climate change will be expressed at these scales" [18: 91] and the GCMs might not exactly simulate the local climate and seasonal trends. Despite the differences between the calculated trends and the compiled model results, the linear regression model was used to project the probable dimension and extent of future climate change for Wulumuqi station.

4.2.2 Precipitation Projection

The measured monthly precipitation values at Wulumuqi station for the last 36 years do not exhibit a normal distribution for nine months of the year (see

Table 4.4 Annual and monthly mean temperature: mean value 1980–1999 from measured data, mean value 2080–2099 from linear trend model and their difference compared to regional averages of temperature projections from 21 global models in the MMD for the A1B scenario

	NOAA	Linear trend model			IPCC A1B scenario model summary statistics				
	1980–99	2080–99	Δ	Δseason	Min	25 %	50 %	75 %	Max
Year	7.1	11.7	4.6	4.6	2.8	3.2	3.8	4.5	6.1
Jan	−12.4	−9.7	2.7	4.7	2.8	3.7	4.1	4.9	6.9
Feb	−9.7	−4.0	5.7						
Mar	−1.3	7.5	8.8	5.3	2.5	2.9	3.6	4.3	6.3
Apr	10.2	15.6	5.3						
May	17.0	18.9	1.9						
Jun	21.7	24.5	2.8	0.2	2.7	3.2	4	4.7	5.4
Jul	24.0	22.7	−1.3						
Aug	22.6	21.8	−0.8						
Sep	16.7	18.9	2.2	7.2	2.7	3.3	3.8	4.6	6.2
Oct	8.0	14.6	6.6						
Nov	−2.1	10.7	12.8						
Dec	−9.2	−3.4	5.8	4.7	2.8	3.7	4.1	4.9	6.9

(*Data source* NOAA NCDC [15], own calculations, [5])

Table 4.2), therefore a linear regression model could not be applied to build a future climate scenario. When tested for monthly precipitation values, it produced negative projected values for several months. It seems that it is not possible to describe precipitation trends with a simple linear model. Therefore, the **precipitation values were extrapolated using mean projections of GCMs for precipitation trends in the region Central Asia and Tibet** [5]. The application of more detailed results from GCMs would have gone beyond the possible work load for this research, but could be implemented to improve the spatial resolution of the effects of climate change. The mean values 1980–1999 for precipitation were extrapolated with the relative increase according to the regional averages of precipitation projections from a set of 21 global models for the A1B scenario presented by Christensen et al. [5].

Because Wulumuqi station exhibited in the past already very high values with regard to precipitation trends [3], **50, 75 %, and maximum precipitation response** were chosen. The monthly projections were calculated using the seasonal precipitation response (Table 4.5). Three future climate scenarios for 2050 with the same linearly projected temperatures (mean, minimum, maximum and dew point) and three different projections of precipitation (50, 75 % and maximum values of modelled precipitation responses, see Sect. 4.1) were compiled (see Fig. 4.5 and Tables A.21–A.23). The calculated values for the climatic variables for the year 2050 were then used in the hydrological model as described before. It is assumed that the distribution of temperature and precipitation does not change and the same extrapolation method as described in Sect. 3.3.2 and TRMM and MODIS data sets are used. When comparing the projected temperature for 2050 to the mean values from 1975–2010 and 2001–2010 (Fig. 4.6), one can see that the temperatures in

Table 4.5 Projection of precipitation values for 2050 using mean observed values from 1980 to 1999 and the modelled 50 %, 75 % and maximum precipitation response for 2080–2099 based on the regional averages of 21 GCMs

Month	Mean 1980–1999 [mm]	50 % difference			75 % difference			Maximum difference		
		[%]	2050 [mm]	Mean 2080–2099 [mm]	[%]	2050 [mm]	Mean 2080–2099 [mm]	[%]	2050 [mm]	Mean 2080–2099 [mm]
1	9.71	19	10.83	11.55	26	11.24	12.23	36	11.82	13.20
2	12.78	19	14.24	15.20	26	14.79	16.10	36	15.56	17.38
3	23.69	10	25.13	26.06	14	25.70	27.01	34	28.57	31.75
4	41.15	10	43.64	45.26	14	44.63	46.91	34	49.61	55.14
5	54.55	10	57.85	60.00	14	59.17	62.18	34	65.77	73.09
6	34.56	4	35.40	35.95	10	36.66	38.02	28	40.42	44.24
7	37.77	4	38.69	39.28	10	40.06	41.55	28	44.17	48.35
8	42.46	4	43.48	44.15	10	45.03	46.70	28	49.65	54.34
9	40.73	8	42.70	43.99	14	44.18	46.43	21	45.90	49.28
10	28.39	8	29.76	30.66	14	30.79	32.36	21	31.99	34.35
11	20.99	8	22.01	22.67	14	22.77	23.93	21	23.66	25.40
12	18.38	19	20.49	21.87	26	21.27	23.15	36	22.38	24.99
Sum	365.15		384.22	396.64		396.29	416.57		429.50	471.51

(*Data source* NOAA NCDC [15], [5], own calculations)

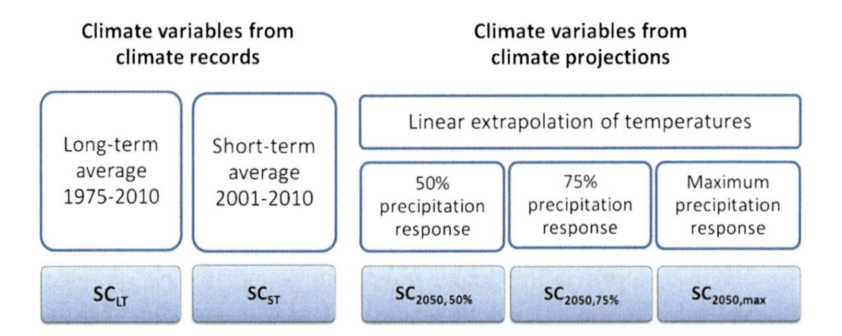

Fig. 4.5 Overview over climate change scenarios used for evaluating effects of climate change on the hydrological system (*Source* own design)

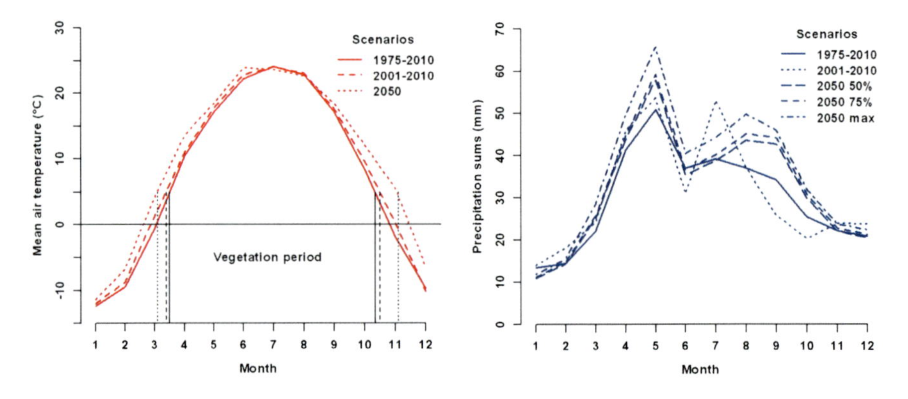

Fig. 4.6 Temperature and precipitation values for 1975–2010 and 2001–2010 for Wulumuqi station from NCDC climate data and projected values for 2050 with the three different projections for precipitation values(max, 75 and 50 %) (*dotted*) (*Data source* NOAA NCDC [15], own calculations)

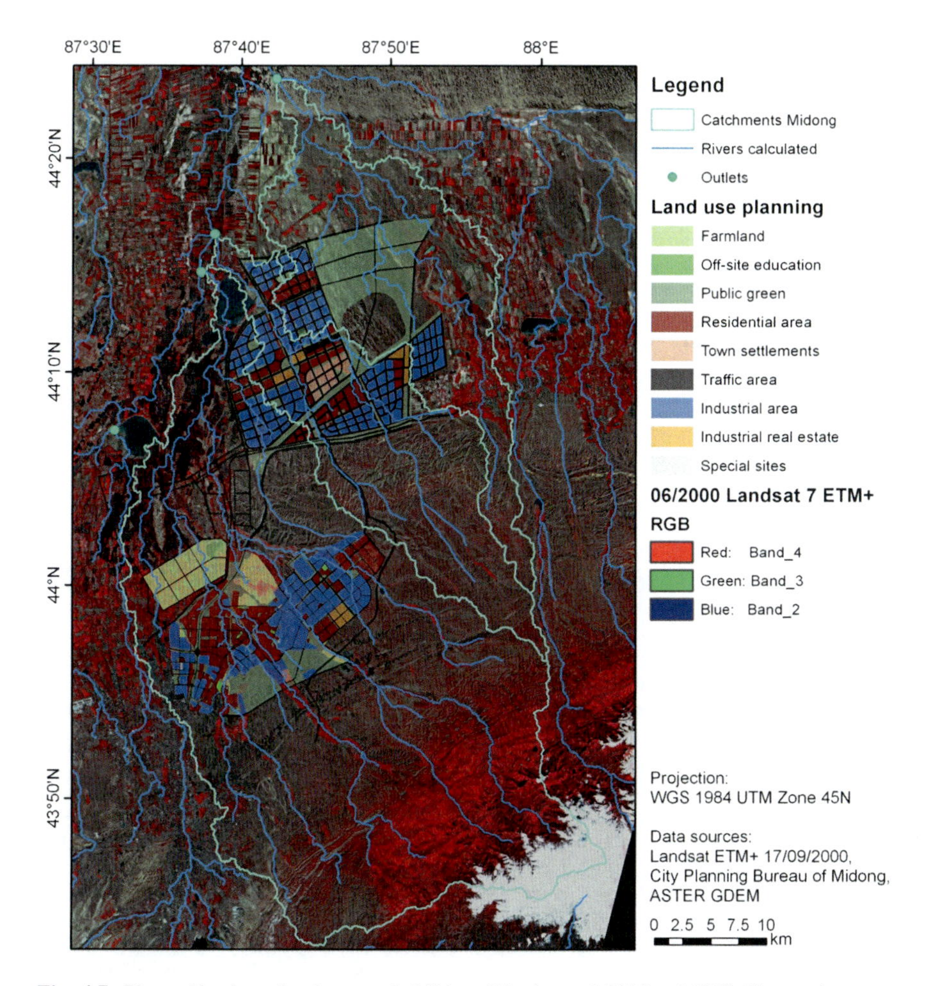

Fig. 4.7 Planned land use development in Midong District until 2020 and 2050. The southern part is planned to be completed until 2020 and the northern part until 2050 (*Source* own design, based on [6], Landsat ETM+ 17/09/2000 ([14]), catchment areas calculated based on ASTER GDEM ([21])

summer will not increase significantly. However, they do increase from autumn to spring, a time period when the temperature influences important hydrological variables such as snow fall and snow melt. Precipitation will mainly increase in the summer months, when the available water is subject to evapotranspiration.

4.3 Land Use Change Scenario

Major land use changes in Urumqi Region are planned for Midong District where an industrial zone and accompanying residential and retail areas will be developed in two steps until 2050 as shown in Fig. 4.7 ([6]). The sealed area will increase from 78 km^2 in 2007 to 130 km^2 in 2020 and 343 km^2 in 2050. The extent of 2007 is taken from a survey conducted in 2007 within a diploma thesis, where Midong New District was mapped and classified according to a field survey and analysis of Google Earth images ([6, 7]). For the calculation of sealed area in 2020 and 2050, a planning map published by the City Planning Bureau of Midong was digitised and analysed according to the land use classification. Residential area, town settlements, traffic area, industrial area, industrial real estate and special sites were considered as sealed area (cf. land use classes from the land use classification in Sect. 3.3.3). The planned development area is situated within the catchments of several smaller rivers that flow into the Urumqi River.

References

1. Abdi, H. (2007). Kendall rank correlation. In N. J. Salkind (Ed.), *Encyclopedia of Measurement and Statistics* (pp. 509–511). Thousand Oaks, Ca: Sage Publications.
2. Arnell, N. (1996). *Global Warming, River Flows and Water Resources* (234 p). Chichester: Wiley.
3. Barth, N. C. (2011). Auswertung der Temperatur- und Niederschlagsdaten von 15 Klimastationen im Umkreis von Urumqi (AR Xinjiang, China) (65 p). Unpublished bachelor thesis, Ruprecht-Karls-Universität Heidelberg.
4. Bolch, T. (2007). Climate change and glacier retreat in northern Tien Shan (Kazakhstan/Kyrgyzstan) using remote sensing data: Climate change impacts on mountain glaciers and permafrost. *Global and Planetary Change, 56*(1–2), 1–120.
5. Christensen, J. H., Hewitson, B., Busuioc, A., Chen, A., Gao, X., Held, I., Jones, R., Kolli, R. K., Kwon, W.-T., Laprise, R., & Magaña Rueda, V. (2007). Regional climate projections. In S. Solomon, D. Qin, M. Manning, Z. Chen, M. Marquis, K. B. Averyt, & H. L. Miller (Eds.), *Climate change 2007: The physical science basis. contribution of working group I to the fourth assessment report of the Intergovernmental Panel on Climate Change* (pp. 847–940). Cambridge, UK: Cambridge University Press.
6. City Planning Bureau of Midong (2007). Midong New District Master Plan 2020 and Midong New District Schematic Map of the overall planning. Urumqi: City Planning Bureau of Midong.
7. Fricke, K. (2008). The development of Midong New District, Urumqi, PR China: Ecological and historical context and environmental consequences (165 p). Unpublished diploma thesis, Heidelberg University.
8. Hoffmann, T., & Rödel, R. (2004). Leitfaden für die statistische Auswertung geographischer Daten, Greifswald (110 p) (Greifswalder Geographische Arbeiten, 33).
9. Intergovernmental Panel on Climate Change (IPCC) (2000) *IPCC special report: emission scenarios summary for policymakers: a special report of IPCC working group III* (20 p). Retrieved 16 Nov 2011, from http://www.ipcc.ch/pdf/special-reports/spm/sres-en.pdf.
10. Intergovernmental Panel on Climate Change (IPCC) (2007) *Climate Change 2007: Synthesis Report* (73 p). Retrieved 16 Nov 2011, from http://www.ipcc.ch/pdf/assessment-report/ar4/syr/ar4_syr.pdf.
11. Mann, M. E. (2004). On smoothing potentially non-stationary climate time series. *Geophysical Research Letters, 31*(7), L07214.
12. Mann, M. E. (2008). Smoothing of climate time series revisited. *Geophysical Research Letters, 35*(16), L1670.

13. Meehl, G. A., Stocker, T. F., Collins, W. D., Friedlingstein, P. G., Gregory, J. M., Kitoh, A., Knutti, R., Murphy, J. M., Noda, A., Raper, S. C., Watterson, I. G., Weaver, A. J., & Zhao, Z. (2007). Global climate projections. In S. Solomon, D. Qin, M. Manning, Z. Chen, M. Marquis, K. B. Averyt & H. L. Miller (Eds.), *Climate change 2007: The physical science basis. contribution of working group I to the fourth assessment report of the Intergovernmental Panel on Climate Change* (pp. 747–845). Cambridge, UK: Cambridge University Press.
14. NASA Landsat Program (2010). Landsat ETM+ scenes L1T, USGS, Sioux Falls. Retrieved from http://glovis.usgs.gov/.
15. National Oceanic and Atmospheric Administration of the U.S. Department of Commerce National Climatic Data Centre (NOAA NCDC) (2011). Global summary of the Day Wulumuqi Station, Station number 514630, 22.08.1956–31.12.2010. Retrieved 16 May, 2011, from http://www7.ncdc.noaa.gov/CDO/cdo.
16. R Development Core Team (2011). R, Version 2.13.0 [computer program]. Retrieved from http://www.r-project.org/.
17. Randall, D. A., Wood, R. A., Bony, S., Colman, R., Fichefet, T., Fyfe, J., Kattsov, V., Pitman, A., Shukla, J., Srinivasan, J., Stouffer, R. J., & Sumi, A. (2007). Climate Models and their Evaluation. In S. Solomon, D. Qin, M. Manning, Z. Chen, M. Marquis, K. B. Averyt, & H. L. Miller (Eds.), *Climate change 2007: The physical science basis. Contribution of working group I to the fourth assessment report of the Intergovernmental Panel on Climate Change* (pp. 589–662). Cambridge, UK: Cambridge University Press.
18. Solomon, S., Qin, D., Manning, M., Alley, R. B., Berntsen, T., Bindoff, N. L., Chen, Z., Chidthaisong, A., Gregory, J. M., Hegerl, G. C., Heimann, M., Hewitson, B., Hoskins, B. J., Joos, F., Jouzel, J., Kattsov, V., Lohmann, U., Matsuno, T., Molina, M., Nicholls, N., Overpeck, J., Raga, G., Ramaswamy, V., Ren, J., Rusticucci, M., Somerville, R., Stocker, T. F., Whetton, P., Wood, R. A., & Wratt, D. (2007). Technical summary. In S. Solomon, D. Qin, M. Manning, Z. Chen, M. Marquis, K. B. Averyt & H. L. Miller (Eds.), *Climate change 2007: The physical science basis. Contribution of working group I to the fourth assessment report of the Intergovernmental Panel on Climate Change* (pp. 19–91). Cambridge, UK: Cambridge University Press.
19. Stadler, D., Bründl, M., Schneebeli, M., Meyer-Grass, M., & Flühler, H. (1998). *Hydrologische Prozesse im subalpinen Wald im Winter* (145 p). Zürich: vdf Hochsch.-Verl. an der ETH.
20. Trenberth, K. E., Jones, P. D., Ambenje, P., Bojariu, R., Easterling, D., Klein Tank, A., Parker, D., Rahimzadeh, F., Renwick, J. A., & Rusticucci, M. (2007). Surface and Atmospheric Climate Change. In S. Solomon, D. Qin, M. Manning, Z. Chen, M. Marquis, K. B. Averyt & H. L. Miller (Eds.), *Climate change 2007: The physical science basis. Contribution of working group I to the fourth assessment report of the Intergovernmental Panel on Climate Change* (pp. 235–336). Cambridge, UK: Cambridge University Press.
21. U.S. Geological Survey & Japan ASTER Programme (2009). ASTER GDEM Version 1. NASA Land Processes Distributed Active Archive Center, Sioux Falls. Retrieved 15 Jul, 2009, from http://earthexplorer.usgs.gov/.

Chapter 5
Simulation Results

After the calibration and validation of the water balance model, it was used to simulate the water balance in Urumqi Region under different scenarios. The hydrological cycle and availability and supply of water resources can be altered both by anthropogenic activities and by changes of the climatic system [1]. Therefore, the water availability was simulated with the water balance model for climate change and land use transformation. To analyse the reaction of the hydrological cycle to climate changes, five different climate data sets were used as input data for the water balance model. First, the water balance in the long and short term past was calculated based on the average climate variables for two time periods in the past (SC_{LT} and SC_{ST} in Sect. 5.1). Additionally, climate conditions were projected for the year 2050 and three climate change scenarios for the year 2050 have been compiled based on the recent climate change in Urumqi Region and regional climate projections by the IPCC Report 2007 in Sect. 4.1. The simulation results of the climate change scenarios are presented in Sect. 5.2. The scenario for land use change was taken from the planned land use development in Midong District until 2020 and 2050 published by the City Planning Bureau of Midong [2]. The effects of the altered land surface characteristics on the hydrological cycle are then evaluated for Midong District in Sect. 5.3. All results of the simulations depend on the assumption made when developing the water balance model.

5.1 Water Balance in the Past

First, the results for the recent situation of the water balance in Urumqi are presented, i.e. the model runs with SC_{LT} and SC_{ST}. ET_p, precipitation, ET_a, melt water, surface runoff, groundwater and total water flows were simulated to obtain information about the distribution of the water balance components and to identify certain spatial characteristics. Then, the water balance components were evaluated on the basis of their absolute mean annual sums in several catchment areas. The units compared were the spatial distribution of mm m^{-2} and for the catchment areas the average mm m^{-2} and total m^3 per year. In Fig. 2.4 or on the last page,

K. Fricke, *Analysis and Modelling of Water Supply and Demand Under Climate Change, Land Use Transformation and Socio-Economic Development*, Springer Theses, DOI: 10.1007/978-3-319-01610-8_5,

the selected catchment areas are shown. The catchments chosen for comparison have been presented in Chap. 2.

According to the simulated distribution, the evapotranspiration potential is lowest in the high mountain areas due to lower temperatures, higher albedo and evaporation resistance terms there. A medium evapotranspiration potential can be observed on the alluvial plain and lower areas. In contrast, the denser vegetated areas on the foothills and on the plain with forest, grassland or agricultural crops show a higher evaporation potential because the vegetation there has lower albedo and resistance terms than the uncovered areas.

The difference between the two simulations is not easy to detect in Fig. 5.1 as the difference is relatively low compared to the total range of ET_p values. However, the average increase of ET_p for the total research area is around 29 mm per year. ET_p is especially high in the vegetated areas in the lower mountains that react strongest to the rising temperature values. For ET_a, the amount of water available for evapotranspiration provided by precipitation is equally important as ET_p as it is a precondition to the spatial distribution of ET_a. When looking at the ET_a for the SC_{LT} or the SC_{ST} as baseline scenarios and distributions, ET_a is relatively low in the mountain areas above the tree line due to the low ET_p values (Fig. 5.2). The little amount of available precipitation in the plain areas also reduces ET_a there. Higher ET_a can only be found in areas with high ET_p and precipitation, i.e. mainly the lower mountain regions and areas with dense vegetation. The distribution of

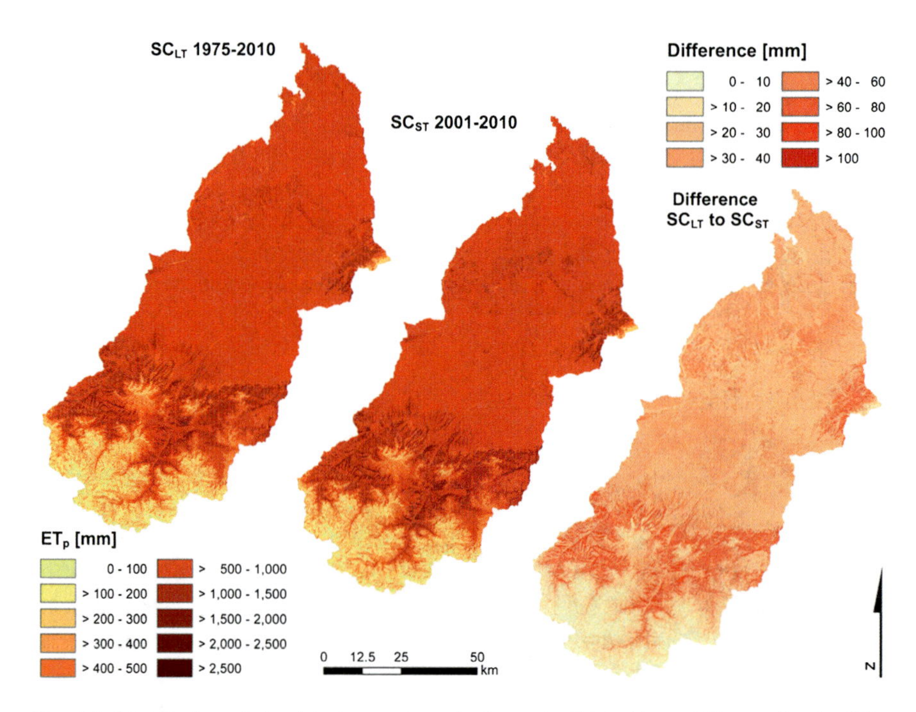

Fig. 5.1 Distribution of simulated evapotranspiration potential in Urumqi Region for the SC_{LT} and SC_{ST} and the difference between the two simulations (*Source* own calculations)

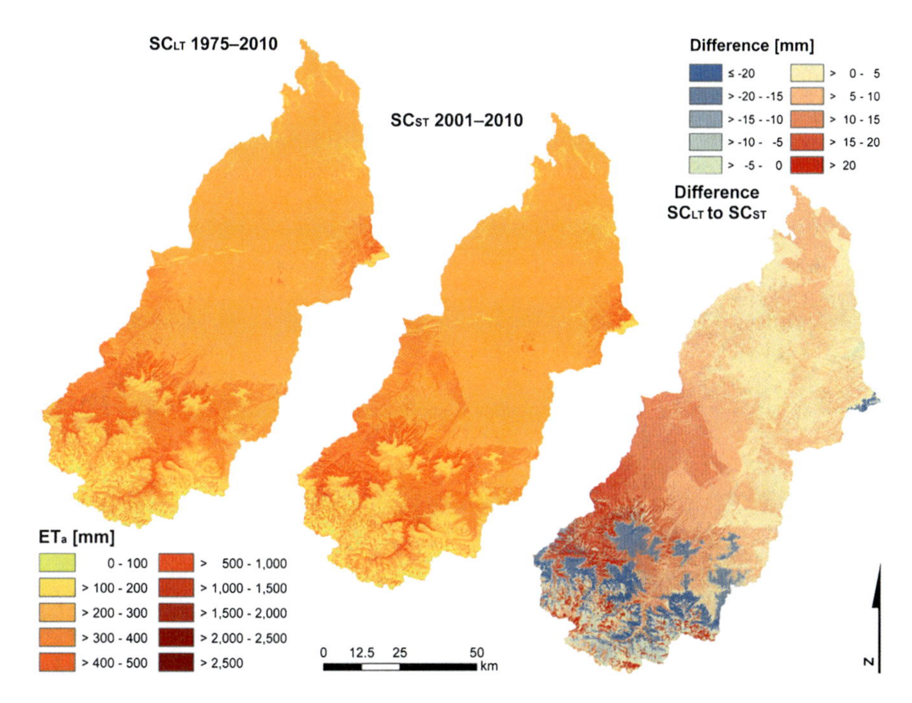

Fig. 5.2 Distribution of simulated actual evapotranspiration in Urumqi Region for the SC_{LT} and SC_{ST} and their difference (*Source* own calculations)

precipitation is not altered from SC_{LT} to SC_{ST}, but the monthly values changed. On the one hand, the **total annual precipitation sum increased in the SC_{ST} compared to the SC_{LT} 4.3 % in Urumqi Region** (spatial results not shown here, refer to Table 5.1). On the other hand, monthly precipitation sums increased in the months November till May and July, but decreased in the months June, September and October. **This lead to a mean increase of about 4 mm actual evapotranspiration** (ET_a) per year, but also to a decrease of ET_a in parts of the high mountain

Table 5.1 Simulated annual precipitation, actual evapotranspiration and net precipitation for catchments in Urumqi Region

	1975–2010 (mm)			2001–2010 (mm)			Difference (%)		
	P	ET_a	P_n	P	ET_a	P_n	P	ET_a	P_n
Urumqi Mountain	482	273	209	500	272	227	3.7	–0.4	8.6
Wulapo Reservoir	426	272	153	442	275	167	3.8	1.1	9.2
Urumqi City	337	251	86	350	257	93	3.9	2.4	8.1
Urumqi River	410	269	141	425	271	154	3.7	0.7	9.2
Toutun Mountain	492	300	192	517	301	216	5.1	0.3	12.5
Toutun River	415	285	130	435	292	144	4.8	2.5	10.8
Midong	339	241	98	350	245	106	3.2	1.7	8.2
Urumqi Region	393	267	127	410	271	138	4.3	1.5	8.7

Urumqi City sub-catchment excluded upstream catchments of Wulapo Reservoir and Urumqi Mountain (*units* mm and %) (*Source* own calculations)

areas (see Fig. 5.2 and Urumqi Mountain catchment in Table 5.1). There the recently higher evapotranspiration potential could not be exhausted because of lower precipitation values in the months stated above. In the other catchments, ET_a increased, but was on average outbalanced by the precipitation increase and lead to

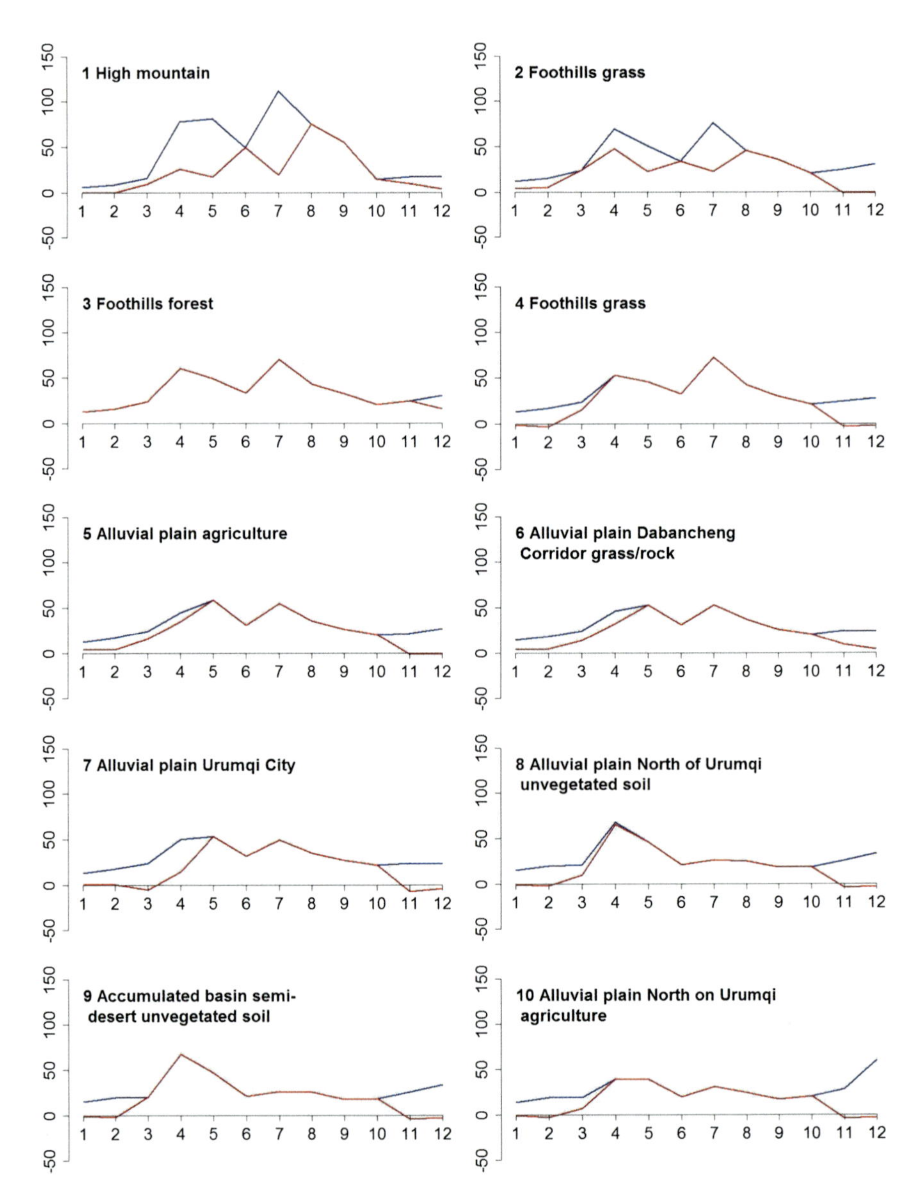

Fig. 5.3 Monthly values of precipitation (*blue line*) and actual evapotranspiration (*red line*) at representative points (*red points* in the map) in the research area, calculated with climate variables from SC_{ST} (units mm, *Source* own calculations)

a higher net precipitation P_n. The proportion of P_n from P for the whole Urumqi Region also increased from 32.1 to 33.9 %.

The relation between precipitation and ET_a is quite different for characteristic sites and regions in Urumqi Region. To illustrate these differences, the monthly values for precipitation and ET_a for several representative points are shown in Figs. 5.3 and 5.4 to visualize the changing situation from the high mountains to the

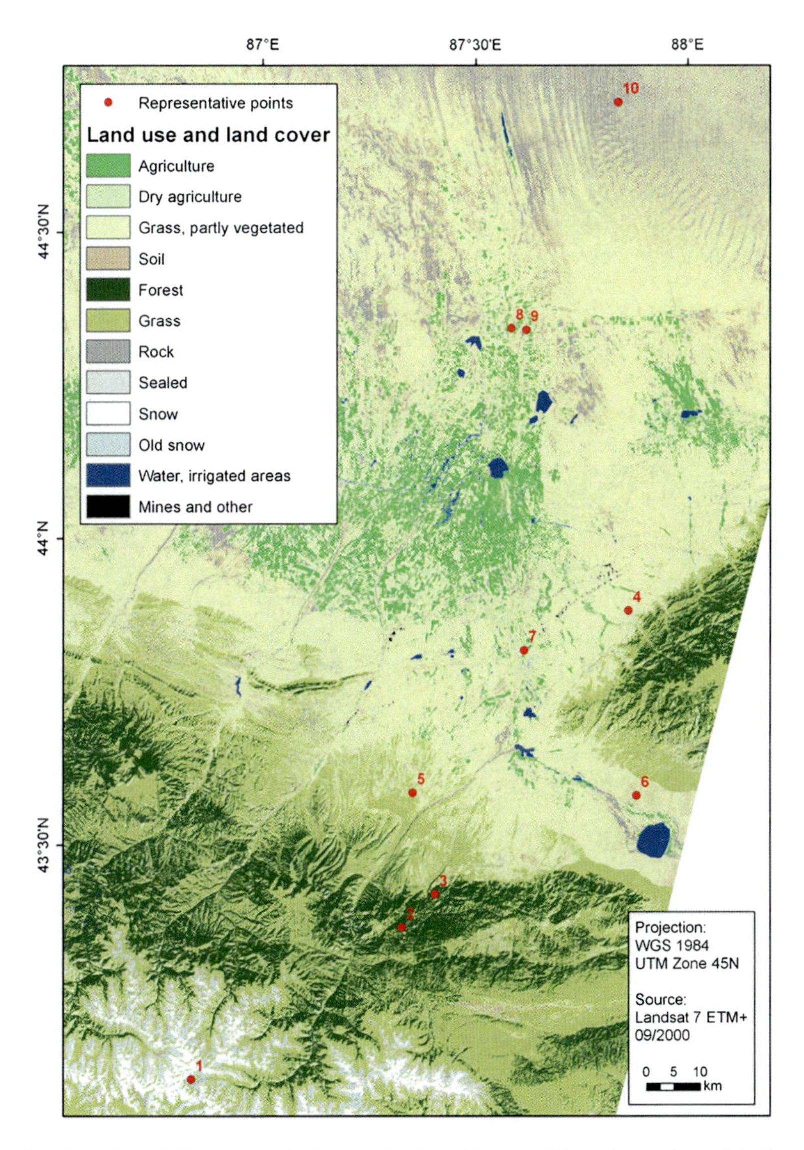

Fig. 5.4 Location of the representative points for which monthly values of precipitation and actual precipitation were presented in Fig. 5.3 (Data source NASA Landsat Program [3], own calculations)

basin. In the high mountain areas with lower temperature and little vegetation ((1) and (2)), precipitation is high enough to exceed the evapotranspiration potential also in the summer months. In the denser vegetated forested areas, however, transpiration consumes almost all available water from precipitation. Descending from the mountain and turning towards the basin in the North, precipitation values decrease as well as ET_a.

Vegetation always leads to higher evapotranspiration (see (3) vs. (2) and (4), (5) vs. (6), and (8) vs. (7) and (9)) which is consuming the available water from precipitation. Compared to areas without or with less vegetation, the date when the precipitation is completely consumed by evapotranspiration sets in earlier for vegetated areas.

Similar to ET_a, the total water flow Q_n depends on the spatial and temporal distribution of evapotranspiration and precipitation as well as melt water which is coupled to temperature. The highest values of Q_n are reached in the mountain areas where precipitation is high and evapotranspiration low. Due to the relatively low values for ET_a in the lower foothills, Q_n is also quite high there. We can observe an overall increase of 34 mm in Urumqi Region when comparing the simulation of Q_n for the SC_{LT} with the SC_{ST} (see Fig. 5.5 and Table 5.2). Over the sealed areas, snow and ice, Q_n did not change very much. Higher precipitation values were offset by the increase in temperature and also evapotranspiration in winter over the

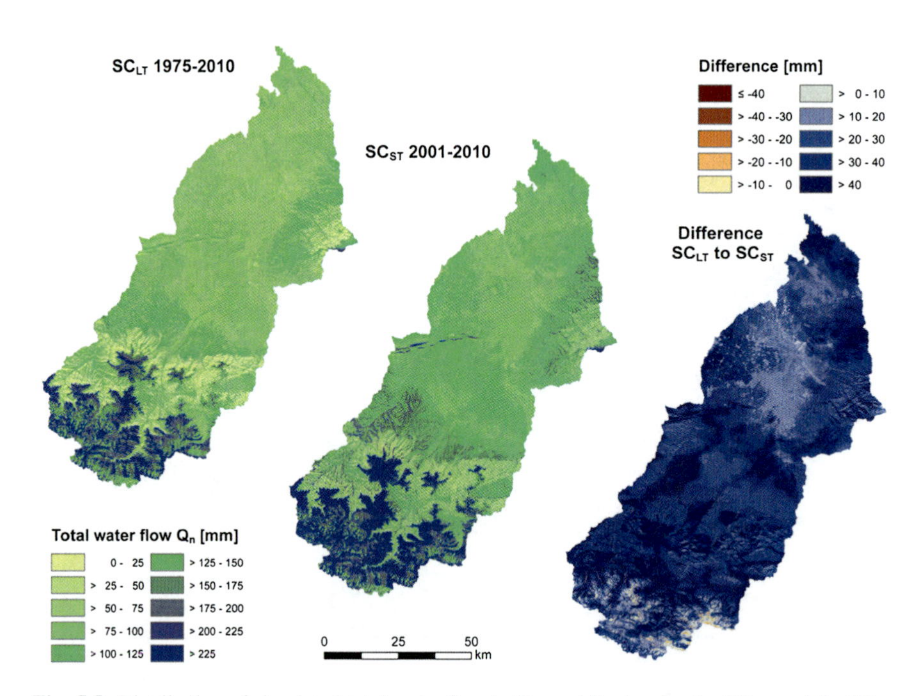

Fig. 5.5 Distribution of simulated total water flow in Urumqi Region for the SC_{LT} and the SC_{ST} and their difference (*Source* own calculations)

Table 5.2 Simulated annual flow values for the SC_{LT} and SC_{ST} for the catchment areas in Urumqi Region. Urumqi City includes only the subcatchment of the city area, all upstream catchments are excluded

	1975–2010 (mm)				2001–2010 (mm)				Difference (%)			
	Q_m	Q_n	Q_{sf}	Q_{gw}	Q_m	Q_n	Q_{sf}	Q_{gw}	Q_m	Q_n	Q_{sf}	Q_{gw}
Urumqi Mountain	76	151	78	74	95	187	95	93	25.2	23.8	21.8	25.8
Wulapo Reservoir	68	114	59	55	90	150	76	74	31.2	31.7	29.8	33.7
Urumqi City	65	71	37	33	77	95	49	46	19.4	34.3	32.4	36.3
Urumqi River	68	106	55	51	88	140	72	69	29.2	32.0	30.1	34.0
Toutun Mountain	71	141	75	67	93	184	96	88	30.7	30.1	29.0	31.4
Toutun River	66	101	54	47	88	138	73	65	32.1	37.3	35.6	39.3
Midong	67	77	37	40	88	106	50	56	30.3	38.3	37.1	39.3
Urumqi Region	67	97	50	47	88	131	67	64	31.3	35.1	34.0	36.2

Q_m melt water flow, Q_n total water flow, Q_{sf} surface water flow, Q_{gw} groundwater flow (*unit* mm and %) (*Source* own calculations)

dark heat islands of the city. In the high mountain areas, the Q_n decreased because of the increased temperatures and ET_a. Over the other areas, however, the increased precipitation in the winter months led to a rise in snow fall and accumulated water equivalent and subsequently to a higher melt water flow Q_m and total water flow Q_n (Figs. 5.6 and 5.7).

As observable from the average surface runoff values in Fig. 5.1, surface flow is dominant in areas either with impervious surfaces such as urban, sealed areas or steep mountain regions, where only a small part of available water can infiltrate (Fig. 5.5). The mountain ranges receive also a very high net precipitation leading to the highest surface runoff in Urumqi Region. Compared to the alluvial plain, the urban areas exhibit also relatively high surface runoff values. Vice versa is average groundwater recharge high in gently sloped areas with permeable underground, mainly in the alluvial plains and the Junggar Basin (Fig. 5.6). These are also identified as the areas with the highest groundwater flow in Urumqi Region with about 15,000 m^3 per 1,000 m^2 and year through loosely layered or porous material [4], refer also to the landscape classification in Sect. 2.1).

Except for the areas where the total water flow changed only very little (high mountains and urban, sealed areas), based on the dominant process and the characteristics of the area either surface or groundwater flow became larger. Surface water flow increased on the lower mountain areas where the slopes are still steep and allow only a small part of the available water to infiltrate, while groundwater flow increased in the areas with more permeable underground such as Dabancheng Corridor and the plain but decreased in the sealed areas.

Furthermore, the results for water flows per catchment are presented and evaluated. Generally, water flow is always highest in the mountain catchments and becomes lower the further downstream the outlet is located (see Table 5.2). This can be seen with the sequence Urumqi Mountain–Wulapo Reservoir–Urumqi

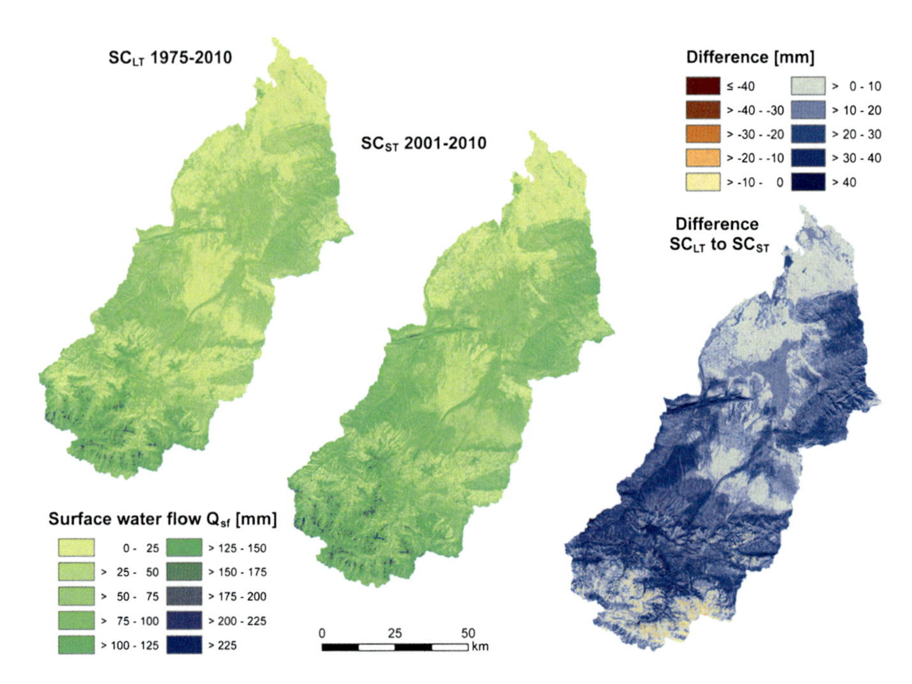

Fig. 5.6 Distribution of surface water flow in Urumqi Region calculated for SC_{LT} and SC_{ST} (*Source* own calculations)

River and Toutun Mountain–Toutun River. The catchment area of Urumqi City was calculated separately from the upstream catchment of Wulapo Reservoir to investigate the characteristics of an urban catchment area. It receives the lowest precipitation and net precipitation, thus all flows show the lowest values. Midong watershed was not divided in sub-catchments and exhibits flow values somewhere between the mountain (Urumqi Mountain and Toutun Mountain) and the city catchment areas (Urumqi City). However, the proportion of groundwater recharge is slightly above 50 % while in the other watersheds it is otherwise. This is due to the high proportion of only gently sloped and low proportion of sealed and impermeable areas in this catchment.

The amount of melt water calculated shows that melt water plays a major role for generating runoff throughout the year. **According to the simulation, 64, 65 and 88 % of the total amount of runoff came from melt water for SC_{LT}** in the catchments of Urumqi River, Toutun River, and Midong, respectively. This is due to the low temperature throughout November till March, high precipitation values in the mountainous and therefore colder areas and the precipitation lost to evapotranspiration in the summer months. The percentage is reduced to 62, 63, and 83 % for SC_{ST}. The proportion of melt water flow compared to total water flow is lowest in the Urumqi Mountain (50 and 51 %) and Toutun Mountain catchment (50 and 51 %), while it is highest in the catchments of Urumqi City (92 and 82 %) and Midong (88 and 83 %). In these lower areas, precipitation is only accumulated

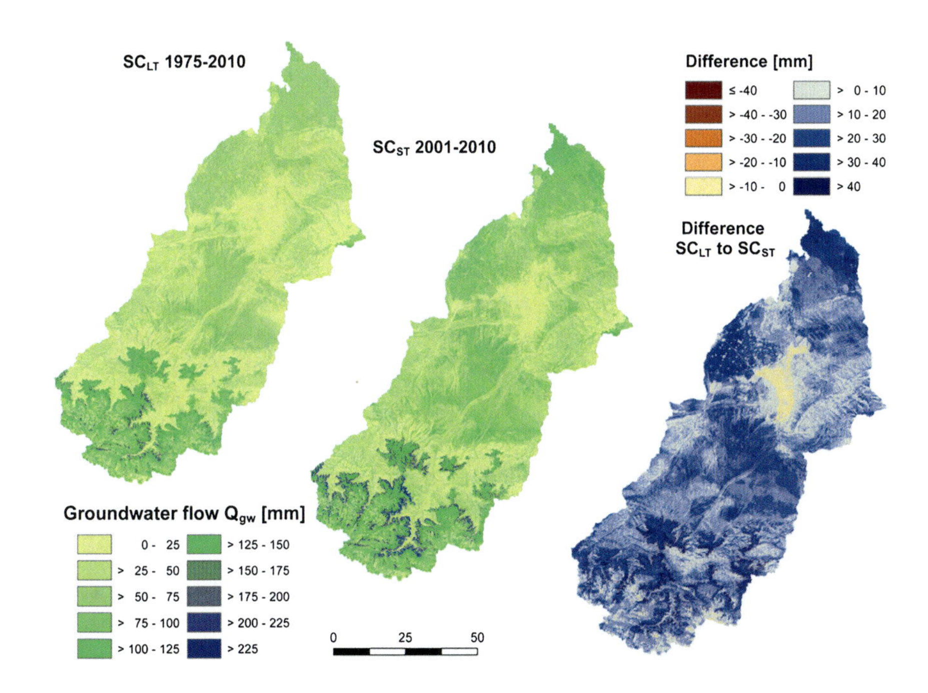

Fig. 5.7 Distribution of annual groundwater flows in Urumqi Region calculated for SC_{LT} and SC_{ST} (*Source* own calculations)

in winter in the form of snow, while it instantaneously evaporates in summer due to the higher temperatures. In the mountain regions, additional runoff from precipitation is possible.

All flows increase from the SC_{LT} to the SC_{ST} (see Table 5.2). While the increase is already significant with over 30 % on average, it is even higher in the downstream catchments. The total Toutun River watershed for example has experienced a 2 % higher increase in melt water flow than its upstream mountain catchment. This is due to high precipitation values in winter that are stored as snow and then released as melt water in spring. But in the urban area of Urumqi City, evapotranspiration becomes higher than in the surrounding areas and the accumulated snow water equivalent and subsequently melt water is lower (only 19 % increase). This is the case because the urban area is darker in winter than the snow covered agricultural fields and grasslands and reflects less radiation. Therefore more water evaporates.

5.2 Results of Climate Change Scenarios

Here, we present the calculated total water balance of Urumqi for the three projected future climate scenarios with linearly projected temperatures and three different projections of precipitation ($SC_{2050,\ 50\ \%}$, $SC_{2050,\ 75\ \%}$ and $SC_{2050,\ max}$, see Sect. 4.1) and compares them to the SC_{LT} and SC_{ST}. Apart from the quantitative results, the spatial distribution of the water balance components potential and actual evapotranspiration, melt water flow, total water flow and surface and groundwater flow now and in the future is described here.

First, the spatial distribution of ET_p, P and ET_a is described and summed up for the selected catchment areas. The change of ET_p for the different climate data sets is relatively small compared to the total range of ET_p (Fig. 5.8). **For the SC_{ST}, an increase of ET_p in the whole region and especially in the lower mountain areas is visible**, probably due to the sensitivity of the slopes and vegetation to the increasing temperatures. This trend is even more distinct when examining the changes from SC_{LT} to SC_{2050} when increases of up to 180 mm were calculated for the mountain slopes. A pronounced increase of ET_p in 2050 is also visible for areas with denser vegetation cover and agricultural fields which could transpire more water than the thinner and drier scrubs and grasses.

The distribution of precipitation is derived from the combination of satellite images and station data as described in Sect. 3.3.2.2 and shows the typical increase of precipitation with altitude in Fig. 5.9. The mountain ranges of the Tianshan in the Southwest and Bogdan Shan in the East of the research area and corresponding higher altitudes receive the most precipitation for all climate scenarios. However,

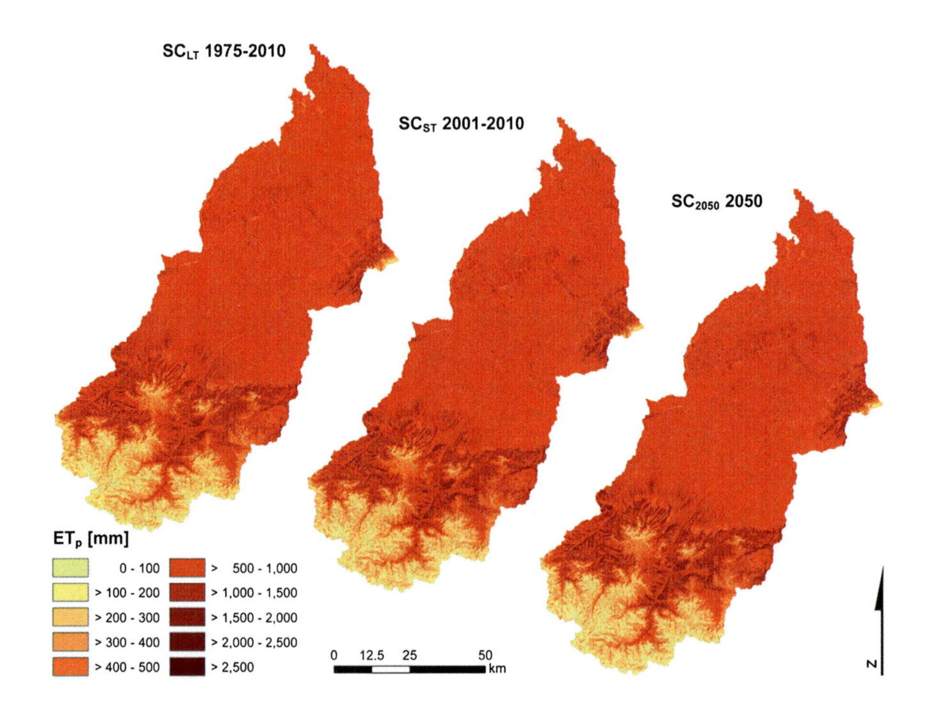

Fig. 5.8 Distribution of potential evapotranspiration in Urumqi Region for the three climate scenarios SC_{LT}, SC_{ST} and SC_{2050} (*Source* own calculations)

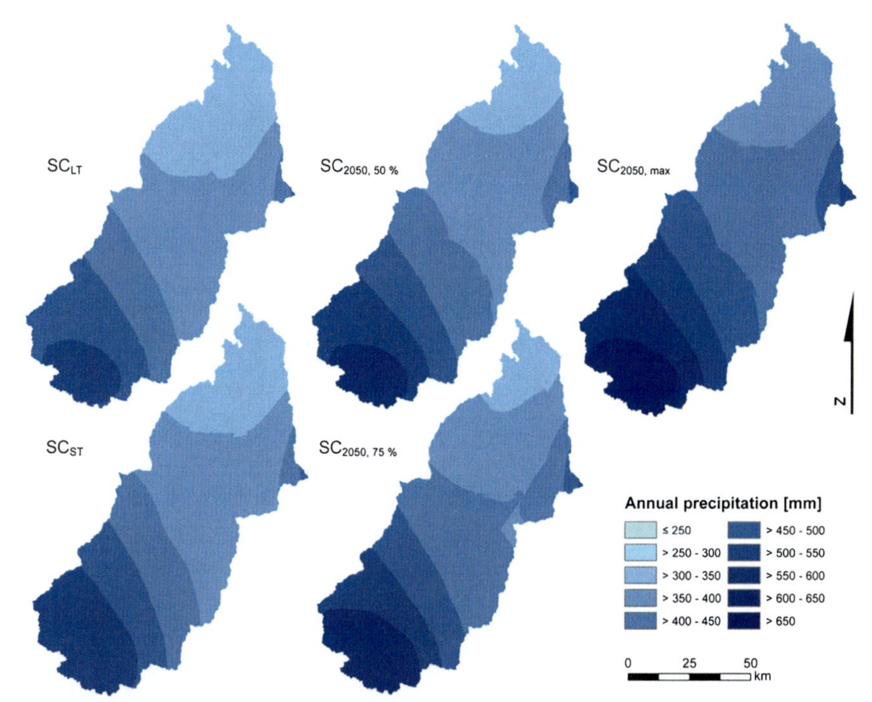

Fig. 5.9 Distribution of annual precipitation in Urumqi Region for the five different climate sets SC_{LT}, SC_{ST}, $SC_{2050,\ 50\ \%}$, $SC_{2050,\ 75\ \%}$ and $SC_{2050,\ max}$ (*Source* own calculations)

the amount of precipitation increased continuously from the SC_{LT} to the SC_{ST} and the three different scenarios for 2050. This could be expected when keeping the annual precipitation values from the climate input data in mind.

Simultaneously with precipitation increases, ET_a augments as well, as it can be seen in the simulated precipitation scenarios for 2050. Only in the high mountain areas, ET_a remains at the same level as before as ET_p is very low there, but in the lower mountains and foothills, ET_a rises as ET_p is higher there and the available precipitation continually increases (Fig. 5.10).

The results for precipitation and evapotranspiration are aggregated on catchment level in Table 5.3: the changes affect in particular mountainous regions. The sums of P, ET_a and P_n are higher in the mountainous catchments despite the differences in ET_a of vegetated and unvegetated mountain areas due to the large amount of locally available water. The further downstream and further to the plain in the North, the lower the precipitation, ET_a and P_n (see the catchments Urumqi City and Midong).

When comparing simulated P and ET_a for the different climate scenarios, the effects of climate change could already be recognised from the results for the SC_{LT} and SC_{ST}. Both precipitation and evapotranspiration sums have risen due to increasing temperature and precipitation that have been documented for the last 36 years (see Sect. 2.4.1). Until now, the increase in precipitation outweighs the

Table 5.3 Simulation of annual precipitation sums for different catchment areas with SC_{ST} and three SC_{2050}

	SC_{ST}			$SC_{2050,\ 50\ \%}$			$SC_{2050,\ 75\ \%}$			$SC_{2050,\ max}$		
Catchment	P	ET_a	P_n	P	ET_a	P_n	P	ET_a	P_n	P	ET_a	P_n
Urumqi Mountain	545	272	227	523	291	232	539	298	242	586	314	272
Wulapo Reservoir	442	275	167	460	297	164	474	304	170	515	326	189
Urumqi City	350	257	93	362	283	78	373	292	82	404	315	89
Urumqi River	425	271	154	442	294	148	456	302	154	495	324	171
Toutun Mountain	517	301	216	531	318	213	547	325	222	596	343	253
Toutun River	435	292	144	446	311	134	460	320	140	499	344	155
Midong	350	245	106	363	268	95	439	303	136	406	299	107
Urumqi Region	410	271	138	423	293	130	437	301	135	474	324	149

(*unit* mm m^{-2}) (*Source* own calculations)

rising evapotranspiration and P_n sums showed a positive trend as well (see Sect. 5.1). The simulation results for 2050 on the other hand suggest that with the rising temperatures an increase in P at least equivalent to the $SC_{2050,\ 75\ \%}$ is necessary to equal the situation of the last 10 years. The proportion of net precipitation left over after ET_a has slightly decreased from the past (32.1 and 33.9 %) to the future climate scenarios (30.7, 31.1 and 31.6 %).

The distributed flow values for melt water, total water flow, surface and groundwater flow are shown in Figs. 5.11, 5.12, 5.13, 5.14, 5.15, 5.16, 5.17, 5.18 and 5.19. The overall increase of flow values from SC_{LT} to SC_{ST} is clearly visible in all watersheds for all flow values. The simulated flow values for 2050 strongly depend on the projected precipitation response of the climate change. Both the simulation with the $SC_{2050,\ 50\ \%}$ and $SC_{2050,\ 75\ \%}$ would not achieve the same P_n as the SC_{ST}.

Not only the amount of P_n, but also the form of precipitation is important for the water balance. The importance of snow melt for the water balance in the past has been shown in Sect. 5.1. However, the proportion of the total water runoff coming from Q_m was 62, 63, and 83 % for the catchments Urumqi River, Toutun River and Midong under SC_{ST} and will be further reduced to 46–53, 47–50, and 60 % for the three scenarios in 2050. This decrease can also be seen in Fig. 5.12. The total amount of melt water in 2050 will even fall below the level of SC_{LT} for all catchments and projected climate scenarios except Toutun Mountain, whose larger part is located high enough to profit from the increased precipitation as snow fall.

As shown in Fig. 5.13, the total water flow Q_n is high in the upper mountain regions due to high precipitation values and low ET_a rates for all climate scenarios. However, these areas are relatively small compared to the whole watershed. **The Q_n trend shows an overall increase from SC_{LT} to SC_{ST} (see also Fig. 5.14). From SC_{ST} to SC_{2050}, Q_n increased only in the higher mountains and shows little positive change on the foothills**, where ET_a is also quite low and Q_n therefore high, no change or decrease on the urban area and the plain. The lower

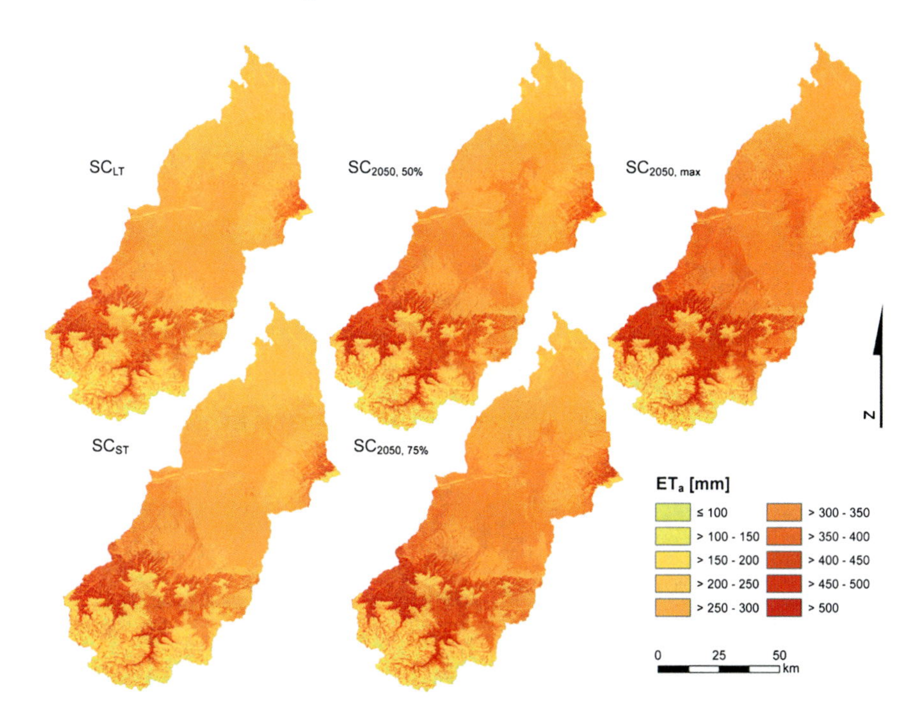

Fig. 5.10 Distribution of calculated actual evapotranspiration in Urumqi Region for the five different climate scenarios SC_{LT}, SC_{ST}, $SC_{2050,\ 50\ \%}$, $SC_{2050,\ 75\ \%}$ and $SC_{2050,\ max}$ (*Source* own calculations)

mountain areas, the plain and the urban areas have lower total water flow due to high ET_a depending on vegetation, low precipitation, and high ET_a over sealed areas in winter, respectively.

The quantitative result of the simulated total water flows confirms the distributed results. In 2050, Q_n is in all catchment areas under the three SC_{2050} higher than simulated for the SC_{LT} (see Fig. 5.15). When focussing on the trend compared to SC_{ST} which is representing the recent hydrological situation, the total water flow will decrease until 2050 except for Urumqi Mountain, Wulapo Reservoir, Urumqi River, and Toutun Mountain catchment under the $SC_{2050max}$. For the catchments of Urumqi City, Toutun River and Midong, Q_n will decrease under all projected scenarios.

A general, slightly positive trend for Q_{sf} in the whole catchment area can be observed from ths SC_{LT} to SC_{ST}. In the simulations for SC_{2050}, however, Q_{sf} increased only in the high altitude mountains, the areas where surface runoff already dominates. The other spatial distribution remains the same with relatively high runoff on the foothills and the urban areas. In the latter, the increase in Q_n was not pronounced enough to considerably increase Q_{sf} there. Similar to the total water flows, simulated Q_{sf} increase in 2050 compared to the SC_{LT} (see Fig. 5.17). However, only Q_{sf} in the Urumqi Mountain catchment and under the $SC_{2050,\ max}$ in

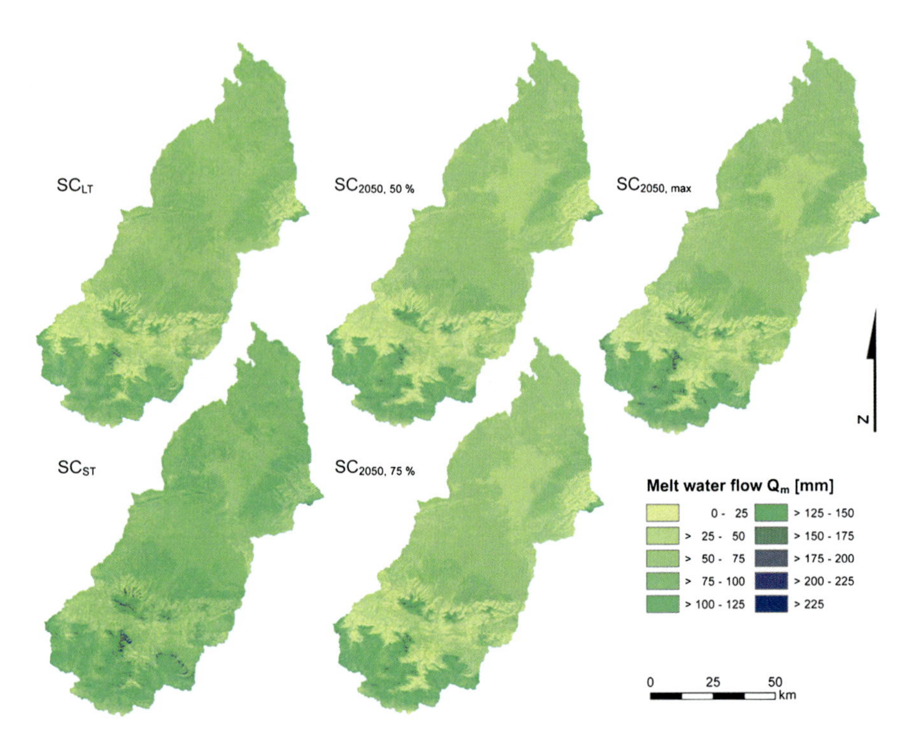

Fig. 5.11 Distribution of melt water flow in Urumqi Region for the five climate scenarios (*Source* own calculations)

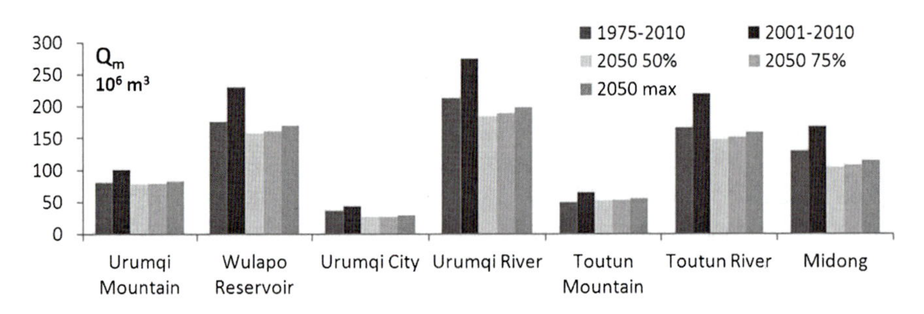

Fig. 5.12 Simulated melt water flows with different climate data sets for the catchments of Urumqi Region (*Source* own calculations)

the catchments of Wulapo Reservoir, Urumqi River, Toutun Mountain, and Toutun River is above the simulated recent surface runoff for the SC_{ST}. Urumqi City and Midong again show decreased surface runoff values.

With regard to the spatial distribution, groundwater infiltration mainly increases in the mountains at the margins of the perennial snow covered and glaciered area and also in the areas with already high groundwater recharge. These are the

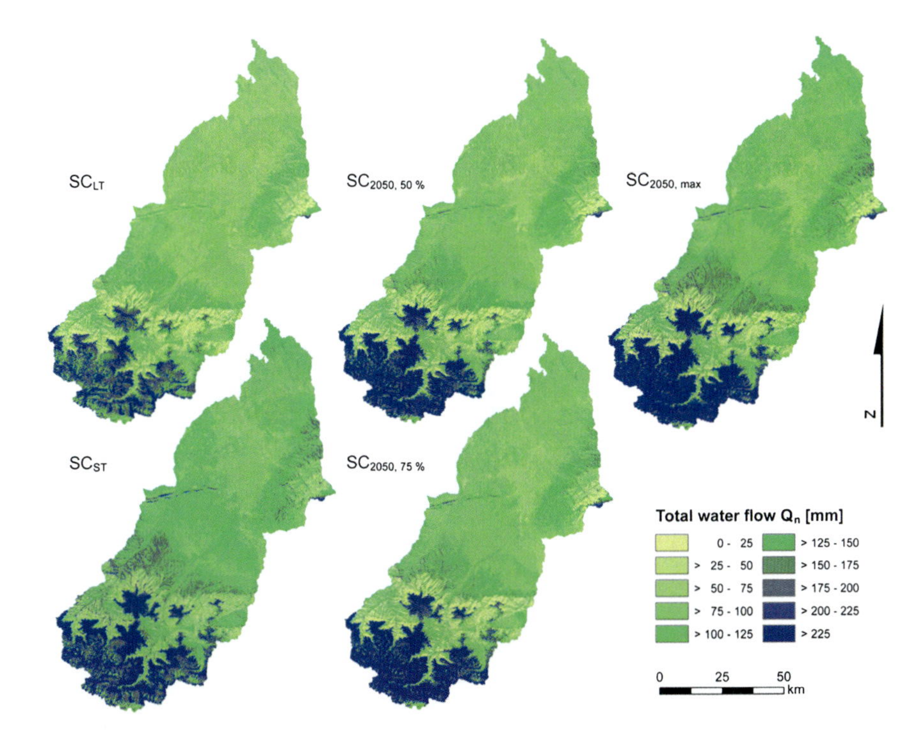

Fig. 5.13 Distribution of total simulated water flow in Urumqi Region (*Source* own calculations)

alluvial fans and the plain, where groundwater recharge is the dominating hydrological process as explained in Sect. 2.3. From the SC_{LT} to the SC_{ST}, Q_{gw} there increased, but the groundwater recharge rates then became lower in all SC_{2050} scenarios. Simulated Q_{gw} in 2050 are higher than the SC_{LT}, but higher than the SC_{ST} only for the maximum precipitation response at Urumqi Mountain, Wulapo Reservoir and Toutun Mountain (see Fig. 5.19). The increase in the mountain catchments is consistent with the spatial pattern.

Apart from the absolute changes and spatial distribution, the ratio of surface and groundwater flows changed as well: **the percentage of simulated surface water flow fell for all catchment areas from the $SC_{LT}$to the $SC_{ST}$and rose again for the projected climate scenarios in 2050** (Fig. 5.20). The increase of water flow in areas predominantly producing Q_{sf} will apparently be larger than in areas with more groundwater recharge. Nonetheless, the percentage of Q_{sf} in Midong catchment will stay significantly lower than in the other catchments.

The changes of the water balance components due to the different climate scenarios are summarised in Table 5.4, for a complete presentation of the absolute results refer to Tables A.25, A.26, A.27 and A.28 in the appendix. When comparing SC_{LT} to SC_{ST}, losses due to evapotranspiration increased only slightly while precipitation, melt water flow and total runoff in all catchment areas was

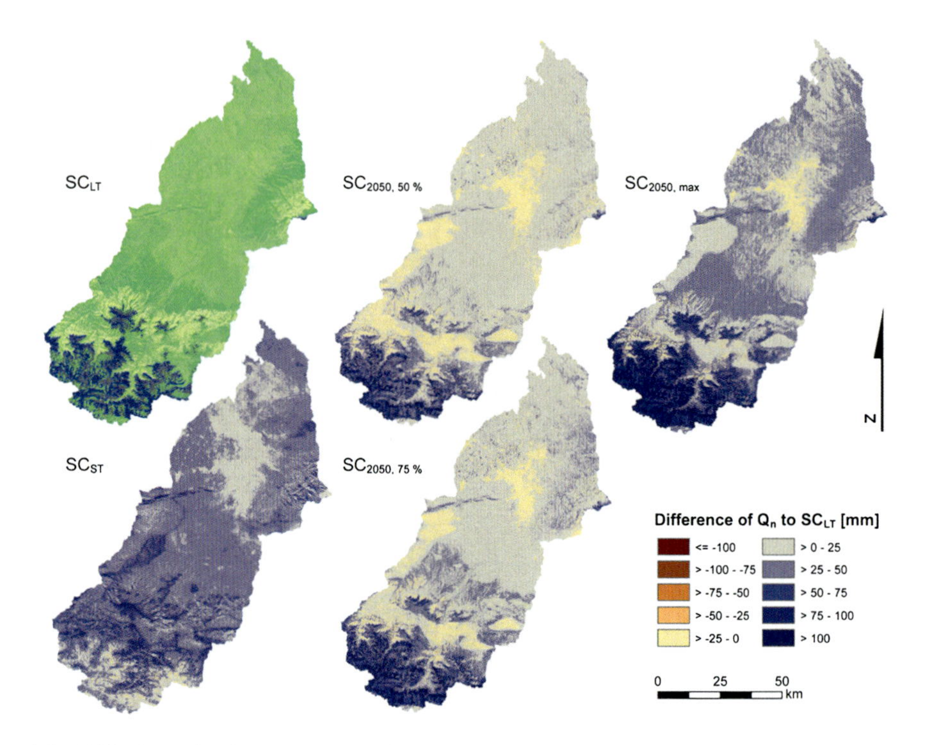

Fig. 5.14 Total simulated water flow in Urumqi Region and differences between simulations (*Source* own calculations)

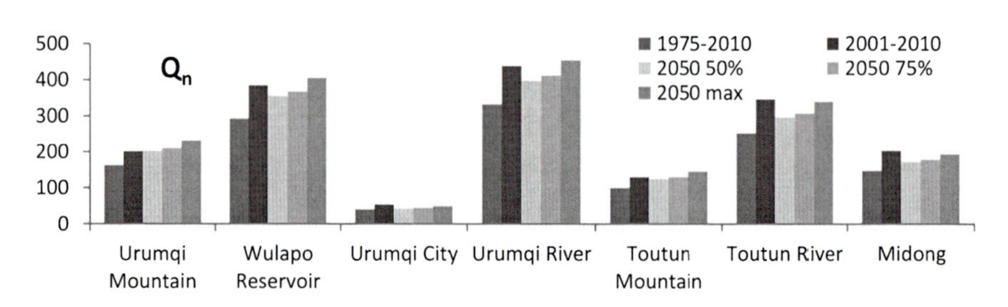

Fig. 5.15 Simulated total water flows with different climate data sets for the catchments of Urumqi Region (*Source* own calculations)

greater than before. However, a rather negative runoff trend is visible from the SC$_{ST}$ to the three projected climate scenarios for 2050 as total water flow and surface runoff in Urumqi Region was diminished or remained the same. A positive trend is only projected in the mountainous catchments of Urumqi Mountain, Wulapo Reservoir and Toutun River in SC$_{2050, max}$ when the precipitation response is at its maximum. The amount of precipitation in the mountain catchments is higher and able to balance the growing evapotranspiration due to rising

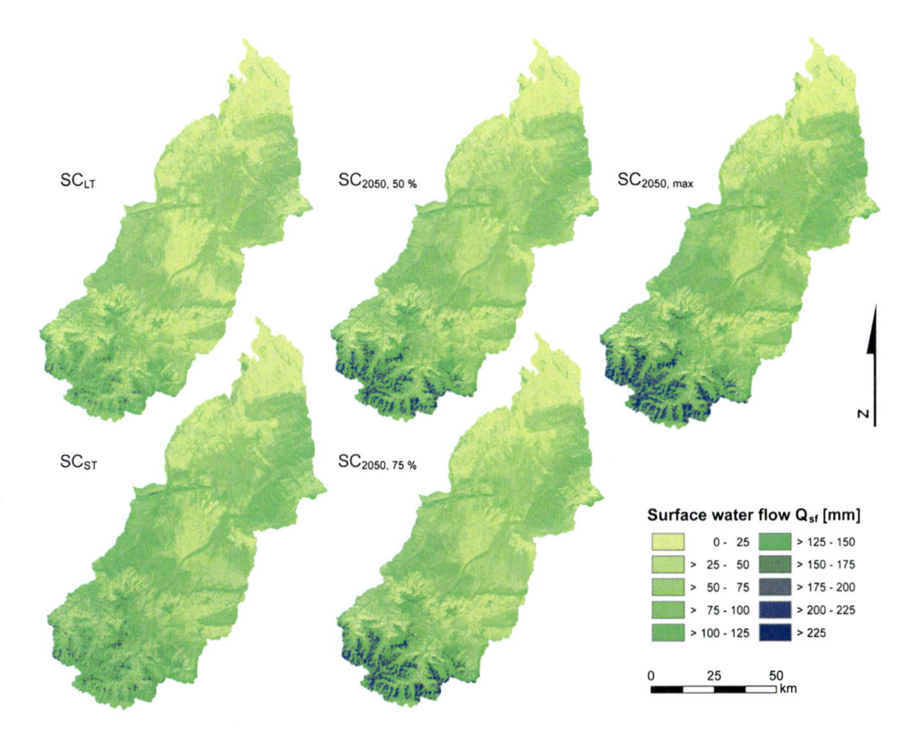

Fig. 5.16 Distribution of simulated surface water flow in Urumqi Region for different climate scenarios (*Source* own calculations)

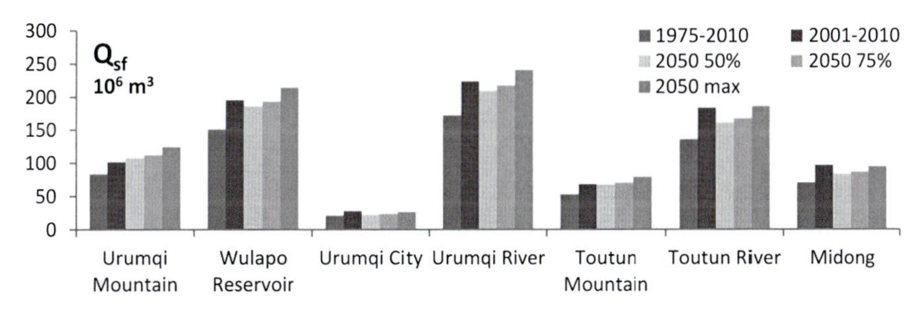

Fig. 5.17 Simulated surface water flows with different climate data sets for the catchments of Urumqi Region (*Source* own calculations)

temperatures (see also Table 5.3). The absolute evapotranspiration grew more in the downstream catchments, namely Urumqi City and Midong catchment. Hence, the highest decrease of water flows could also be observed there. Most prominent when comparing the recent to the future climate scenarios is the **reduced runoff from melt water of –15 to almost –40 % in some catchments under the SC$_{2050,\ 50\%}$as significantly less precipitation falls as snow and accumulates in**

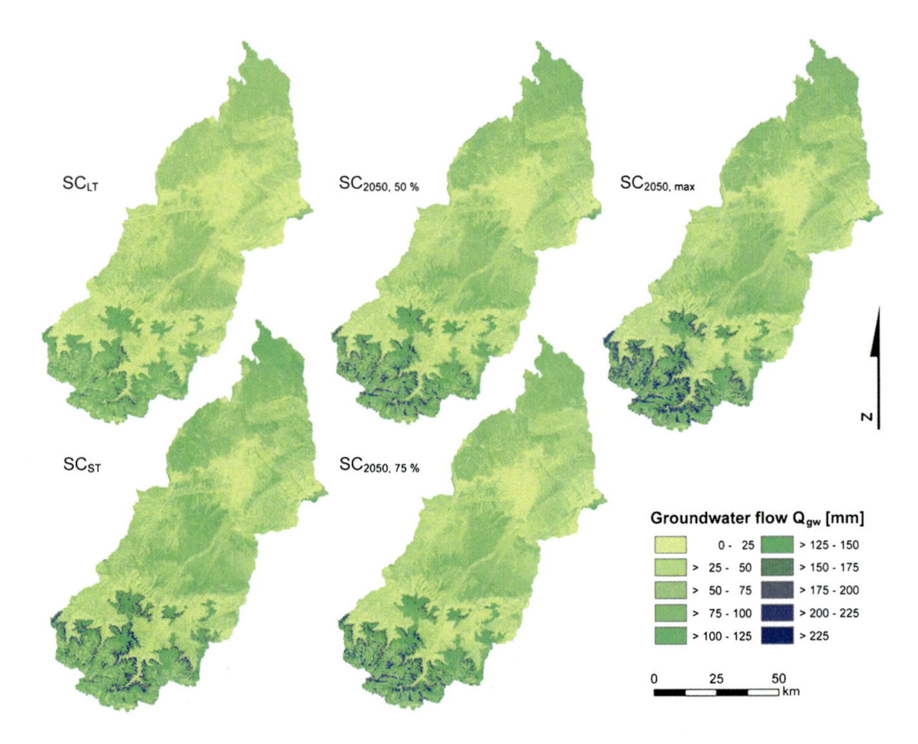

Fig. 5.18 Distribution of groundwater recharge in Urumqi Region simulated for different climate scenarios (*Source* own calculations)

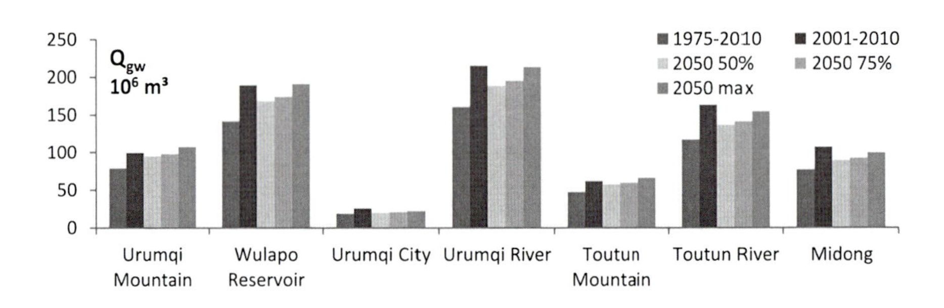

Fig. 5.19 Simulated groundwater flows with different climate data sets for the catchments of Urumqi Region (*Source* own calculations)

winter. Furthermore, the groundwater flow benefits less from increasing total water runoff and in the case of reduced total water flow less water is available for groundwater recharge than for surface runoff.

Due to the drawbacks of the water balance model with regard to monthly results and the modelling of runoff dynamics (see Sect. 3.4.3), the analysis focussed on

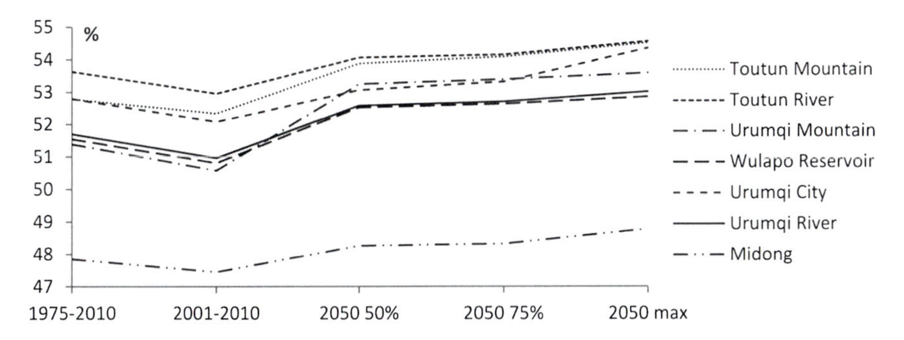

Fig. 5.20 Percentage of surface water flow in several catchment areas under different climate scenarios (*Source* own calculations)

Table 5.4 Difference in % of simulation results for annual precipitation, actual evapotranspiration, melt water, total water flow and surface and groundwater runoff in the catchment areas of Urumqi Region, calculated with the SC_{LT}, SC_{ST} and three different climate projections for the year 2050 ($SC_{2050,\ 50\ \%}$, $SC_{2050,\ 75\ \%}$, and $SC_{2050,\ max}$) (*blue* less evapotranspiration, more runoff; *red* more evapotranspiration, less runoff; *source* own calculations)

	Catchment area	Urumqi Mountain	Wulapo Reservoir	Urumqi city	Urumqi River	Toutun Mountain	Toutun River	Midong	Urumqi Region		Urumqi Mountain	Wulapo Reservoir	Urumqi city	Urumqi River	Toutun Mountain	Toutun River	Midong	Urumqi Region
		1975–2010									2001–2010							
P	2001–2010	3.7	3.7	3.8	3.7	5.1	5.0	3.5	4.1									
ET_a		−0.2	0.8	2.3	1.0	0.5	2.5	1.6	1.7									
Q_m		25.2	31.2	19.4	29.2	30.7	32.1	30.3	30.4									
Q_n		23.8	31.7	34.3	32.0	30.1	37.3	38.3	35.1									
Q_{sf}		21.8	29.8	32.4	30.1	29.0	35.6	37.1	33.4									
Q_{gw}		25.8	33.7	36.3	34.0	31.4	39.3	39.3	36.9									
P	2050 50%	8.4	8.0	7.3	7.9	7.9	7.4	7.3	7.6	2050 50%	4.6	4.2	3.4	4.0	2.6	2.4	3.7	3.4
ET_a		6.6	8.8	12.7	9.5	6.1	9.4	11.4	9.9		6.8	8.0	10.2	8.4	5.5	6.7	9.6	8.1
Q_m		−3.5	−10.2	−27.4	−13.1	3.1	−11.2	−19.8	−14.2		−22.9	−31.5	−39.2	−32.8	−21.2	−32.8	−38.5	−34.2
Q_n		24.7	21.2	6.9	19.5	24.9	17.6	17.0	18.3		0.8	−7.9	−20.4	−9.5	−4.0	−14.4	−15.4	−12.4
Q_{sf}		29.2	23.5	7.4	21.6	27.5	18.6	18.0	19.8		6.1	−4.8	−18.9	−6.6	−1.2	−12.6	−13.9	−10.2
Q_{gw}		19.9	18.8	6.2	17.3	22.0	16.5	16.0	16.8		−4.7	−11.2	−22.0	−12.5	−7.1	−16.4	−16.7	−14.7
P	2050 75%	11.8	11.4	10.7	11.3	11.2	10.8	10.6	11.0	2050 75%	7.9	7.4	6.6	7.3	5.8	5.6	6.9	6.6
ET_a		9.0	11.7	16.0	12.5	8.4	12.5	14.7	13.0		9.2	10.9	13.4	11.3	7.9	9.7	12.8	11.1
Q_m		−1.8	−8.0	−24.8	−10.9	5.0	−8.8	−16.9	−11.7		−21.6	−29.9	−37.0	−31.0	−19.7	−31.0	−36.2	−32.3
Q_n		29.3	25.7	11.4	24.0	29.9	22.0	21.3	22.8		4.5	−4.5	−17.0	−6.1	−0.2	−11.2	−12.3	−9.1
Q_{sf}		34.4	28.4	12.5	26.4	33.1	23.2	22.5	24.5		10.3	−1.1	−15.0	−2.8	3.1	−9.2	−10.7	−6.6
Q_{gw}		24.0	22.9	10.2	21.4	26.3	20.6	20.2	20.9		−1.4	−8.1	−19.1	−9.4	−3.8	−13.4	−13.7	−11.7
P	2050 max	21.5	20.9	19.9	20.7	21.1	20.4	19.9	20.4	2050 max	17.2	16.6	15.6	16.4	15.2	14.7	15.8	15.7
ET_a		15.1	19.5	25.4	20.5	14.4	20.9	24.3	21.5		15.3	18.6	22.6	19.3	13.8	18.0	22.3	19.5
Q_m		3.0	−2.9	−20.8	−6.0	11.1	−3.7	−11.2	−6.6		−17.8	−26.0	−33.6	−27.2	−15.0	−27.1	−31.9	−28.4
Q_n		43.0	38.6	22.4	36.7	45.6	34.9	31.7	35.0		15.5	5.3	−8.9	3.5	11.9	−1.8	−4.8	0.0
Q_{sf}		49.1	42.2	26.0	40.2	50.4	37.2	34.2	38.0		22.4	9.5	−4.9	7.7	16.5	1.2	−2.1	3.5
Q_{gw}		36.5	34.9	18.3	32.9	40.3	32.2	29.3	31.9		8.5	0.9	−13.2	−0.8	6.8	−5.2	−7.2	−3.7

the annual results. However, the monthly values of surface water flow for the five different scenarios are compared here. The monthly results are not investigated in detail and only Q_n is shortly evaluated here to show the seasonal effects of the climate scenarios. In order to asses also the spatial differences, the monthly results were calculated for the catchments Urumqi Mountain, Wulapo Reservoir, Urumqi City and whole Urumqi Region (see Fig. 5.21).

In the summer months, precipitation in the mountain catchments contributes more to runoff than in the lower catchments of Urumqi City or the whole Urumqi Region. In the lower catchments, precipitation in the summer months is lost to evapotranspiration and those areas rely more on the melt water flow in spring. However, the water flow in the lower catchments did not become significantly larger in spring, but rather in autumn probably due to increased precipitation.

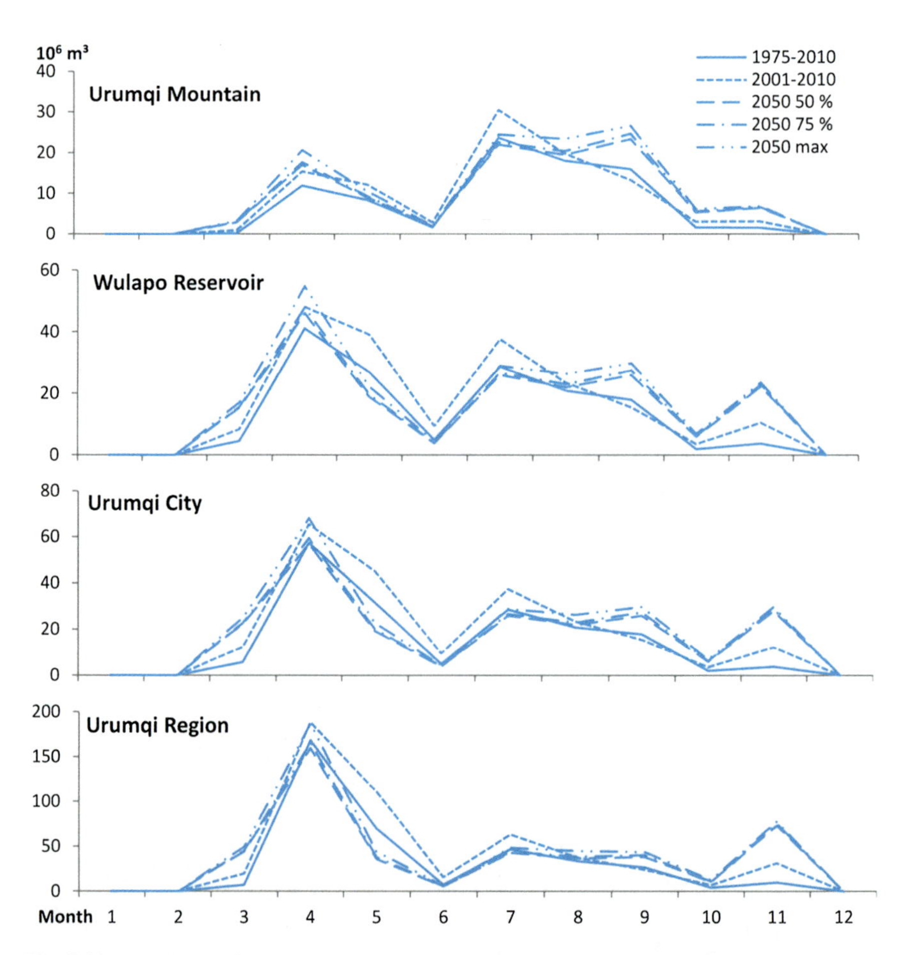

Fig. 5.21 Monthly surface water flow at Urumqi Mountain, Wulapo Reservoir, Urumqi City, and Urumqi Basin for SC_{LT}, SC_{ST}, and all three scenarios SC_{2050} (*Source* own calculations)

These results coincide with the distribution and change of melt water flows evaluated before. The SC_{ST} led to an increase of total runoff in all catchments and all months except September compared to SC_{LT}, similar to the even spatial distribution of runoff increase (see Figs. 5.13 and 5.14). For the three SC_{2050}, an increase in runoff can be expected for September to March influenced by the increase in precipitation. During the summer months May to August, however, the directly available runoff is projected to decrease because the evapotranspiration losses also become higher. It is remarkable that the runoff increased in November which is probably due to the increase in temperature and reduced snow fall.

5.3 Results for Land Use Change

Apart from climate change, changes initiated by anthropogenic activities are expected for Urumqi Region. These changes mainly involve alterations of land use and cover which in turn affect characteristics of the land surface and hydrological processes. For example evapotranspiration is regulated by albedo, surface resistance, and heat conductivity of the land surface, while soil type, degree of imperviousness and land cover type effect infiltration and surface runoff. Whereas changes of the soil properties can also happen due to climate change as described by Arnell [5], they are not considered here, as they are overshadowed by the much faster happening effects of human activities. The planned changes in land use (see Sect. 4.3) were integrated into the land use maps and derived parameters such as albedo, roughness height, stomatal resistance terms, degree-day-factor and proportion of runoff. The simulation of the distributed water balance was repeated with the average climate data of the SC_{ST} and new surface parameters and then compared to the original run for the period.

The spatial changes of potential and actual evapotranspiration, melt water, total water, and surface and groundwater flow are presented in Figs. 5.22, 5.23, 5.24, 5.25 and 5.26. The reduced ET_p is clearly visible for the planned development areas. This is due to the lower evapotranspiration potential of sealed urban surfaces in general compared to areas with vegetation cover, integrated in the model through the stomatal resistance factor r_s (see also Sect. 3.1.1). The importance of stomatal resistance for the calculation of ET_p was also demonstrated in the sensitivity analysis (Sect. 3.2.1). In contrast to the unvegetated surrounding areas covered with snow in winter, the evapotranspiration potential of the urban areas is higher due to lower albedo values and aerodynamic resistance r_a.

The seasonal differences are also shown in Fig. 5.23 where ET_p for January is shown on the left side and for July on the right side. The difference between urban areas and the surroundings is reversed from winter to summer, but the situation in summer affects the annual evapotranspiration potential more as the values for ET_p in summer are a multiple of the winter values. In summer, however, the largest part of precipitation is lost to evapotranspiration anyway. Changes from 105 to 85 mm

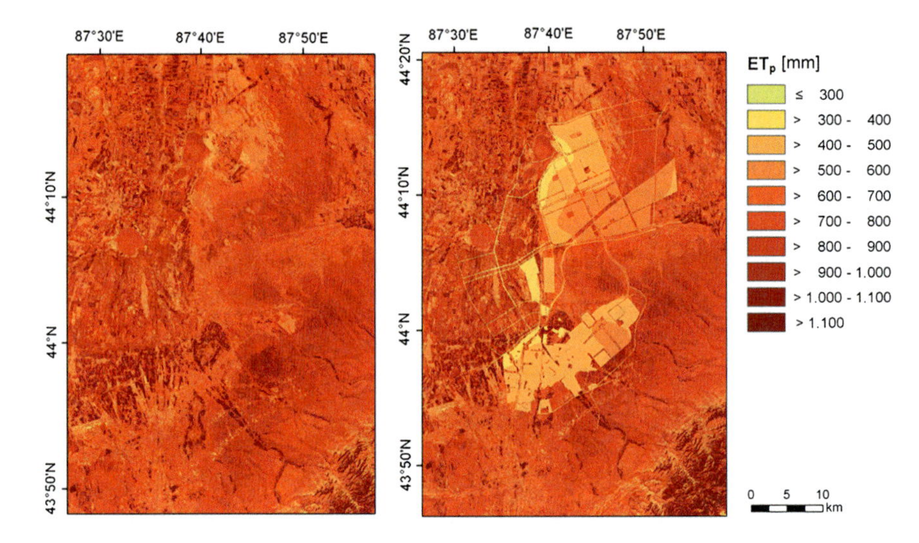

Fig. 5.22 Distribution of ET_p in Midong before and after the planned land use change calculated with SC_{ST} The total annual evaporation potential is lower for sealed areas (*Source* own calculations)

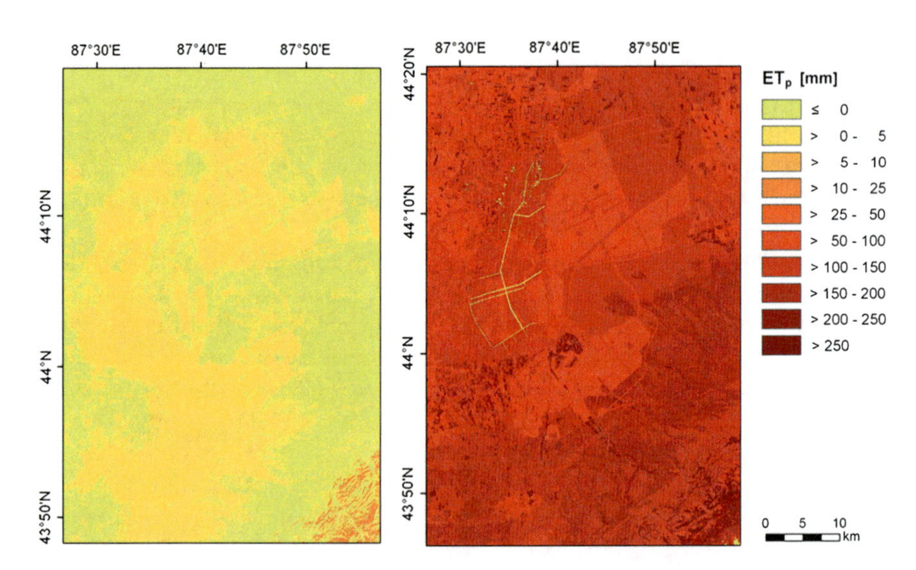

Fig. 5.23 Distribution of ET_p in Midong under the planned land use scenario for January and July calculated with SC_{ST}. ET_p is higher for sealed areas in winter but lower in summer when the overall potential evapotranspiration is a multiple of the values for ET_p in winter (*Source* own calculations)

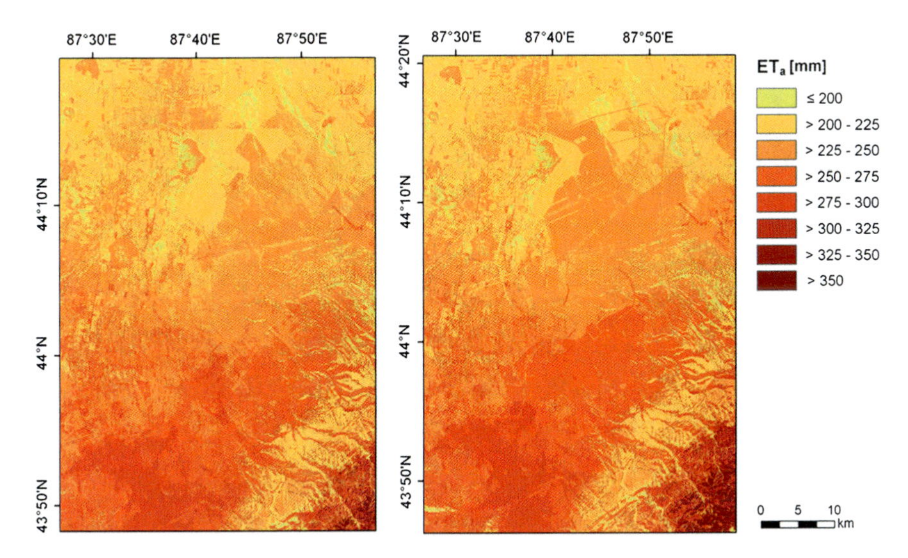

Fig. 5.24 Distribution of ET_a in Midong under different land use scenarios calculated with SC_{ST} (*Source* own calculations)

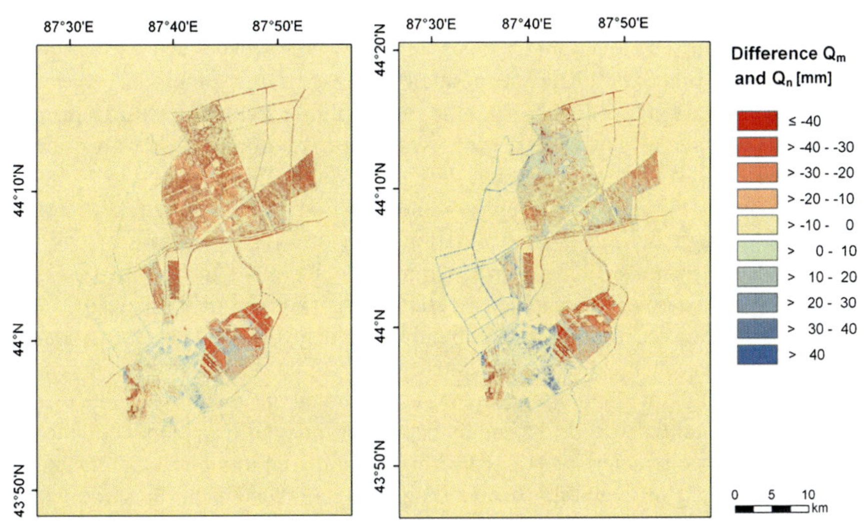

Fig. 5.25 Difference in melt water flow and total water flow in Midong after the land use change calculated with SC_{ST} (*Source* own calculations)

evapotranspiration potential compared to 30 mm rainfall in July do not change the monthly water balance while in winter a lower ET_p directly increases accumulated snow and subsequently snow melt water. Therefore, it is not surprising that the calculation of ET_a shows a different distribution of high and low values. The distribution of precipitation did not change, therefore it is not presented here.

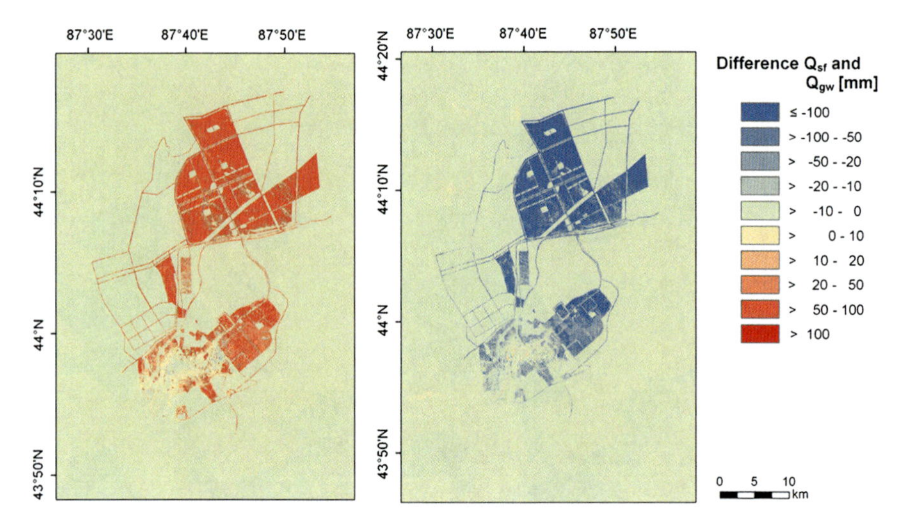

Fig. 5.26 Difference in surface and groundwater water flow for Midong development area calculated with SC_{ST} (*Source* own calculations)

However, it is important when precipitation occurs and it seems that the combination of higher ET_p over urban areas in the winter season and the temporal distribution of precipitation led to a small increase of the annual ET_a over the planned sealed areas (see Fig. 5.24). **Due to the higher evapotranspiration over urban areas especially in winter, snow water equivalent and melt water flow is decreased as well as the total water flow** (see Fig. 5.25).

Additionally, the degree of imperviousness and proportion of surface runoff is increased through buildings, transportation lines and other sealed surfaces and leads to a significantly higher surface runoff of about 40 to 75 mm (see also Fig. 5.26). On the other hand, **groundwater flow is reduced in these areas**. Little changes are observed for the areas already sealed in the land use classification from 2000. The water balance aggregated for the three outlets of Midong catchment affected by the land use changes is presented in Table 5.5. The evapotranspiration potential was decreased by 6.5 % for the whole catchment due to the changed surface characteristics of sealed areas compared to natural and vegetated ones. The amount of available precipitation was calculated with the same values for both simulations and remained the same.

Actual evapotranspiration however increased slightly by 0.7 % probably due to changes in the temporal distribution of ET_p in relation to precipitation (see Fig. 5.23). This reduction in net available water also has a small impact on the total water flow which is reduced by 0.9 % as well. Apparently, melt water flow is also affected by the increased evapotranspiration of precipitation and reduced by 2.1 %. This also indicates that actual evapotranspiration increased mainly in the winter and thus reduced snow fall and subsequently the melt water runoff. Not surprisingly, the development of sealed areas has a significant effect on the

Table 5.5 Water balance components in Midong catchment before and after planned land use development calculated with SC_{ST}

		Before		After		Diff	
		10^6 m^3	mm	10^6 m^3	mm	mm	%
Potential evapotranspiration	Total catchment	1,471	767	1,376	717	–50	–6.5
	Sealed area		744		466	–277	–37.2
Precipitation	Total catchment	672	350	672	350	0	0.0
	Sealed area		333		333	0	0.0
Actual evapotranspiration	Total catchment	469	245	473	246	1	0.7
	Sealed area		238		247	8	3.8
Melt water	Total catchment	169	88	165	86	–2	–2.1
	Sealed area		83		73	–10	–12.0
Surface runoff	Total catchment	97	50	112	58	8	15.9
	Sealed area		29		75	46	258.6
Groundwater recharge	Total catchment	107	56	90	47	–9	–16.1
	Sealed area		63		12	–51	–81.0
Total water flow	Total catchment	203	106	201	105	–1	–0.9
	Sealed area		92		87	–5	–5.2

(*Source* own calculations)

partition of surface and groundwater flows. Due to the increased degree of imperviousness, surface water flow is increased by 15.9 % and groundwater flow is significantly reduced by 16.1 % (for a summary refer to Table 5.5).

References

1. Bastiaanssen, W. G., & Harshadeep, N. R. (2005). Managing scarce water resources in Asia: the nature of the problem and can remote sensing help? *Irrigation and Drainage Systems, 19*, 269–284.
2. City Planning Bureau of Midong (2007). Midong New District Master Plan 2020 and Midong New District Schematic Map of the Overall Planning. Urumqi: City Planning Bureau of Midong.
3. NASA Landsat Program (2010). Landsat ETM+ scenes L1T, USGS, Sioux Falls. Retrieved from http://glovis.usgs.gov/.
4. Zhang, Z., & Li, L. (2004). *Zhongguo dixiashui ziyuan yu huanjing tuji. The Atlas of groundwater resources and environment of China* (262 p). Beijing: Zhongguo ditu chubanshe.
5. Arnell, N. (1996). Global warming, river flows and water resources (234 p). Chichester: Wiley.

Chapter 6
Projection of Water Consumption

The growth of population and economic output in newly industrialising regions, including agriculture and industry, normally leads to a growing water demand. Additionally, the demand is driven by consumption quantity and consumption patterns of the society populating the according spatial unit. With further industrial development and growing per capita GDP decreases the agricultural water consumption in a non-linear manner, simultaneously rises the industrial water demand and per capita water consumption for living purposes [1]. The rising living standard leads to a higher water consumption of a growing number of households, while more efficient utilities and infrastructure, waste water treatment and recycling are not yet fully implemented [31]. The purpose of this chapter is to model the relationship of development and water demand for Urumqi City to determine the absolute quantitative future demand and the relative composition of water demand by the different user groups in agriculture, industry and population. The water demand projection does not take into account possible changes in consumption due to climate change but only due to the development of population and economy.

The wide interest in determining the relationship between water consumption and the development of population and economy has been manifested in many publications and research papers on this topic. Du et al. [8] already tried to describe the relationship between population development, per capita GDP, gross industrial output, government revenue and water consumption for Urumqi using polynomial functions. Unfortunately poly-nomial functions are not useful for extrapolation as they tend to soar up or down at the margins [10]. Therefore, another but also simple method was chosen: basic econometric modelling with algebraic functions on the city level. The development of water demand in Urumqi Region was estimated using a method developed by Trieb and Müller-Steinhagen [29] and also used by Trieb [28]. Only sectoral water demand, growth rates of economy and population as well as several efficiency coefficients were necessary. Although not as detailed as the water balance model describing the distribution of water resources, it is used to analyse the development of the water demand as well. This data was available from published reports and statistical data.

K. Fricke, *Analysis and Modelling of Water Supply and Demand Under Climate Change, Land Use Transformation and Socio-Economic Development*, Springer Theses, DOI: 10.1007/978-3-319-01610-8_6,

6.1 Methodology for Econometric Modelling

The water demand ω of each population (municipal water consumption), agriculture (mainly irrigation), and industry is calculated separately as a function of the time t:

$$\omega(t) = \omega(t-1) * (1 + \gamma(t)) * \frac{\eta(t-1)}{\eta(t)} * (1 - \mu) \tag{6.1}$$

The driving factors are growth rate of population or GDP $\gamma(t)$, the efficiency of the water distribution system $\eta(t)$, the end use efficiency enhancement $\mu(t)$.

When comparing the actual growth rates, the relationship between population and GDP growth rate was not stable meaning they could not be freely exchanged as drivers of water consumption. For the calibration period, agricultural water demand showed a slightly higher correlation with population growth and industrial water demand with GDP growth. Hence, they were modelled with the respective drivers. However, municipal water demand did not show a significant correlation to neither population nor GDP growth. Therefore, an adapted formula was chosen to account for both driving factors for municipal water demand $\omega_{mun}(t)$:

$$\omega_{mun}(t) = \omega_{mun}(t-1)\left(1 + \left(0.5\gamma_{pop}(t) + 0.5\gamma_{GDP}(t)\right)\right)\frac{\eta_{mun}(t-1)}{\eta_{mun}(t)}(1 - \mu_{mun}) \tag{6.2}$$

with the growth rate of population $\gamma_{pop}(t)$ and the growth rate of GDP $\gamma_{GDP}(t)$. The distribution efficiency $\eta(t)$ is calculated according to

$$\eta(t) = \eta(t_E) * \varepsilon(t) + \eta(t_S) * (1 - \varepsilon(t)) \tag{6.3}$$

where $\eta(t_S)$ is the efficiency in the starting year and $\eta(t_E)$ efficiency of the end year, using progress factor α and best practice β:

$$\eta(t_E) = \eta(t_S) + \alpha * (\beta - \eta(t_S)) \tag{6.4}$$

and weighing factor ε calculated with the time t, starting time t_S and end time t_E:

$$\varepsilon = \frac{t - t_S}{t_E - t_S} \tag{6.5}$$

6.2 Driving Factors

For the projection of future population and GDP growth, internationally published studies and prognoses were consulted. First, the population development of Urumqi is compared to China as a whole and two possible scenarios are chosen. In the following subsection, three studies about the economic development of China are compared and evaluated.

6.2.1 Population Growth

In general, the projection of population is difficult as reversing of demographic or migration trends cannot be predicted [5]. Hence, the aim was to find a simple method to project the population growth and to outline a realistic population development for Urumqi until 2050. A simple extrapolation using linear regression and the prediction of the UN World Population Prospects for overall China were compared [30] (Fig. 6.2).

The UN Report on the World Population Prospects [30] offers three different scenarios for the population development of PR China, low, medium and high. The population de-velopment in Urumqi in the past 60 years with an average increase of 5.32 % p.a. (1949–2009, [26]) exceeded by far the national population development with 1.53 % growth p.a. (1950–2010, [30]). During the last 30 years, the growth rate was quite variable, but especially for the last 15 years (2.35 % p.a.) higher than the population growth rate of China (1.03 % p.a.) or a fast growing city such as Shanghai (1.76 % p.a., see Fig. 6.1). Hence, the United Nations high growth scenario was considered suitable for further calculations. This is realistic as Xinjiang and Urumqi experienced large in-migration from people seeking job opportunities [4, 19]. The projection of Urumqi's population was calculated with the number of inhabitants in 2009 which was extrapolated using the growth rates stated by the high growth scenario of the UN Report [30].

To account for possible errors and variance in the underlying data, the mean deviation of the population data from 1949 to 2009 for Urumqi was taken as the error of the projection and also propagated (see Fig. 6.2). The second extrapolation method uses a linear regression model for the population from 1949 until 2009 [26]. To account for data errors and inaccuracies, the partial regression line of the linear regression were taken as possible range of the future development. However, the deviance was small compared to the absolute figures and variations of other projected scenarios and therefore neglected for further calculations. As the linear regression line is fited into the whole data period from 1949 until 2009, the

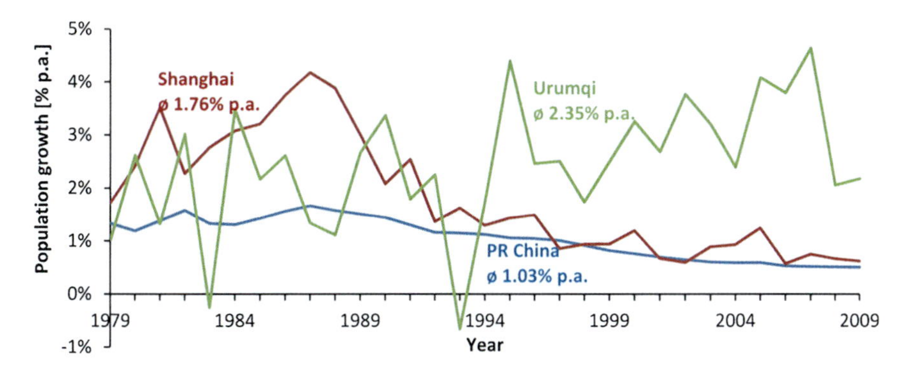

Fig. 6.1 Population growth rates 1979–2009 for PR China, Shanghai and Urumqi (*Source* [17, 25, 26])

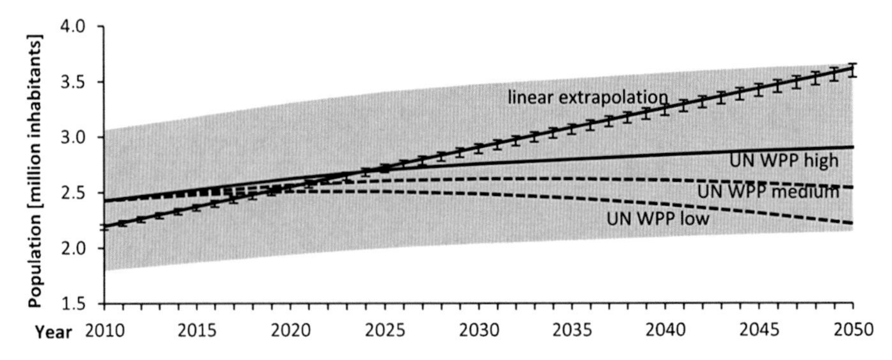

Fig. 6.2 Projected population of Urumqi 2010–2050 using a linear regression model and the three scenarios of the UN World population prospects. Error bars of the linear regression calculated using the errors of slope and axis intercept of the line of best fit, error band (*grey*) of the high growth scenario calculated using the mean deviation of the underlying population figures. 1949–2009 (*Source* [26, 30], own calculations)

calculated value for 2009 deviates from the observed value. Hence, the population projection starts at a lower value than the projection of the UN WPP. The results for population development calculated with the linear regression model was considered as the high growth scenario for Urumqi City, while the United Nations high growth scenario was used as the low growth scenario. As shown in Fig. 6.2, the mathematical error of the projected growth scenarios is quite variable and the differences between the scenarios actually larger than the calculated error. The propagation of the error values are omitted as they would simulate a degree of accuracy for the results which is not justified.

6.2.2 Gross Domestic Product

For the projection of the gross domestic product (GDP) and corresponding per capita GDP in Urumqi, several studies dealing with the probable economic development of China have been evaluated [9, 13, 36] as realistic projections for Urumqi or Xinjiang were not available. The GDP growth rates of Urumqi City and China are comparable for the last 20 years (see Fig. 6.4) and therefore the national projections could be transferred to Urumqi. The predicted GDP growth rates differ quite widely and cause diverging future GDP values when applying them to Urumqi (see Fig. 6.3). Hawksworth [9] applies only a constant GDP growth rate, while the other two reports chose a stepwise degression of a much higher starting rate (9.5 and 7.7 % respectively). To represent a wide range of possible developments, the projections of Wilson and Stupnytska [36] and Keidel [13] were used for the calculation of water demand as they show a realistic expert judgement of the future development. Not used was the lower projection I by Hawksworth [9], as the growth rates were considered too low when comparing them to the ones of Urumqi and China for the last 30 years (see Fig. 6.4).

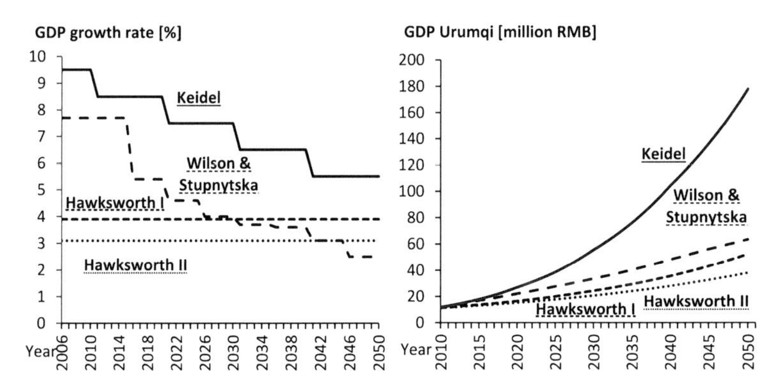

Fig. 6.3 Projected GDP growth rates and GDP of Urumqi 2010–2050 (*Source* [9, 13, 26, 36], own calculations)

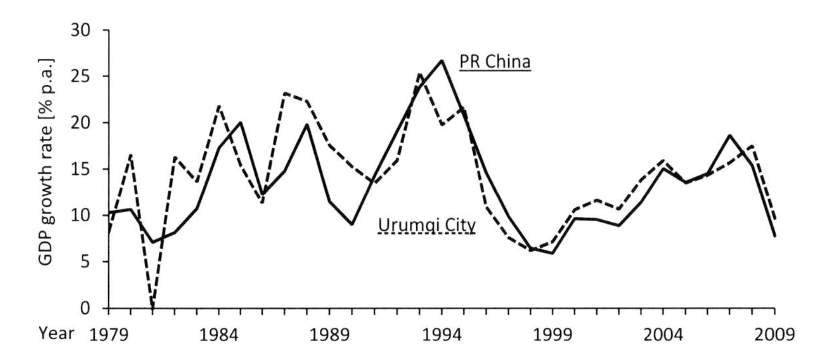

Fig. 6.4 GDP growth rates of PR China and Urumqi 1979–2009 (*Source* [17, 26])

6.2.3 Water Distribution Efficiency, Best Practice and Progress Factors

The level of operational and utility efficiency in urban water distribution networks is measured with indicators such as non-revenue water (NRW) or unaccounted-for water. The International Water Association recommends the use of NRW [15]. Non-revenue water is defined as the water losses including physical water losses due to leakage etc. and unbilled consumption. It can also be calculated as the total production volume (water supplied) minus the total consumption volume which produces revenue (water sold) divided by the total production volume (net water supplied) in percent [16, 11]. Unaccounted-for water (UFW) is often used in the same connotation, but not identical. UFW does not include legal usage that is not paid for (e.g. water from hydrants used for firefighting), and is usually measured in m^3/conn/day. The difference is usually small and here only non revenue water is used due to better data availability [12].

The International Benchmarking Network for Water and Sanitation Utilities estimated the non revenue water for a sample of Chinese water utilities at 27 % (2005) and 22 % (2009) [11] (e.g. Chengdu 18 % (for 2001 in [2], Beijing 8 %, Hong Kong 36 %, Shanghai 14 % and Tianjin 11 % (for 1997 in [16]). One explanation for the comparatively low NRW may be the high population density in larger Chinese cities. The AWWA Leak Detection and Water Accountability Committee recommends because of increasing demand and higher operational costs, the goal for NRW should be less than 10 % [3]. The NRW is therefore estimated to be only around 20 % on average and around 10 % for the best utilities.

Based on the references from other Chinese cities, the value of 80 % was used as the factual efficiency of distribution in the starting year $\eta(t_S)$ in Urumqi. The best practice percentage of distribution efficiency, which could be achieved in advanced high density networks, was set at 95 %. The parameters α and β imply that α % of the gap between the present efficiency and best practice β % will be closed until the end of the estimation period. With a progress factor α of 65 %, a distribution efficiency of 90 % would be achieved by 2050. As most of the industrial areas and companies in Urumqi are connected to the municipal water system, the factors of municipal water demand are adopted for industrial water demand as well.

The irrigation water use efficiency in China is generally regarded as low, several authors cite numbers between 38–46 % [14]: 46 %, [7]: circa 40 %, [37]: approximately 45 %, [27]: 38–45 %). A more comprehensive analysis was conducted by Cornish [6], who compared and benchmarked several irrigation systems all over China. The main system water delivery efficiency was between 41–85 %. As the irrigation system efficiency around Urumqi, the provincial centre of development, should not be the worst, an actual efficiency of distribution of 45 % and a best practice efficiency of 90 % were assumed (see Table 6.1).

Available data about sectoral water demand 2003–2008 taken from Water Reports and Statistical Yearbooks served as the basis for the calculation and were used to evaluate and calibrate the factors above (see Fig. 6.5). The values of the general end use efficiency enhancement were first adopted from Trieb [28] as they were assumed to be reasonable also for Urumqi. However, when comparing calculated estimations about sectoral water demand with available data for 2003–2008 taken from Water Reports and Statistical Yearbooks, the general end

Table 6.1 Factors for estimating future water demand

Sector	Agriculture	Municipal	Industrial
Driving force	γ_{pop}	γ_{GDP}	γ_{ind}
Efficiency of distribution	$\eta_{irr} = 45$ %	$\eta_{mun} = 80$ %	$\eta_{ind} = 80$ %
Best practice factor	$\beta_{irr} = 90$ %	$\beta_{mun} = 95$ %	$\beta_{ind} = 95$ %
Progress factor	$\alpha_{irr} = 50$ %	$\alpha_{mun} = 65$ %	$\alpha_{ind} = 65$ %
General end use eff. enhancement	$\mu_{irr} = 1$ % y^{-1}	$\mu_{mun} = 2$ % y^{-1}	$\mu_{ind} = 2$ % y^{-1}

(modified according to [28])

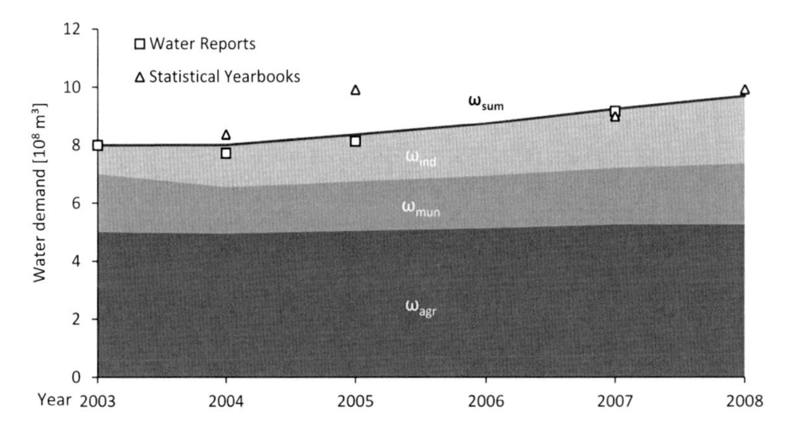

Fig. 6.5 Water demand of Urumqi 2003–2008 from Water Reports and Xinjiang Statistical Yearbooks (data points) combined with the calibrated modelled water demand ω_{agr}, ω_{mun}, ω_{ind} and ω_{sum} (*Source* [20], [22–24] and [32–35], own calculations)

use efficiency enhancements proved to be larger than the proposed values by Trieb [28] and were therefore calibrated and increased accordingly.

As with other methods for modelling and projections, the results are only as good as the available input parameters and calibration data. The calibration data from the Water Reports and the Statistical Yearbooks however was not consistent and showed some outliers. Additionally, the model results could not be validated because data was only available for a calibration period of 5 years. Considering also the uncertainties of the projections of population and GDP growth as mentioned in Sects. 6.2.1 and 6.2.2, the estimation results have to be regarded as a possible scenario of water demand in 2050 that was developed to be as realistic as possible.

6.3 Water Demand Results

The water demand was calculated with six different scenarios for population and GDP development from 2003 until 2050 (for an overview of the scenarios see Fig. 6.6). The calculations were implemented in a Microsoft Excel file and the results presented in Fig. 6.7. As one can see from the wide range of values for the different scenarios, the results are highly dependent on the reliability of the input data and factors and on the subjective assessment of the investigator. The different scenarios for the population growth proved to have only little effect on the total water supply. This observation is reasonable as the GDP is the driving factor of two of the water demanding sectors and growing at a higher rate than the population. Almost every value could be achieved, therefore the most reasonable one is chosen. The constrained projection for GDP growth by Hawksworth [9] seems to underestimate

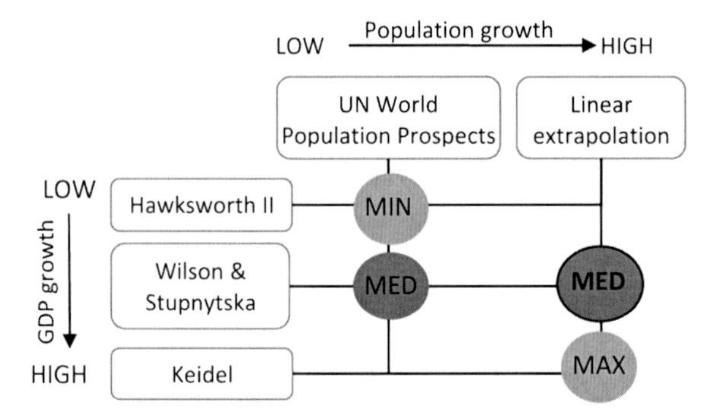

Fig. 6.6 Combination of scenarios for population and GDP development (*Source* own calculations)

at least the short term growth of GDP in this model as it causes an immediate decrease of water demand from 2009 on. Using the GDP projection by Keidel [13], the GDP growth and accompanying municipal and industrial water demand would cancel the increasing distribution and end usage efficiency. Together with the linear extrapolation of population growth it would send up the water demand to a maximum amount that could not be met neither by the available water resources nor the production and distribution system (cf. maximum water demand in Figs. 6.6 and 6.7). On the other hand, the GDP scenario by Hawksworth [9] would imply an immediate decrease of water consumption. Combined with the population growth as suggested by the scenario taken from the UN WPP [30] it can be considered as the minimum water demand scenario, but regarded as unrealistic and been omitted. As the projected population growth adapted from UN WPP [30] actually underrepresented the actual development until 2010, the scenarios based on this projection were also not considered for further evaluation. According to expert judgment, an increase of 50 % of the water demand is the most extreme amount that could be met by the water supply in Urumqi region. Assuming self-regulating processes in the

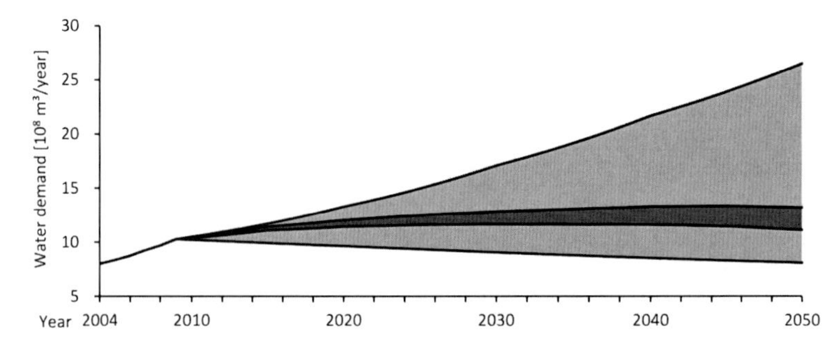

Fig. 6.7 Water demand 2003–2008 and estimated water demand 2003–2050 for the minimum, maximum and two medium scenarios (*black lines*) as introduced in Fig. 6.6 (for the exact figures refer to Table A-29 in the appendix) (*Source* own calculations)

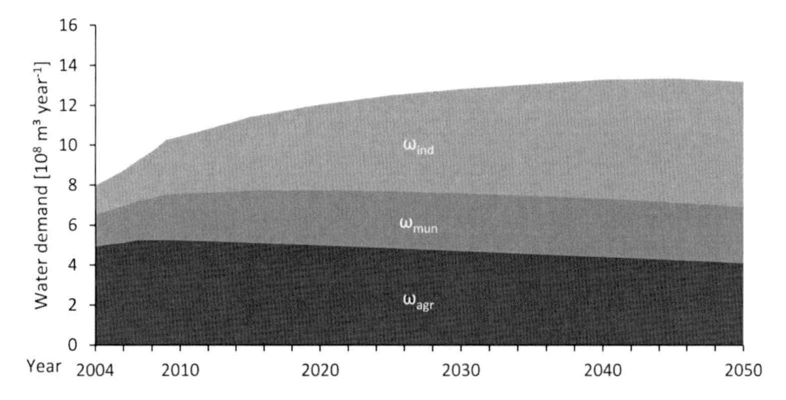

Fig. 6.8 Composition of water demand 2003–2050 using the scenario of [36] for GDP growth and a linear extrapolation of the population growth (higher medium scenario combination as presented in Fig. 6.7, *Source* own calculations)

region, the GDP scenario by Wilson and Stupnytska [36] combined with the linear regression model for the population growth were chosen as the most reasonable water demand drivers.

According to the estimation results of the most probable scenario (GDP growth scenario by Wilson and Stupnytska [36] and linear extrapolation of the population growth), water demand would increase from 800 to $1{,}000 \cdot 10^6$ m^3 to about $1{,}300 \cdot 10^6$ m^3. This is a projected rise of 30 to 62.5 %. The composition of the water demand will change as well (see Fig. 6.8 and Table 6.2). Mainly the municipal water demand will increase due to the relationship with GDP growth and a high starting share, while the proportion of the irrigation water demand will decrease. This could be realistic, as the irrigation sector offers the highest potential for improvement and with the transition to a more industry-oriented economy, the areas assigned to agricultural production will decrease as well. The economic output per water used is higher in the industrial sector, therefore industrial water use has a higher cost-benefit ration. The estimation of water demand in 2050 would also be comparable to the average ratio in OECD countries, where 65 % of

Table 6.2 Actual (2004) and projected water demand of Urumqi City 2010–2050

Years	Water demand [10^6 m^3]						
	ω_{agr}		ω_{mun}		ω_{ind}		ω_{sum}
2004	495	61.9 %	160	20.0 %	145	18.1 %	800
2010	524	50.2 %	232	22.3 %	287	27.5 %	1,044
2020	500	41.4 %	275	22.8 %	431	35.8 %	1,206
2030	471	36.7 %	289	22.5 %	524	40.8 %	1,283
2040	441	33.1 %	292	21.9 %	597	44.9 %	1,330
2050	410	31.1 %	282	21.3 %	628	47.6 %	1,320

(*Source* 2004: estimated from Water Affairs Bureau Urumqi [33] and Statistics Bureau of Urumqi [21], 2010–2050: own calculations)

the water is used for industrial purposes and only 30 % for agriculture [18]. The results for the projected water demand as presented in Table 6.2 are compared with the water availability under the climate change scenarios in Sect. 7.1.

References

1. Anand, P. B. (2007). *Scarcity, entitlements and the economics of water in developing countries* (310 p). Cheltenham: Edward Elgar.
2. Andrews, C. T., & Yñgiuez, C. E. (2004). *Water in Asian cities: Utilities' performance and civil society views* (97 p). ADB: Manila.
3. AWWA Leak Detection and Water Accountability Committee. (1996). Committee report—water accountability. *Journal AWWA, 88*(7), 108–111.
4. Becquelin, N. (2000). Xinjiang in the nineties. *The China Journal, 44*, 65–90.
5. Bretz, M. (2001). Zur Treffsicherheit von Bevölkerungsprognosen. *Wirtschaft und Statistik, 11*, 906–921.
6. Cornish, G. A. (2005). Performance benchmarking in the irrigation and drainage sector: Experience to date and conclusions, Department for International Development and HR Wallingford (66 p). Retrieved 18 Oct, 2011, from http://www.dfid.gov.uk/r4d/pdf/outputs/R8164.pdf.
7. Deng, X.-P., Shan, L., Zhang, H., & Turner, N. C. (2004). Improving agricultural water use efficiency in arid and semi-arid areas of China. In T. Fischer (Ed.), *Proceedings of the 4th International Crop Science Congress in conjunction with 5th Asian Crop Science Conference, 12th Australian Agronomy Conference 2004, in Brisbane, Queensland, Australia 26* Sept - 1 Oct, 2004 (pp. 1–14). Brisbane, Australia.
8. Du, H., Zhang, X., & Wang, B. (2006). Co-adaptation between modern oasis urbanisation and water resources exploitation: A case of Urumqi. *Chinese Science Bulletin, 51* (Supp. I) 189–195.
9. Hawksworth, J. (2006). *The world in 2050: Implications of global growth for carbon emissions and climate change policy* (67 p). London: PricewaterhouseCoopers.
10. Hoffmann, T., & Rödel, R. (2004). Leitfaden für die statistische Auswertung geographischer Daten (110 p). Greifswald: (Greifswalder Geographische Arbeiten, 33).
11. International Benchmarking Network (2011a). Country report China, The international benchmarking network for water and sanitation utilities (2 p). Retrieved Oct 24, 2011, from http://www.ib-net.org/production/?action=country.
12. International Benchmarking Network (2011b). Water and sanitation utilities: Context information and indicators—definitions, international benchmarking network (22 p). Retrieved May 23, 2012, from http://www.ib-net.org/en/documents/IBNETindicator definitionsSept04_000.doc.
13. Keidel, A. (2008). China's Economic Rise: Fact and Fiction (15 p). (Policy Brief, 61).
14. Khan, S., Tariq, R., Cui, Y., & Blackwell, J. (2004). Can irrigation be sustainable? In: T. Fischer (Ed.), *Proceedings 4th International Crop Science Congress in conjunction with 5th Asian Crop Science Conference, 12th Australian Agronomy Conference Sunday 26 Sept—1 Oct, 2004* (pp. 1–10). Brisbane, Australia.
15. Lambert, A. (2003). Assessing non-revenue water and its components: a practical approach. Water21 08/2003: 50–51.
16. McIntosh, A. C., Yñiguez, C. E., & Asian Development Bank (Eds.) (1997). *Second water utilities data book: Asian and Pacific region*, (210 p). Manila Philippines: Asian Development Bank.
17. National Bureau of Statistics China (2010). *China Statistical Yearbook 2010*. CD-ROM, Beijing: China Statistics Press

18. OECD (1999). *The price of water. Trends in OECD countries*, (174 p). OECD: Paris.
19. Shen, Y., & Lein, H. (2005). Land and water resource management problems in Xinjiang Uygur Autonomous Region China. *Norwegian Journal of Geography, 59*(3), 237–245.
20. Statistics Bureau of Urumqi (2005). *Urumqi Statistical Yearbook 2005*, (714 p). Beijing: China Statistics Press.
21. Statistics Bureau of Urumqi (2005). Urumqi Statistical Yearbook 2005 (366 p). Beijing: China Statistics Press.
22. Statistics Bureau of Urumqi (2006). Urumqi Statistical Yearbook 2006 (439 p). Beijing: China Statistics Press
23. Statistics Bureau of Urumqi (2008). *Urumqi Statistical Yearbook 2008*, (455 p). Beijing: China Statistics Press.
24. Statistics Bureau of Urumqi (2009). *Urumqi Statistical Yearbook 2009* (506 p). Beijing: China Statistics Press.
25. Statistics Bureau of Shanghai (2010). *Shanghai Statistical Yearbook 2010*. CD-ROM, Beijing: China Statistics Press.
26. Statistics Bureau of Urumqi (2010). *Urumqi Statistical Yearbook 2010* (456 p). Beijing: China Statistics Press.
27. Tian, Y. (1986). Tugayi in the delta and lower reaches of the Kerya river—A natural complex reflecting ecological degradation. *Journal of Desert Research, 6*(2), 1–24.
28. Trieb, F. (2007). *AQUA-CSP: Concentrating Solar Power for Seawater Desalination* (279 p). DLR (Final Report, Bundesministerium für Umwelt, Naturschutz und Reaktorsicherheit (BMU)).
29. Trieb, F., & Müller-Steinhagen, H. (2008). Concentrating solar power for seawater desalination in the Middle East and North Africa. Desalination *220*, 165–183.
30. UN Population Division of the Department of Economic and Social Affairs (2009). World Population Prospects: The 2008 Revision (801 p). United Nations, New York.
31. van Edig, A., & van Edig, H. (2005). Integriertes Wasserressourcen-Management: Schlüssel zur nachhaltigen Entwicklung. In: S. Neubert (Ed.), *Integriertes Wasserressourcen-Management (IWRM). Ein Konzept in die Praxis überführen* pp. 135–157. Nomos Verl.-Ges: Baden–Baden.
32. Water Affairs Bureau Urumqi (2003). Water Report 2003. Water Affairs Bureau Urumqi City, Urumqi, digital.
33. Water Affairs Bureau Urumqi (2004). Water Report 2004 (27 p). Water Affairs Bureau Urumqi City, Urumqi.
34. Water Affairs Bureau Urumqi (2005). Water Report 2005 (27 p). Urumqi: Water Affairs Bureau Urumqi City.
35. Water Affairs Bureau Urumqi (2007). Water Report 2007 (24 p). Water Affairs Bureau Urumqi City, Urumqi.
36. Wilson, D., & Stupnytska, A. (2007). The N-11: More Than an Acronym (23 p) (Global Economics Paper, 153).
37. Xinhuanet (2006): *China's 11th five-year (2006–2010) social and economic development plan* (14 p). Retrieved 24 Oct, 2011, from http://www.sandpconsulting.com/admin/upload/ChinaParliamentendorses11thFive-YearPlan.pdf.

Chapter 7
Discussion

In this chapter, the modelling results will be summarised and discussed. This includes a comparison of the projected future water resources and predicted consumption in Urumqi Region. Then, strategies for an improved water management are suggested and evaluated within the local context of Urumqi Region. Rather than classifying them according to the part of the water system they belong to (e.g., water supply, water saving usage, water treatment, water quality, and water reuse), the consequences of climate and land use change are discussed for the different uses and the strategies suggested.

7.1 Future Water Resources and Predicted Consumption

In the previous chapters, the results of the water demand and water supply in Urumqi Region now and projected for the year 2050 with three different scenarios were presented. The changes of the available water resources was simulated with a hydrological water balance model while the future consumption in Urumqi City is the result of the econometric modelling. Nevertheless, a direct comparison is not possible, as water supply is modelled, calibrated and validated on the basis of river watersheds, while the water demand is based on and projected for the whole administrative region. Despite these constraints, the relevant results can be analysed and compared to assess the severity and challenges of the future situation.

As both temperature and precipitation are expected to increase, the resulting trend for the available runoff is not necessarily clear. It might increase due to more available water from precipitation, but also decrease due to the rising potential evapotranspiration. Aizen et al. [1: 1402] expect the role of evapotranspiration to "be more significant in regions with summer maxima of precipitation than in regions with winter maxima". Urumqi shows the maximum precipitation in summer, however, precipitation is predicted to increase from January to May and July and August. This may lead to more snowfall and melt water in spring, but also to more actual evapotranspiration because more water is available. The simulation results show that the timing was indeed important for the resulting runoff, as the

K. Fricke, *Analysis and Modelling of Water Supply and Demand Under Climate Change, Land Use Transformation and Socio-Economic Development*, Springer Theses, DOI: 10.1007/978-3-319-01610-8_7,

precipitation increase in the summer months will show only little effect on the total runoff then (see Fig. 5.20). But unlike the prognosis by Aizen et al. [1] the high evapotranspiration potential in the urban and lower areas reduces the snow accumulation in winter and melt water flow in spring, different to other parts of the research area (see Table 5.4).

Aizen et al. [1] reported for the Tianshan rivers no significant changes or a decrease in annual runoff over the past 52 years. Although increases in precipitation were measured, they were not associated with a synchronous rise in river runoff. The authors suggest that the precipitation must be lost as evaporation or as percolation to deep groundwater circulation systems in the foothills [1]. Rising air temperatures are cited as the main factor controlling changes in river runoff as it determines the fraction of liquid precipitation compared to snow fall and the evaporation potential. For the Urumqi Basin however, Casassa et al. [2] found from several local studies that runoff from glacier melt had increased for different periods 1958–2003, some linked to a warming trend and stronger glacier melt, but in the lower catchments strongly related to the precipitation increase. These findings were also confirmed by the result of this study for recent climate change that showed an increase of 35 % of total runoff between the short- and the long-term scenario. For the short-term climate scenario (SC_{ST}), temperatures are rising compared to the long-term climate scenario. Therefore, ET_p is increasing, but the deviation is small compared to the total range. More important are the effects of climate change on ET_a which is also depending on the water availability. Thus, the increase in precipitation in the scenarios is also accompanied by an increase of ET_a (see Table 5.1 in Sect. 5.1). However, the increase in precipitation of the SC_{ST} exceeded the rise of ET_a and lead to a higher net precipitation, available for runoff. The amount of melt water is larger due to the larger precipitation values in winter, as well as the amount of surface runoff and groundwater recharge. A slightly larger proportion of total runoff also percolated into the groundwater in the more permeable areas as suggested by Aizen et al. [1].

For the evaluation of the projected climate scenarios, the comparison of the future water balance with the recent climate scenario SC_{ST} is more interesting. The water infrastructure, distribution and consumption are adjusted to the recent water availability. Therefore, aberrations from this situation have a severe impact on the region. As the climate scenarios show an equal rise of temperature, the increase of ET_p is the same for all three scenarios. The increase is especially pronounced in the lower mountain areas and other regions with denser vegetation such as agricultural areas. Despite their different canopy structure grassland and coniferous forest have similar maximum hourly evapotranspiration rates and maximum surface conductances [3]. Simultaneously with the projected precipitation increase, ET_a rises as well. The increase is most pronounced in the lower, vegetated mountain areas, because ET_p is highest there and the amount of available precipitation is large. **But only for the scenario with the highest precipitation values (maximum precipitation response $SC_{2050,\ max}$) the same net precipitation value as for the SC_{ST} is reached in whole Urumqi Region**. Only with a heavy increase of precipitation, rising temperatures and evapotranspiration potentials water losses can be compensated. However, melt water flow is decreasing even compared to the SC_{LT} for all three projected scenarios due to the higher temperatures except Toutun Mountain. This catchment includes a larger part of high altitudes that exhibits low ET_p and still benefits from increased precipitation in the form of snow in winter. In the lower catchments, the amount of snow fall in winter and snow melt in spring is reduced and is only partly balanced by increased precipitation in autumn. Unfortunately, these changes will lead to lower water availability during the vegetation period in spring and summer. Fuchs [4] showed that the vegetation period based on thermal criteria has been preponed by over 5 days from 1973 to 2011. Although the thermal preconditions for agriculture might further improve in the future and enable a longer growing period, the water supply does not follow this trend. Similar are the results for total water runoff Qn under the future climate scenarios, it will increase only in the mountain areas. Based on the results for the catchments in Urumqi Region, the total water flow will decrease until 2050 except for Urumqi Mountain, Wulapo Reservoir, Urumqi River, and Toutun Mountain under the $SC_{2050,\ max}$. The increase of Qn caused mainly an increase of surface runoff as in the above mentioned areas surface runoff is the dominating hydrological process. The proportion of groundwater flow is reduced in all catchment areas (see also Fig. 5.19).

Climate change effects also the spatial distribution of the water balance components. Based on the spatial structure of the hydrological system as presented in Fig. 2.10, the consequences of climate change for the hydrological system in Urumqi are illustrated in Fig. 7.1. ET_a rises mainly in areas with denser vegetation cover, i.e., vegetated lower mountains and agricultural fields on the alluvial plain, as the evapotranspiration potential in these areas increases more. The amount of melt water is overall reduced due to the increased temperatures. Only in the mountain areas where increasing temperatures change the snow fall and melting period only marginally, the amount of melt water is stable. Precipitation led to higher total water flow values in the mountain catchments, but due to rising ET_a and decreasing net precipitation, it is reduced downstream. Because of the reduced water availability existing water problems as described in Sect. 2.5 are likely to increase. The recession of river flows to the downstream basins will worsen as the

Table 7.1 Simulated total water flows for catchments and nested sub-catchments (*italic*) in Urumqi Region and five climate scenarios

Catchment	km^2	SC_{LT}	SC_{ST}	$SC_{2050,\ 50\ \%}$	$SC_{2050,\ 75\ \%}$	$SC_{2050,\ max}$
Urumqi mountain	1,071	162	201	202	210	232
Wulapo	2,560	292	385	354	368	405
Urumqi River	3,126	323	538	428	446	490
Toutun mountain	705	100	130	125	129	145
Toutun River	2,507	253	347	297	308	341
Midong	1,918	147	203	172	178	194
Σ Urumqi Region	7,552	732	989	866	899	989

(*unit* $10^6\ m^3$)
Source Own calculations

		High mountains	Low mountains/ foothills	Piedmont	Settlements	Alluvial plain	Accumulation basin
ET_p	SC_{LT}	–	++	+	+	+	+
	SC_{ST}	→	↑↑	↑	↑	↑	↑
	SC_{2050}	↑	↑	↑	↑	↑	↑
ET_a	SC_{LT}	–	+	+	–	–	–
	SC_{ST}	→	→	↑	→	→	→
	SC_{2050}	↑	↑	↑	↑	↑	↑
Q_m	SC_{LT}	++	–	+	+	+	+
	SC_{ST}	↑	↑	↑	↑	↑	↑
	SC_{2050}	↑	↓	↓	↓	↓	↓
Q_n	SC_{LT}	++	–	+	–	–	–
	SC_{ST}	↑	→	↑	→	↑	→
	SC_{2050}	↑	↓	↓	↓	→	↓
Q_{sf}	SC_{LT}	+	–	+	+	–	–
	SC_{ST}	→	↑	↑	↑	→	→
	SC_{2050}	↑	↓	↑	↓	→	→
Q_{gw}	SC_{LT}	+	–	–	–	+	+
	SC_{ST}	↑	→	→	→	↑	↑
	SC_{2050}	↑	→	→	↓	↓	↓

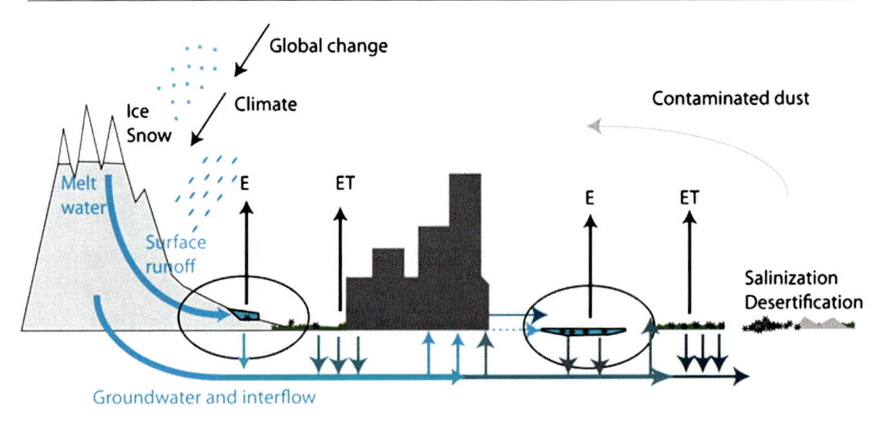

Fig. 7.1 Distribution of the water balance components over the different landscape units of Urumqi Region simulated under climate change (+ higher than average, – lower than average) and their trend for recent (2001–2010 average) and future (2050 scenarios) climate change (↑ increase, → no trend, ↓ decrease). *Source* Fricke et al. [5], own calculations

increase of total surface runoff will be consumed before reaching downstream catchments. The surface runoff in the downstream catchments itself will be reduced e.g., in the Urumqi City catchment from 49 mm surface runoff for SC_{ST} to 40–47 mm for the three SC_{2050} or in the Midong catchments from 50 to 43–49 mm. Due to the higher water consumption, river water quality is expected to decrease as well, putting even more pressure on downstream ecosystems.

The results of land use change and climate change in the urban areas partly confirm the prognosis made by Meßer [6]: transpiration by vegetation is reduced, but evaporation increased due to a changed radiation budget, and surface runoff will increase. Only the overall rise in total water flow projected for urban areas in Europe cannot be found in Urumqi as the increase in ET_a and sublimation of snow in winter offsets the reduced transpiration in summer as demonstrated for the land use change in Midong District (see Sect. 5.3). Groundwater recharge will decline under the urbanisation process as well as the amount of water stored in groundwater bodies (storage problem). Additional water quality problems and floods due to the surface runoff over sealed areas can add to this situation, but have not been modelled or investigated here (cf. [6]). On the alluvial plain in Dabancheng Corridor and the Gobi zone, the decrease of groundwater recharge will continue as ET_a in this area will increase and lead to a reduction of total water flow and water availability for groundwater recharge. Combined with a growing groundwater demand as highly sought after water resource and substitute of less available surface water, groundwater tables will continue to fall. The larger ET_a in the agricultural areas towards the plain has to be met by more irrigation which in turn causes further soil salinization and rising groundwater tables in the farmlands. Increased extraction and lower quality of water available for recharge will also worsen the groundwater quality (see Fig. 7.1).

Subsequently, the projections of water availability and demand are compared. Projections of water demand are very difficult, already the assumptions about economic and population development differ widely (see Chap. 6). According to the different scenarios used for the projection of population and economic growth, water consumption could remain on the water demand level of the 2000s (around 800–900 $\times$ 10^6 m^3) and below the water demand in 2008 (993 $\times$ 10^6 m^3), or it could triple until 2050 for the highest scenarios (see Table A.29). Most realistic and consistent with the recent development on the basis of the available data are the trends and results of the medium socio-economic scenarios. According to these results however, water demand would increase 11–32 % compared to the last available figures from 2008.

In contrast, the amount of water available from surface runoff and groundwater recharge would in the best case remain the same and in the worse cases be reduced by 9.1–12.4 % (Table 7.1). With a consumption rate of 74 % in 2007 (see Fig. 2.7) this projected development would severely enforce the water related problems. Urban and economic development would have to stagnate to cope with the limited local water resources. Otherwise, the technological development and efficiency coefficients were dramatically underestimated. Deng et al. [7: 289] evaluated the situation of water resources for China as a whole and also warned that this "contradiction between supply and demand of water resources is continually aggravating". In Urumqi, the water resources are actually relatively abundant due to the location at the foot zone of the Tianshan Mountains. However, due to the large number of inhabitants in this region, Urumqi Region has been identified as region with real and severe water scarcity (see Sect. 2.5 and [8]).

The increasing demand connected to the social and economic development leads to a 'bottleneck' situation. But not only the supply of economy and population is important. Before the situation aggravates for the supply of these user groups, environmental and ecological problems can be observed. For the provision of ecosystem services such as dilution and transport of polluted effluents, waste water, and irrigation water, an overhead of available water resources is necessary. In the case of Urumqi, the water available for the environment was already very little in the last years. In the year 2007, the consumption rate of surface water, which is more easily available for this purpose, was 89 % while it was 62 % for groundwater resources [9]. The consumption rate is likely to rise as water demand will probably increase and water supply will decrease until 2050 as described above. As the groundwater recharge is projected to decrease, natural treatment processes of former surface water, when water is percolating through soil and underground, will diminish and further reduce the available amount of better quality water. Urumqi would still have to rely mainly on the surface water resources, further increasing the consumption rate and the problem of surface water quality.

7.2 Sectoral Evaluation and Strategies for Improvement

The simulated results for the effects of climate and land use change in Urumqi Region on the availability and distribution of water resources are now used to develop possible response strategies. The general considerations behind the suggested water conservation strategy are explained here.

The causes and mitigation of climate change are largely related to energy production and consumption. Although the latter is also connected to water production and treatment, a comprehensive examination of this topic could not be conducted and will not be discussed here. This chapter will instead focus on adaptation measures, i.e., how to modify the existing water system to adapt to the changing circumstances. Indeed, this is a very important topic for agglomeration areas as the areas within cities "that are highly vulnerable to the consequences (of climate change), with no defense against them and no means to reduce that vulnerability" [10: 185]. The aim is to avoid water scarcity, pollution and wasting. Every region has its natural, economic and socio-cultural preconditions and requires different reasonable measures in order to establish a sustainable and integrated water resource management. Regional boundary conditions such as distribution of precipitation, characteristics of the main water bodies as simulated for the different scenarios in Chap. 5, runoff regulation and drinking water abstraction and consumption have to be taken into account [11]. It is equally important to concentrate the measures on the main consumer groups and the sector with the largest potential for water savings.

The reasonable usage of water resources is opposed by two system immanent problems. First, water is an almost ubiquitary resource and common property and is used accordingly. The **tragedy of the commons** is that every acting object, whether it is a single person, an enterprise or a state, attempts to maximise its benefit by using the common goods, including water. The individual advantage of this decision is usually larger than the negative consequences as the latter are distributed over the whole society and every object has to pay only for a small part [12]. This impact is also called a 'negative externality' of individual decisions [13]. This leads to a decoupling of decisions and consequences and complicates the rational evaluation and comprehensive calculation of costs and benefits. The introduction of social arrangements is necessary to enforce responsibility and sustainable usage without exceeding the carrying capacity (cf. [12]). Second, the paradox of continuous population growth and limited resource availability leads to a reduction of the per capita resources. Free-market theories assume that it can be eased on a regional scale by increased regional water trade and transfer [14]. Economic theory indicates that reduced availability of natural resources will lead to increasing prices and support a more efficient usage, substitution with other resources, as well as technological improvements. Due to the limited global water resources and continuously growing population, this process is notwithstanding finite [12, 15]).

To use the available water resources without reducing their quantity and quality permanently, the consumption rate has to be within the **carrying capacity** of the regional ecosystem. A solution for this problem of decreasing per capita resources is necessary, otherwise the consequence is "the necessity of abandoning the freedom to breed" [12: 1248]. With regard to water resources, this would imply turning away from the freedom of unrestricted development and consumption. Simultaneously, distribution equity and a participatory water management on the level of 'governance' and political executive have to be created to avoid distribution and development conflicts of water consumption [14].

In 2008, the amount of water resources per capita of Urumqi City was 413 m^3 per capita, which is well below the provincial average of 3,798 m^3 per capita and indicates absolute water scarcity (<500 m^3 per capita) [16]. Thus, the endowment of water for production in Urumqi Region is not sufficient and represents a comparative disadvantage for further development. Apart from the absolute values of available water, the factors for production and the comparative advantage are also influenced by environmental standards implemented and environmental costs connected to the production of certain goods which become more and more important [17]: decreasing water availability due to water abstraction and implementation of environmental standards as promoted recently in China both increase (environmental) production costs. Although resource availability speaks against Urumqi Region as a centre of provincial development, the industrial production is growing and attracting more economic investment. Production patterns do not always follow the concept of comparative advantage or take advantage of the endowment factors, also called the 'Leontief paradoxon' [17]. The continuous attraction, growth and production in Urumqi Region can be attributed to other factors such as existing infrastructure and production, transportation, centrality, policies and primarity.

These aspects have to be considered when evaluating possible measurements and strategies, because the fact that a strategy is possible does not mean that it is probable when it goes against economic considerations or policies. Kahlenborn and Kraemer [18] define an ecologic, social-cultural, and economic dimension of water resources management. As a measure for the economic aspect, the absolute economic output and the relative economic output per water used, which are measured by the gross domestic product (GDP) and the GDP per m^3 water used respectively, and the amount of water consumption per output unit of certain goods and products have to be accounted for. The value of a product with regard to water and hereby considered as its ecological value is assessed using the concept of 'virtual water'. 'Virtual water' is defined by Allan [19] as the water necessary and used to produce goods. The exchange of water through the trade and transport of goods between regions and countries is called 'virtual water transfer'. '**Virtual water transfer**' can even substitute real water transfers: Produced goods normally weigh less than the amount of water used for their production, so the transportation of traded good is more efficient than transportation of water [20]. The consumption of water is accompanied by the production of waste water, in this case 'virtual wastewater'. It is the amount of polluted water discharged into the ecosystem and exported through the import of produced goods [17]. The concept can be applied to all kinds of products and services, including agricultural and industrial production. Their yield can be evaluated with regard to the amount of water used for the production, thus representing their 'virtual water price'. However, the paid price of water does not include the economic value of the environmental resource and ecosystem 'water' as the value of the functions provided by water resources (production, dilution, transportation, buffering) are difficult to assess and rarely monitored. Turner et al. [15] state that environmental valuation is more "a social discourse process relying upon social agreement and [...] only loosely tied [...] to technical valuation methods" [15: 20]. Nonetheless water is a natural resource, necessary for the sustainability of environmental systems, and an economic resource, important for human livelihoods [15].

7.2.1 General Measures

Water management is comprised of multiple very diverse activities and procedures to the distribution and regulation of water that can be included in an adaptation strategy as intended in this chapter [21] and corresponding to the concept of integrated water resource management, all parts of the catchment area, the most important factors, different levels, concerned stakeholders and actors as well as available and useful methods, data and information should be integrated [22]. General implementation measures have already been suggested for Midong New District taking the different stakeholders and actor groups concerned with water supply and demand into account including the dimensions of water supply, water saving and improvement of usage efficiency, water treatment and reuse [23]. Municipal and governmental actors, population, agriculture and industry have to be incorporated including their economic, institutional, administrative and social possibilities. Possible strategies and measures for an integrated water resource management in Midong New District are presented in Table 7.2. Many technical possibilities exist nowadays to improve water saving, usage efficiency, treatment and reuse, but their implementation has to be adapted to the local circumstances.

Table 7.2 Possible strategies for an integrated water resource management (translated according to Fricke [23])

Sector Actors	Water supply	Water saving	Water treatment	Water reuse
Municipal/ provincial government	Protection of water resources (reduction of evaporation and leakage, groundwater protection)	Water saving campaign, public buildings as role model and adapted urban design	Expansion of canalisation and number of connections, increased capacity utilisation of water treatment plants, improved water treatment, cleaning-up of existing waste and sewage deposits	Process and grey water reuse in public buildings or for the irrigation of public green
Population		Usage-bound charging, water-saving appliances	Installation of pit drainages and primary sedimentation basins	Grey-water usage, garden irrigation with treated waste water
Agriculture	Sufficient irrigation and drainage, encourage percolation and groundwater recharge	Adapted and improved irrigation technology and management	Pit drainage	Usage of treated waste water
Industry		Optimisation of production processes and reduction of process water	Internal pre-treatment (especially in smaller enterprises), monitoring of discharge	Internal water cycles, water reuse

Additionally to the strategic orientation of the measures on the key aspects, the connection and requirements of economic and social structural conditions have to be considered. This also includes national and international specification for the protection of soil and water resources, which have to be incorporated into the services for the public [24, 25]. However, economic, social and judicial aspects can only be touched briefly in this chapter, as it would require a shift in the research focus. Nonetheless, **all suggested technical measures should be supported by policies and instruments for sustainable water resources management** as suggested by OECD [26] or Abderrahman [27]: modification of water use patterns through water pricing and tariff reform, education and information of users and the introduction of improved technology and infrastructure. These instruments are based on institutional and regulatory reforms where the role of the government has to be reconsidered as well. The respective government has the responsibility of water supply and sustainability, but now it has to shift from the sole decision-maker to a decision-enabler. The growing number of instances concerned with water resources management ideally requires a multi-stakeholder decision making where the government has to make the necessary regulatory decisions and offer a transparent regulatory framework [27]. Together with a comprehensive information based on water resources and use, the integrated and flexible allocation of water between competing uses could be achieved. Improved monitoring and evaluation systems similar to the hydrological and socio-economic model presented and used in this research can support and are actually necessary for an integrated water resources management and short- and long-term water plans [27].

7.2.2 *Agriculture*

In the administrative unit of Urumqi City, the largest water consumer is the agricultural sector. In 2007, it was responsible for 65 % of the water consumption in Urumqi City, while population and industry used 23 and 12 % respectively [9]. Thus, the main focus of the water conservation strategy will be on water savings in the agricultural sector.

The Chinese State council passed out the aim to build a "new socialist countryside" with directions for the 11th Five Year Plan in 2006 to make significant commitments to China's agri-food sector and rural industries with the major focus on improving the well-being of China's farmers and redressing the imbalance in living condition between urban and rural citizens [28: 39]. Thus, improvements of the livelihood and productivity of the farming system are expected. The agriculture in Urumqi Region, for which the water balance was modelled, is mainly located in Dabancheng Corridor between the two mountain ridges of the Tianshan Mountains and North of the city towards the Junggar Basin (see also the land use classification in Sect. 3.3.3.4).

The larger part of the agricultural production depends on irrigation, in 2006, 95.5 % of the cultivated area was sown with crops and 85 % was irrigated [29]. During the last 10 years, runoff and available water resources increased, but are expected to decrease in the lower catchment areas again until 2050. This is mainly due to higher ET_a which will increase the transpiration of agricultural crops and evaporation of irrigation water thus increasing the irrigation water demand in two ways. The beginning of the vegetation period and the snow melt water flow have been preponed around the same time period, but the total amount of melt water and water flow will decline in the future. At the same time, the competition with industry and population over water supply will grow (see water demand projections in Sect. 6.3), mainly for the groundwater resources that provide most of the water consumed by industry and population (in 2007, 44 and 69 % according to [9]). Also within the agricultural sector, competition will increase. Animal husbandry produced 46 % of the agricultural GDP in 2008 and increased twice as much as the output of crops [30]. Animal products have much higher 'virtual water content' than cultivated products, but generate a larger economic benefit [31]. Unfortunately, the specific water demands of crop cultivation and animal husbandry are not available and cannot be deduced from the existing figures. Measures for both agricultural sectors are therefore considered as equally important and most of the suggested measures apply to both of them anyway.

Not only the total water flow will probably decrease and the competition for the scarce resources will increase between the user groups, but also the spatial availability will change. Due to the large water demand, agriculture will suffer most when the water resources are not locally available anymore. The difference between the total water flow under the short-term climate scenario to the projected climate scenarios for 2050 can be regarded as a 'risk map' of increased water scarcity in the future (see Fig. 7.2). Unfortunately, most agricultural areas will experience reduced local water availability.

Not surprisingly, as the available water resources have recently increased, the total sown area in Urumqi Region increased by 20 % over the last 10 years (1998–2008 [32]). The proportion of area sown with grain crops has continuously decreased, while the area with vegetables increased. Of the total agricultural sown area, 36.78 % are cultivated with grain crops, 23.35 % with vegetables, 10.41 % with oil-bearing crops, 6.41 % with cotton, and 23.21 % with other crops (tuber, sugar beet, alfalfa) (see also Fig. 7.3). The area of irrigation intensive crops such as cotton is relatively low. However, the far above the water supply available by annual precipitation ($\sim$100 mm per year as shown in Fig. 7.4). Additionally, most of the agricultural areas are situated in areas with net precipitation values even below the regional average, e.g., North of Urumqi City. Irrigation is necessary, thus 85 % of the cultivated land was irrigated in 2006 (area of cultivated land (1,000 ha): 69.75; irrigated areas (1,000 ha): 59.28; sown area to farm crops (1,000 ha): 66.62 ([33], Tables 11.4, 11.9, 11.14)) and agricultural cultivation of this region with field crops is not sustainable.

As demonstrated in Chap. 6, the population in Urumqi is projected to increase further for the next 40 years and the local food demand will rise. At the same time,

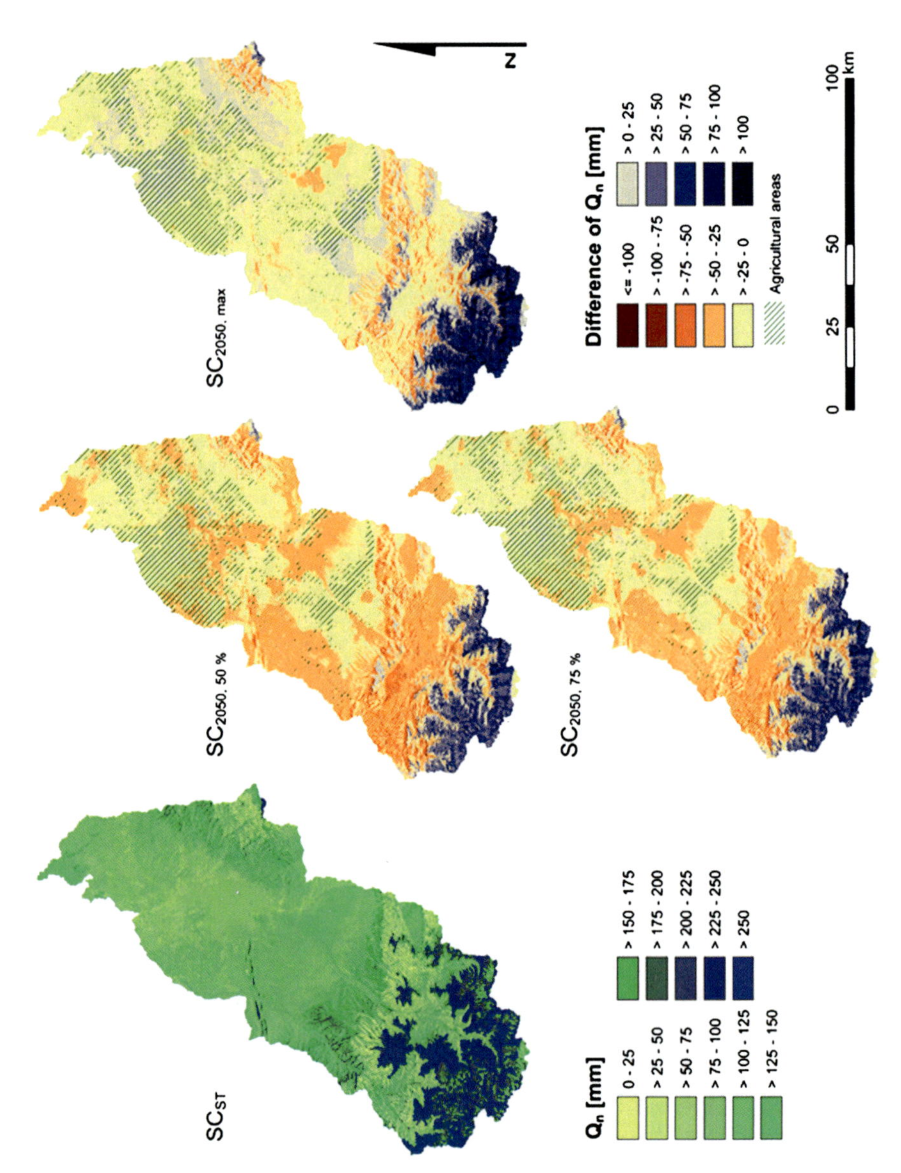

Fig. 7.2 Simulated total water flow for the short-term climate scenario and the difference to the three climate scenarios for 2050 with agricultural areas. Under the two climate scenarios $SC_{2050, 50\%}$ and $SC_{2050, 75\%}$, total water availability will decrease in most of the agricultural areas (*yellow* and *orange* coloured in the map). *Source* LULC classification, own calculations

Xinjiang is supposed to produce food for the denser populated areas in Eastern China, where the population is growing as well. The agricultural development was also part of the "one black, one white" strategy of the central government for the Northwestern province, focussing on both industrial (oil extraction and

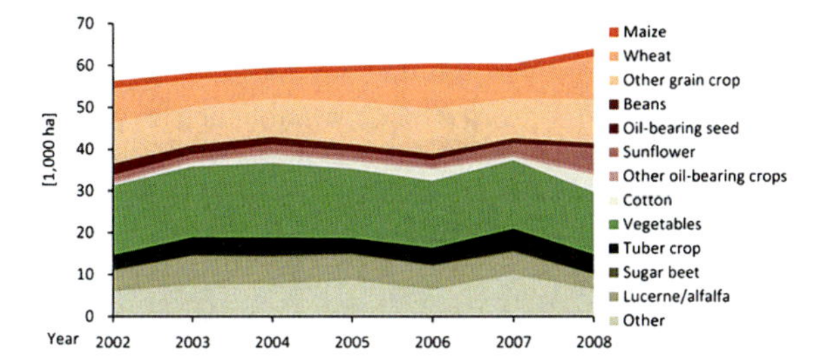

Fig. 7.3 Sown areas of main crops in Urumqi City 1978–2008 (*unit* 1,000 ha). *Source* Statistics Bureau of Urumqi [34]

petrochemical production) and agricultural development (mainly cotton as a cash crop for whole Xinjiang). The expansion and reclamation of agricultural areas was advanced first at the fringes of the Junggar and Tarim Basins and then pushed further to areas in the basins suitable for mechanical cultivation, but unsuitable with regard to the climatic conditions and water supply. For this purpose, the Xinjiang Production and Construction Corps, an economic and semi-military governmental organisation, and state farms were established in Xinjiang Province.

In Urumqi Region, where the industrial development and productivity of Xinjiang is concentrated, these two goals of industrial and agricultural development seem to compete against each other for water and land. Liu and Savenije [35] found that the per capita water requirement for food (CWRF) in China has increased from 255 m^3 cap^{-1} yr^{-1} in 1961 to 860 m^3 cap^{-1} yr^{-1} in 2003, "largely due to an increase in the consumption of animal products in recent decades", but the CWRF of China is still much lower than that of developed countries. For example the USA have a CWRF of 1,820 m^3 cap^{-1} yr^{-1} or the 15 member countries of the EU in 2003 have a CWRF of approximately 1,500 m^3 cap^{-1} yr^{-1} [35: 887]. With the continuous population growth, the total water requirement for food is likely to increase in the next three decades [35]. In 2007, agriculture accounted for 65 % of the total water consumption, while the share of industrial

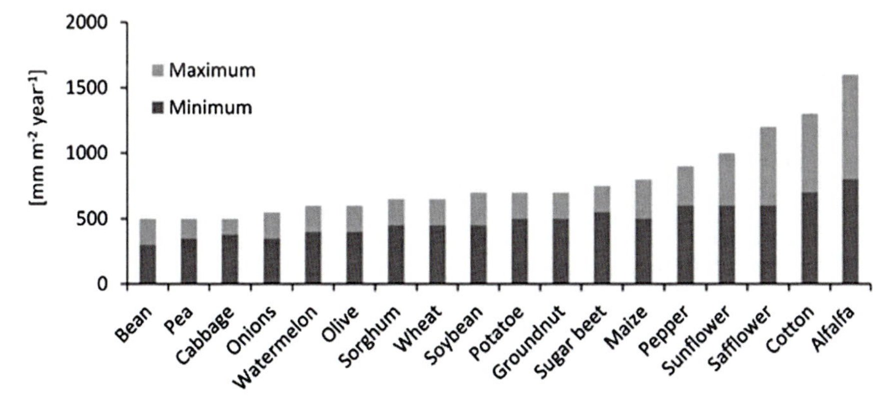

Fig. 7.4 Water requirements of agricultural crops (minimum to maximum values, *unit*: mm m^{-2} $year^{-1}$). *Source* [58, 59]

consumption was only 12 % [9]. By contrast, the proportion of GDP generated by agricultural yield was only 1.98 % in 2007 compared to 41.57 % of industry [32]. However, the food price indices published by the FAO show an ascending trend for food, oil and cereal prices and suggest rising revenue from agriculture in the future (Fig. 7.5).

Still, the revenue per water consumed from industry (28.36 RMB m^{-3}) is 10 times the revenue from agricultural water consumption (2.76 RMB m^{-3}) (calculated from Statistics Bureau of Urumqi [32] and Water Affairs Bureau Urumqi [9]) while agricultural users pay one-third or less of the price paid for water supply by enterprises and put agricultural water demand into an unfavourable negotiation position.

Therefore, the strategies for adaption to the changes of climate and consumption are based on the assumption that the agricultural water consumption will decrease. Agriculture will have to reduce either its relative water consumption per output by improving the usage efficiency or absolute water demand by reducing agricultural production. It is important to note that the suggested strategies are specifically adapted and restricted to Urumqi Region and the situation there. Due to its importance as a provincial capital and regional centre the strategic focus of development shifts towards industrial production and the service sector. Outside Urumqi Region the importance of agricultural production will continue.

Before focussing the consumptive side and increasing the water usage efficiency through water reuse, intensification of agriculture and shifting agricultural areas in and outside of the research region, the potentials and possible strategies for water supply are discussed, e.g., **water supply infrastructure**, storage and transportation facilities, or treatment and reuse of wastewater (also suggested by Hoff et al. [36]). The aim of these measures is to reduce the amount of 'white water', the water lost to transpiration into the atmosphere [35]. This unproductive

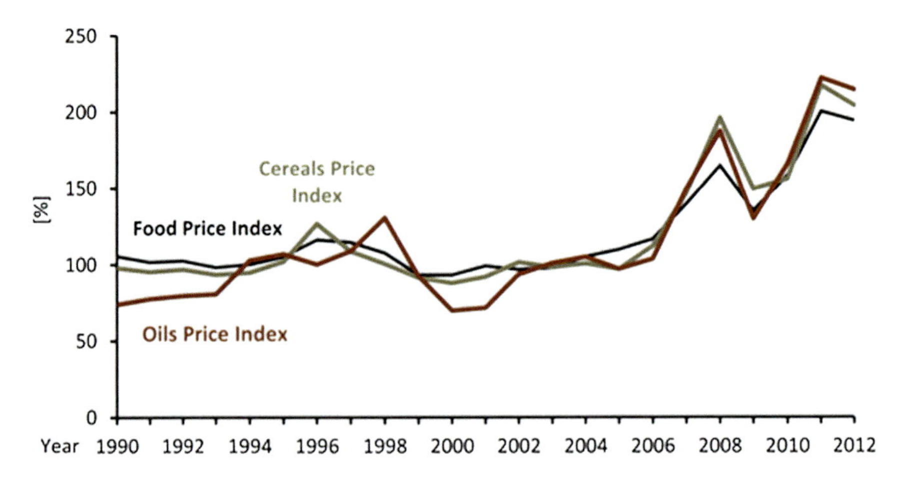

Fig. 7.5 Annual real food price indices (2002–2004 = 100 %). *Source* [60]

water can make up a large part of the water balance [35] as it is the case in Urumqi Region (see actual evapotranspiration in Sect. 5.1). Here, the methods for collecting and storing water resources could be improved and losses during storage and transportation due to leakage or transpiration could be reduced. Transportation channels and reservoirs should be sealed against leakage and as far as possible flows and floods occurring in this area should be directly diverted into the desired infrastructure systems. For the latter purpose, artificial structures and dams already exist in the Dabancheng Corridor. Further engineering has to be done to reduce transpiration from the open water bodies (Fig. 7.6).

Reservoirs and dams, when they are considered necessary, should be located in areas with low evapotranspiration potential, preferred in the mountain regions with considerably lower temperatures, or covered and underground. The same applies to the water channel network, which should be covered up by construction or natural material. As estimated for the projection of agricultural water demand in Sect. 6.2.3, the irrigation efficiency in Urumqi is assumed to be around 45 %. This means 55 % of the water fed into the distribution system is lost. Underground pipes and lined canals could lead to estimated 30–40 % water savings [37]. Thereby not only the quantity, but also the quality of water could be improved.

Another possibility to increase water supply is the **reuse of treated waste water**. The reuse of treated waste water for agriculture is often cited as an example of intersectoral water usage and regarded as a central element of integrated water management in water scarce regions. The practice has several additional advantages such as fertilizer substitution and sewage disposal. Nonetheless, the usage of treated wastewater is an alternative to the intensification of freshwater usage, although not risk-free. With the increased Industrialisation higher pollution loads and standards for the reuse of wastewater prevented further usage [37].

Fig. 7.6 Pictures of open concrete and shadowed channels in Urumqi Region. *Source left* K. Fricke, *right* P. Klenk

Unregulated use of raw sewage can transmit worm diseases, bacteria or virus infections. The systematic use depends on the degree of purification and risk assessment of user and consumer (cf. [37, 38]). Often, the costs for treated waste water are lower than for freshwater and the farmers have to spend less on fertilizer due to the nutrient load of the water. However, the treated waste water is often not suitable for dripping irrigation and increases the irrigation water demand significantly as well as the work demand, acreage restrictions and abdication of highly rentable cultivation of vegetables suitable for consumption [39].

When certain standards for the treated waste water are met, it can be used to irrigate crops not intended for direct consumption. Nutrients in the waste water can lead to a higher yield than the combination of tap water and fertilizers. The advantages of waste water reuse include the reduction of nutrient discharge into surface and groundwater, less demand for freshwater, and potentially decreased costs for waste water treatment when nutrients do not have to be removed [40]. However, it is only possible when the water treatment facilities meet the necessary standards reliably, when the water price is not higher than of freshwater and the crop yields are as high as with conventional farming. It has to be profitable and the users and farmers have to agree with the regional water management strategy [40].

A more efficient water usage is not automatically an economically integrated, non-hazardous and ecologically sustainable measure. In Tunisia for example, the waste water reuse rate stagnates at a rate of 20 % as the advantage of lower water prices and substitution of fertilizers are offset by the opportunity costs, caused by the missed benefit due to acreage restrictions for vegetable cultivation. Cost benefits because of fertilizer substitution are often idle as the exact nutrient load of the treated waste water is unknown and the farmers instead risk over-fertilisation and groundwater contamination [38]. Due to the acreage restrictions when substituting fresh water with treated waste water, the income per farm declines, the minimum size of farms to survive increases and the number of farms drops. The profitability of freshwater substitution also strongly depends on the regional price differences between products produced with different water qualities. Taken as a whole the increase in usable water resources would lead to growing income [39]. The promotion of the strategy via price incentives and subventions, limits for all problematic substances and acreage restrictions are most important [40]. The success is furthermore relying on an adapted waste water treatment system, possibly additional disinfection, deregulation of acreage restrictions, special agricultural management, and institutionalisation of an appropriate judicial and regulatory framework. With regard to the ecological aspect of water savings, the quantitative results that can be realistically realised are limited. In Tunis, only 7.5 % of the water consumption can be substituted, thus demanding additional water saving measures. Qualitatively does the waste water usage relief the nutrient and pollutant load of the receiving stream, but the substantial burden on the agricultural soils and the groundwater bodies below with increasing salt and nutrient concentrations have to be watched closely [38].

As reported in Sect. 2.5, the average treatment rate in Urumqi 1996–2007 was 52.37 % [41]. Since the construction of a new treatment plant in 2009, only an average of 25 % of the total treatment capacity with a national discharge standard 'II' of pollutants for municipal wastewater treatment plants (GB18918–2002) is actually used. Treated wastewater according to national discharge standard 'II' is suitable for the irrigation of fiber plants, grain and oil crops, paddy field grain and right on the limits for open ground vegetables, corresponding to the quality requirements of farmland irrigation water (GB20922–2007). It is not suitable for urban water use as the treatment plant discharge would require additional treatment for disinfection to remove coliforms after the secondary treatment. Indeed, a part of the treated waste water was already used for agricultural irrigation, greening and groundwater recharge: According to the Water Report 2007, 42.07×10^6 m^3 'other' water resources were consumed by agriculture [9]. Yao [41] suggested that the potential of waste water reuse was not completely used due to several problems: the sewer and drainage system was deficient and lagged behind urban development, the wastewater treatment plants build from governmental or municipal funds were lacking economic profitability, the industrial waste water exceeded discharge standards of the waste water treatment plants, and the possibilities of reuse is so far limited to agricultural irrigation. The suggestions to improve the waste water reuse rate in Urumqi included financial incentives (increase of water tariff and reduced charge for treated water), advancement of the treatment process to meet urban water use standards, introduce managerial responsibility to the contracting system of the water treatment plants, heighten the enterprise's sense for environmental protection, strengthen the supervision, administration, and close monitoring of the discharge system [41].

As a measure how to react to the effects of climate change on the **agricultural water demand** side, possible strategies for the **intensification of agriculture** have to be considered. The challenge is to produce more agricultural output per area and water used (crop yield and crop water productivity), i.e., to achieve "more crop per drop" which has also been presented by the Food Agricultural Organisation [42] as a strategy to resolve the world's water problems. With regard to animal husbandry, this would mean the increased introduction of an industrial production system, increased numbers of head per area and shorter growing periods. This is only possible with measures such as more fodder and additives supporting growth and health of the animals. From the grazing to the mixed production system to the industrial production system the total water footprint per unit of product declines except for dairy products. The feed conversion efficiency of the more intensive system (i.e., industrial production system) is better and less water is needed to produce the feed [43]. To achieve an intensification of crop cultivation, irrigation, soil, and crop management were considered. The improvement of irrigation techniques and soil and water management aims at increasing the yield per water unit used and reducing evapotranspiration by plants and soil. It can be measured by increased crop water productivity (CWP) and inversely by a reduced 'virtual water content' of the crop (VWC). At the same time, crop yield per area should not decrease as the cultivation has to remain profitable for the farmers. Adding more irrigation water normally leads to an increase of yield, but not to an effect on the ratio of crop yield to water consumption [36]. With better irrigation management adapted to the imminent water stress and demand of the plants or concentrated on the reproductive stage of the crops, a reduction in irrigation of up to 40 % "would not adversely affect crop production under the appropriate timing and irrigation dose" (less than 10 % drop in yield) for example in wheat production [44: 229]. According to Blanke et al. [37], the main reason for not adopting water saving measures by the farmers is the lack of incentives to save water. Water saving can be encouraged by supporting institutions, providing incentives and information, financial assistance and coordination.

Also, improved soil management including fertilisation, soil conservation, tillage or mulching can reduce water demand for biomass production and evapotranspiration [36]. It should be aimed at reducing the amount of 'white water' by increasing the amount of rain and melt water immediately infiltrating into the soil. This water is then available to the plants as 'green water' or percolates into groundwater storage and flows. According to the definition of Rockström et al. [45], "green water is the soil water held in the unsaturated zone, formed by precipitation and available to plants, while blue water refers to liquid water in rivers, lakes, wetlands and aquifers, which can be withdrawn for irrigation and other human uses" (cited in [36: 178]). One goal of better water, soil and crop management would be to increase the amount of precipitation directly used or infiltrated into to the soil ('green water'), thus reducing the amount of 'white water' lost to non-reproductive transpiration and need for irrigation, ergo 'blue water' (see also [36]). Even when the absolute shortage of blue water indicated by a threshold of 1,000 $m^3\ cap^{-1}\ yr^{-1}$ is reached, many regions still have the potential to develop or improve the use of their green water resources (see [8, 36, 45]). The downside of increased use of fertilizers as stated by Liu and Savenije [35] is the high dependency on probably imported phosphates and the discharge of unused nutrients as pollutants into surface and groundwater. Under certain circumstances, crop yield could increase due to increasing atmospheric CO_2 concentrations without any other measures as noted by Hoff et al. [36]. These possible developments aside is it quite unsure whether the increased crop water productivity would really lead to a reduced water consumption and not instead to an intensification of agriculture on existing fields and increased water demand. Several driving factors have to be taken into account, the necessary investment in agricultural management and technological development, the demand for agricultural crops and water price and availability, i.e., the profitability of the measures. All of these factors could be supported by financial or political incentives.

Besides water and soil management, crop management has to be taken into account. Although the strategy of cotton and oil was issued for the development of Xinjiang, only a relative small area in Urumqi Region is cultivated with the water intensive cash crop (see Figs. 7.3, 7.4). However, the proportion of more water intensive crops such as vegetables, sunflower and maize could be reduced further in favour of less water intensive plants. This could only be achieved by political support as financial incentives through subsidies, manipulation of crop prices or increased water prices are necessary.

Another strategy requiring political support is the idea of **shifting existing agricultural areas** to areas with more water resources, either in the form of precipitation or in form of less distance to water production and storage areas. In Urumqi Region, this would mean expanding the agricultural areas in the Dabancheng Corridor, or—instead of moving to the flat plain in the North—cultivation of the sloped piedmont and foothill areas. These areas have recently been used as meadows for cattle husbandry by local shepherds and farmers. Hence, this measure would require extensive preparation and inclusion of the affected population and farmers to avoid the eruption of conflicts. The expansion onto already used non-agricultural/urban land is even more unlikely as it would require de-construction of already build-up area and displacement of uses that produce more economic output than agricultural fields. Recent trends rather show a reallocation the other way, namely the conversion of agricultural into urban or industrial land (see Sect. 2.2).

Another possibility to reduce water consumption by agriculture in Urumqi Region despite rising food demands is the import of food and 'virtual water trade' (cf. [36]). With regard to agricultural production, it can also be referred to as the sum of 'white' and 'green water' representing the amount of water used for terrestrial biomass production including water taken up by the plants and transpirated into the atmosphere or evaporated from the soil moisture [36]. Agricultural crops that are rainfed and produced in Urumqi Region do not have to be imported as the "rainwater imbedded in agricultural products would not be readily available for any other economic production even if crops were not grown on this land" [17: 165]. Urumqi Region should reduce the exports of irrigated agricultural products to make the saved water available to other production purposes. The limited regional endowment of the factor agricultural water supply makes the shift to other regions and the re-import through 'virtual water transfer' useful.

The net 'virtual water' export at the provincial level in 2002 calculated by Zhang et al. [46] was $395 \times 10^6\ m^3\ yr^{-1}$ for Xinjiang. In this publication, the ratio of net 'virtual water' export to water use of Xinjiang for final demand in individual provinces was relatively low, highest ranked were the provinces with strong industries and major contributors of the national economy and export in China (Tianjin, Shanghai, Jiangsu, Zhejiang, Fujian, Shandong and Guangdong). However, this result might not be representative for Urumqi Region, which is the industrial centre of Xinjiang and much denser populated than the rest of the province.

7.2.3 Urban Areas with Non-Agricultural Production and Population

In the urban areas, the water demand is increasing based on the growth of population and industrial production while available water supply is expected to stagnate or decrease (see Sect. 7.1). As the urban supply is covered up to 59.6 % by groundwater resources [9], the demand for groundwater will increase, too. The rising temperatures and evapotranspiration will probably lead to an increased water demand by population (households), urban green areas and other social amenities, although the relationship was not modelled in this research and not specifically accounted for in the water demand projections. The expanding urban areas will also decrease the accumulation of snow water equivalent, the total water flow and especially groundwater recharge. They could also profit from measures improving infiltration and groundwater recharge by reducing their degree of impermeability. The groundwater consumption rate (consumption divided by available water resources) is 'only' at about 62 %, therefore further exploitation is expected. The advantages of a stable and reliable supply and the good water quality render groundwater into a potential source for further exploitation [47]. However, groundwater recharge is projected to decline more than surface runoff. Groundwater could also be substituted by treated surface water, but this approach is not feasible at the moment due to the treatment costs and the already high consumption rate of surface water (89 % in 2007). Still, both types of water resources should be combined by using groundwater to meet the shortfall of surface water surplus to replenish the groundwater storage through artificial enhanced recharge. As suggested by Zhou et al. [47], increased abstraction from new well fields in the Dabancheng Basin will lead to a reduction in spring discharge, outflow and discharge into lakes and reservoirs, but avoids evaporation losses. Additionally, excess water and water saved from water saving technologies has to be used to increase groundwater recharge to compensate for the growing abstraction [47]. In urban areas additional measures to ensure an adequate quality of the infiltrating and percolating water would have to be implemented.

The concept of 'virtual water' can also be applied to industrial goods that require a large amount of water for production, causing discharge of considerable pollutants into the environment and could be imported into Urumqi Region. Guan and Hubacek [17] calculated an extended regional input–output model for the hydro-economic regions in China and observed that the possibilities to import 'virtual water' have not been exhausted by North China. However, this concept seems difficult to implement and is politically not desired as the focus in Urumqi Region and Xinjiang lies on (petro-) chemical products and it is the declared goal of the provincial and national government to further strengthen industrial development in Northwest China. The export-oriented economic strategy is not compatible with water-saving policies. The recent socio-economic system needs a large input of water, which as an environmental resource is difficult to assess and not valued appropriately in the recent economic calculations and decision-making processes [17].

Therefore, **water saving potential in urban areas** lies in the redevelopment of the distribution network and improved maintenance to reduce leakage, as well as the installation of more efficient devices and technologies for water usage. Additionally, changes in the consumer's behaviour are possible. Within the public water supply system, leakage losses correspond to the water consumption of industry. As researched for the distribution efficiency index in the econometric model, the leakage rate of the distribution system in Urumqi is estimated to be about 10–20 % (see the distribution efficiency in Sect. 6.2.3). The largest water saving potential is the reduction of water losses to leakage and the limitation of network and filter flushing. However, investments in the infrastructure have to be balanced by the revenue of the saved water, which is difficult with the recently very low water fees. And as long as water resources are available for extraction and the water demand can be met, the water provider is not forced to increase the transmission rate of the network.

Water saving appliances in households such as water saving valves, aerators, and toilet flush could save about 50 l per day or 1/3 of the specific drinking water consumption without reducing the living quality. Additional savings could be achieved by introducing water saving washing and dish washing machines and the supplementary installation of water meters in households [48]. Measures that need more investment are the increased application of urinals, rain and grey water recycling. The targeted and systematic information about the advantages of these investments and new acquisitions is necessary as most of the consumers and investors do not act economically rational as long as they do not know the equipment and appliances (see [49] and the 'tragedy of the commons' in Sect. 7.2). However, the substitution of drinking water by rain and grey water is restricted to certain applications and has some problems. Both water types should not be mixed with freshwater and make a secondary piping system necessary. Due to the microbiological pollution and bacterial re-growth in storage containers rainwater is only useful for toilet flush and garden irrigation. Grey water from domestic activities such as washing and bathing and after treatment is also only suitable for toilet flushes and landscape irrigation. Both systems have to be integrated into the building an the investments for grey water usage are higher than for rainwater usage. Rainwater usage is more useful for buildings with large roof areas, while grey water usage is more profitable in large building complexes or multi-storey buildings [48]. The amount and frequency of precipitation is not very high in Urumqi City, thus making rainwater usage rather unlikely. However, grey water usage would be suitable in particular for the multi-storey buildings which make up most of the building in the urban areas of Urumqi City. **Water saving measures in public and office buildings or businesses providing accommodation** can use similar technologies as private households. The implementation in these buildings is even more effective than in households as the utilisation rate is significantly higher, leading to an economic advantage. They can also serve as role models and test area for the comprehensive implementation of selected techniques.

The general reuse of treated waste for urban purposes should also be discussed. So far, treated waste water is only used for irrigating agricultural fields and greenery (see also Sect. 7.2.2). Yao [41] suggested to invest in advanced treatment of the waste water and to enable other possibilities of reuse. These could include toilet flushing, fire fighting, landscape irrigation, car washing and water use for construction when the pollutant loads in the discharge correspond to the national standard 'I B' or better for wastewater treatment plants.

In the industrial production, already existing separated water networks for cooling, process and drinking water offer large advantages and possibilities for water saving measures, e.g., through water reuse and circuits. The control and regulation of water consumption and waste water discharge is mainly possible through economic benefits and incentives, i.e., water price and waste water fees as well as subventions, or waste water and water quality standards for industrial waste water [48]. The decision to invest into water saving measures depends on the economic advantage as the investments have to pay off. In this case, a system flow analysis or life cycle assessment of selected products and processes of Urumqi's industries is helpful to identify possibilities for water saving and to assess relationships to other resources.

The concept and methods of life cycle assessment (LCA) are officially included in the ISO 14000 series (ISO 14040) as a tool for environmental management decision support (ISO 14040 2006). It can be used as a decision support system as it "provides a profile as complete as possible of the interactions of an activity (product, process or service) with the environment, contributes to the understanding of the overall and independent nature of the environmental consequences of human activities and provides decision makers with information, which quantifies the potential environmental impacts of activities and identifies opportunities for environmental improvements" (Allen et al. 1997 in [50: 249]). LCA offers not only advantages for improving the environmental performance, but also the possibility to optimize production processes [51] thus emphasising also the economic advantages of improving the environmental performance of an enterprise. An LCA can furthermore be used as a decision making tool or part of it: an LCA enables the stakeholders to choose the best practicable option and best available technique not entailing excessive costs. This approach provides rather a set of alternatives than a single prescriptive solution which "may be optimal but not necessarily appropriate for a particular situation" [51: 1509f]. Especially in the fast developing and expanding industrial areas of Urumqi City, Midong District, the "High Tech Development Zone" and the "Economic and Technical Development Zone", the introduction of LCA and optimisation of production and recycling processes could reduce the demand and contamination of water resources.

The connections between water and life cycle assessment are generally manifold. A complete life cycle (of a product) includes raw material extraction (including water), processing, transportation, manufacturing, distribution, use, reuse, maintenance, recycling, or final waste disposal (Consoli et al. 1993 in [50]). Water usage, such as process or cooling water, is normally included in life cycle inventories (LCI) for economic products and the consequences for water resources are again included in the life cycle impact assessment (LCIA) and the documentation of environmental impacts. Additionally, water service such as water treatment or water supply can be evaluated using the LCA approach: LCA can also assess the consequences of water extraction and the methods and techniques used to treat sewage and waste water [52]. The environmental burdens associated with water supply beyond extraction include non-renewable energy use, material use, land use, and pollution of air, soil and water resources [50]. LCA can also be applied for the evaluation of additional infrastructure and intensification of production, purification or reuse a. o., because these measures require energy and material and can in turn induce additional burdens on the existing water resources and also contribute the causes of climate change via the energy consumption.

Despite its wide application, the LCA tools exhibit several shortcomings with regard to water resources: Raw materials except energy are in general included but with severe limitations. Water is in most cases not included as the data availability is limited due to confidentiality issues and data gaps. Also, the time-constraints reduce the completeness of the LCI data and water is often less important than energy consumption and other resources (raw materials, minerals, chemicals etc.) due to its low costs and neglected in the LCA. Toxicological impacts on humans and on ecosystems are often incomplete because of the sheer number of chemicals used in society and the lack of knowledge on their behaviour. Also, data on eutrophication of aquatic systems is usually incomplete (due to insufficient data for water emissions), and data for organic compounds contributing towards photo-oxidant formation is expressed as a general parameter (e.g., particulate emission) making differentiation impossible. Therefore it is difficult to forecast consequences for water resources intensively used for fishery, extraction of potable water and emission of wastes (cf. [53]). In the same way, land use, habitat alterations and impacts on biodiversity are in most cases not included. These categories will continue to pose a methodological problem, since there is no agreement on how to consider them in an inventory analysis.

Nonetheless, the reduction of water consumption can also lead to technical problems, e.g., through the lack of network flushing and tractive power in the drainage system, microbial contamination of drinking water pipes or rising groundwater tables. Despite decreasing water consumption, prices would have to rise as 80 % of the drinking water costs are due to infrastructure maintenance, which cannot be easily removed [51]. The infrastructure has also to be available for extreme events, for example for fire-fighting service or the drainage of rainstorm events. Short-term reduction of water fees through water saving is only reasonable when the savings are above-average. But on the long-term, downsizing of the infrastructure can pay off and should be considered in the long-term planning or when large infrastructure investments are pending. In a continuously growing and rebuild city such as Urumqi, there is still room to optimize the infrastructure according to the average and extreme requirements of water demand and discharge even if the suggested measures for water conservation would be implemented.

7.2.4 Ecosystem and Environment

The ecosystem and the environment in Urumqi Region were so far the first to experience the consequences of increased water demand and will be suffering also from the future decrease of water supply. A functioning ecosystem has a large ecological capital, certain environmental factors that are essential components used by individuals and species including nutrition, energy and spatial structures [54]. It can also provide service functions, which can be assessed using methods of the economic value of service functions of ecosystems [55]. With regard to water resources ecosystem offer functions such as buffering of substances emitted into the environment through adsorption and absorption, purification of water, water storage and withdrawal. The question is to which degree ecosystems can cope with anthropogenic influences under the changing circumstances and at what point the carrying capacity of the ecosystems will be exceeded and the ecosystem functions are not working anymore ('beyond the limits' in [56]). Possible consequences are the degradation of regenerative capacities, a dramatic decrease of freshwater resources, increased costs for freshwater abstractions and the transport of sewage from industrial production, agriculture and households. It is questionable whether man-made capital, financial investments, innovative technologies and management concepts are able to compensate the services provided by complex ecosystems (probably at higher costs) when the 'critical' natural capital is used up and the border to reversibility is exceeded [15]. Therefore, protection of water resources and the water environment is definitely necessary [57]. The goals of a strategy for adaptation to climate and land use change are first to prevent the loss of functionality and second to maintain the diversity of the ecosystem [47]. All measures increasing the environmental flow in rivers and green water of natural landscapes or reducing the contamination and increasing the quality of water resources can be regarded as reasonable. To improve the chance of implementation, projects increasing the environmental water flow have to be combined with measures beneficial also for other user groups.

A linkage of the infrastructure measures suggested before with specific and adapted ecosystem protection would be useful, for example renaturation of now artificial river beds and groundwater recharge structures. The aim would be to amplify the natural process of runoff infiltration in and outflow out of the alluvial plain at the transition to the accumulation basin. This cycle could be repeated by groundwater extraction via pumps, repeated infiltration and artificial enhanced recharge to achieve a conjunctive use of surface and groundwater and to use the groundwater bodies as underground storage to avoid the evaporation losses of reservoirs [47]. These quantitative measures and the sharing of scarce water resources should be accompanied by soil and water conservation as well as protection of water quality, wastewater reuse and desalinisation. Simultaneously, the enlargement of settlements and sealed surfaces should be limited or allowed only under the condition to retain the ecological functions. But for the targeted implementation, a hydrological model for the evaluation of measures, an efficient system for the dissemination of information and raising public awareness on water issues is necessary (cf. [47]).

With the results from the water balance model, an overall decrease or in the best case a continuous level of available water resources have been predicted. The spatial differences in the distribution will increase while the proportion of melt water and groundwater flow will decrease. At the same time, water demand by agriculture, population and industry will grow with a decreasing agricultural share. Thus, the suggested measures include in general small-scale measures in all sectors for the reduction of water consumption, but also a shift of water-intensive agricultural production to areas with favourable hydrological conditions and water supply but less competitors for the water resources. On the water supply side, the effects of climate change and the reduced storage in snow and groundwater have to be compensated. This can be achieved by infrastructure measures, for example the construction of additional reservoirs in the mountain areas, and enhancing the natural processes of surface water infiltration and groundwater recharge while benefitting from the filtering and treatment functions of soil and underground.

References

1. Aizen, V. B., Aizen, E. M., Melack, J. M., & Dozier, J. (1997). Climatic and hydrologic changes in the Tien Shan, Central Asia. *Journal of Climate, 10*, 1393–1403.
2. Casassa, G., López, P., Pouyaud, B., & Escobar, F. (2009). Detection of changes in glacial run-off in alpine basins: examples from North America, the Alps, central Asia and the Andes. *Hydrological Processes, 23*, 31–41.
3. Kelliher, F. M., Leuning, R., & Schulze, E. D. (1993). Evaporation and canopy characteristics of coniferous forests and grasslands. *Oecologia, 95*, 153–163.
4. Fuchs, J. (2011). *Multitemporale Detektion der Gletscherveränderung im östlichen Tian Shan (AR Xinjiang, China) im Kontext des Klimawandels: Untersuchungen am Beispiel der Flusseinzugsgebiete von Toutun, Shuixi und Urumqi* (69 p). Ruprecht-Karls-Universität Heidelberg.
5. Fricke, K., Sterr, T., Bubenzer, O., & Eitel, B. (2009). The oasis as a mega city: Urumqi's fast urbanisation in a semi-arid environment. *Die Erde, 140*(4), 449–463.
6. Meßer, J. (1997). *Auswirkungen der Urbanisierung auf die Grundwasserneubildung im Ruhrgebiet unter besonderer Berücksichtigung der Castroper Hochfläche und des Stadtgebietes Herne* (Vol. 58). Essen: Deutsche Montan Technologie GmbH (DMT-Berichte aus Forschung und Entwicklung).
7. Deng, W., Bai, J., & Yan, M. (2002). Problems and countermeasures of water resources for sustainable utilisation in China. *Chinese Geographical Science, 12*(4), 289–293.
8. Falkenmark, M., Berntell, A., Jägerskog, A., Lundqvits, J., Matz, M., & Tropp, H. (2007). *On the verge of a new water scarcity: A call for good governance and human ingenuity* (19 p). Stockholm: SIWI.
9. Water Affairs Bureau Urumqi. (2007). *Water Report 2007* (24 p). Urumqi: Water Affairs Bureau Urumqi City.
10. Zimmerman, R., & Faris, C. (2011). Climate change mitigation and adaptation in North American cities. *Current Opinion in Environmental Sustainability, 3*(3), 181–187.
11. Steinberg, C., Weigert, B., Möller, K., & Jekel, M. (Eds.) (2002). *Nachhaltige Wasserwirtschaft: Entwicklung eines Bewertungs- und Prüfsystems* (311 p). Berlin: Erich Schmidt.
12. Hardin, G. (1968). The tragedy of the commons: The population problem has no technical solution; it requires a fundamental extension in morality. *Science, 162*, 1243–1248.
13. Gao, C. (2009). An analysis of externality economy of Xinjiang water resource development. *Journal of Sustainable Development, 2*(2), 143–147.
14. Shiva, V. (2005). *Der Kampf um das blaue Gold: Ursachen und Folgen der Wasserverknappung* (215 p). Zürich: Rotpunkt-Verlag.
15. Turner, R. K., Bateman, I. J., & Adger, W. N. (2001). Ecological economics and coastal zone ecosystems' values: An overview. In R. K. Turner, I. J. Bateman, & W. N. Adger (Eds.), *Economics of coastal and water resources. Valuing environmental functions* (pp. 1–44). Dordrecht: Kluwer.

16. Statistics Bureau of Xinjiang Uygur Autonomous Region. (2010). *Xinjiang Statistical Yearbook 2010, CD-ROM*. Beijing: China Statistics Press.
17. Guan, D., & Hubacek, K. (2007). Assessment of regional trade and virtual water flows in China. *Ecological Economics, 61*, 159–170.
18. Kahlenborn, W., & Kraemer, R. A. (1999). *Nachhaltige Wasserwirtschaft in Deutschland* (244 p). Berlin: Springer.
19. Allan, J. A. (1998). Virtual water: A strategic resource: Global solutions to regional deficits. *Groundwater, 36*(4), 545–546.
20. Oki, T., & Kanae, S. (2006). Global hydrological cycles and world water resources. *Science, 313*(5790), 1068–1072.
21. Stewart, J. B. (1996). Extrapolation of evaporation at time of satellite overpass to daily totals. In J. B. Stewart, E. T. Engman, R. A. Feddes, & Y. Kerr (Eds.), *Scaling up in hydrology using remote sensing* (pp. 245–255). Chichester, NY: Wiley.
22. Jakeman, A. J., Giupponi, C., Karssenberg, D., Hare, M. P., Fassio, A., & Letcher, R. A. (2006). Integrated management of water resources: Concepts, approaches and challenges. In C. Giupponi (Ed.), *Sustainable management of water resources. An integrated approach* (pp. 3–26). Cheltenham: Elgar.
23. Fricke, K. (2009). Integriertes Wassermanagement—Strategien für das Industriegebiet Midong in Urumqi, NW-China. *UmweltWirtschaftsForum, 17*(3), 291–298.
24. Jakobitz, K. (1994). *Wassergütewirtschaft und Raumplanung: Probleme der Zusammenarbeit und Lösungsansätze* (Vol. 192, 249 p). Hannover: Verlag der ARL (Forschungs- und Sitzungsberichte Akademie für Raumforschung und Landesplanung).
25. Möller, H.-W. (2002). *Trinkwassergefährdung und Trinkwasserpolitik: Eine marktwirtschaftliche Konzeption des Trinkwasserschutzes* (Vol. 42, 458 p). Baden–Baden: Nomos-Verl.-Ges (Verwaltungsorganisation, Staatsaufgaben und öffentlicher Dienst).
26. OECD. (1998). *Water consumption and sustainable water resources management* (64 p). Paris: OECD (OECD Proceedings).
27. Abderrahman, W. A. (2000). Urban water management in developing arid countries. *Water Resource Development, 16*(1), 7–20.
28. OECD (Ed.). (2006). *China in the global economy. Environment, water resources and agricultural policies. Lessons from China and OECD countries* (287 p). Paris: OECD.
29. Statistics Bureau of Xinjiang Uygur Autonomous Region. (2007). *Xinjiang Statistical Yearbook 2007, CD-ROM*. Beijing: China Statistics Press.
30. Statistics Bureau of Urumqi. (2010). *Urumqi Statistical Yearbook 2010* (456 p). Beijing: China Statistics Press.
31. Chapagain, A. K., & Hoekstra, A. Y. (2003). Virtual water trade: A quantification of virtual water flows between nations in relation to international trade of livestock and livestock products. In A. Y. Hoekstra (Ed.), *Virtual water trade. Proceedings of the international expert meeting on virtual water trade* (pp. 49–76). Delft: IHE Delft.
32. Statistics Bureau of Urumqi. (2009). *Urumqi Statistical Yearbook 2009* (506 p). Beijing: China Statistics Press.
33. Statistics Bureau of Xinjiang Uygur Autonomous Region. (2007). *Xinjiang Statistical Yearbook 2007, CD-ROM*. Beijing: China Statistics Press.
34. Statistics Bureau of Urumqi. (2008). *Urumqi Statistical Yearbook 2008* (455 p). Beijing: China Statistics Press.
35. Liu, J., & Savenije, H. H. (2008). Food consumption patterns and their effect on water requirement in China. *Hydrology and Earth System Sciences, 12*, 887–898.
36. Hoff, H., Falkenmark, M., Gerten, D., Gordon, L., Karlberg, L., & Rockström, J. (2010). Greening the global water system: Green-blue water initiative (GBI). *Journal of Hydrology, 384*(3–4), 177–186.
37. Blanke, A., Rozelle, S., Lohmar, B., Wang, J., & Huang, J. (2007). Water saving technology and saving water in China. *Agricultural Water Management, 87*(2), 139–150.

38. Neubert, S. (2005). Abwassernutzung in der Landwirtschaft: ein 'integriertes' und ökologisch nachhaltiges Verfahren? In S. Neubert (Ed.), *Integriertes Wasserressourcen-Management (IWRM). Ein Konzept in die Praxis überführen* (pp. 239–257). Baden–Baden: Nomos Verl.-Ges.
39. Wolff, H.-P., Doppler, W., & Nabulsi, A. (2005). Potenzial und Folgen der Verwendung von Abwasser in ruralen Räumen: das Beispiel des Wassereinzugsgebietes des Jordan. In S. Neubert (Ed.), *Integriertes Wasserressourcen-Management (IWRM). Ein Konzept in die Praxis überführen* (pp. 259–270). Baden–Baden: Nomos Verl.-Ges.
40. Murray, A., & Ray, I. (2010). Wastewater for agriculture: A reuse-oriented planning model and its application in peri-urban China. *Water Research, 44*(5), 1667–1679.
41. Yao, Y. (2011). Water reuse: a case study of Urumqi, China. In IWA (Ed.), *1st Central Asian Regional Young and Senior Water Professionals Conference, CD-ROM,* Almaty.
42. Food Agricultural Organisation (FAO). (2003). *Unlocking the water potential of agriculture* (62 p). Rome: FAO. Retrieved 6 Aug, 2012, from ftp.fao.org/agl/aglw/docs/unlocking_e.pdf.
43. Mekonnen, M. M., & Hoekstra, A. Y. (2012). A global assessment of the water footprint of farm animal products. *Ecosystems, 15*, 401–415.
44. Hu, Y., Moiwo, J. P., Yang, Y., Han, S., & Yang, Y. (2010). Agricultural water-saving and sustainable groundwater management in Shijiazhuang Irrigation District, North China Plain. *Journal of Hydrology, 393*(3–4), 219–232.
45. Rockström, J., Falkenmark, M., Karlberg, L., Hoff, H., Rost, S., & Gerten, D. (2009). Future water availability for global food production: The potential of green water for increasing resilience to global change. *Water Resources Research 45* (W00A12).
46. Zhang, Z. Y., Yang, H., Shi, M. J., Zehnder, A. J., & Abbaspour, K. C. (2011). Analyses of impacts of China's international trade on its water resources and uses. *Hydrology and Earth System Sciences, 15*, 2871–2880.
47. Zhou, Y., Nonner, J. C., Li, W. et al. (2007). *Strategies and techniques for groundwater resources development in Northwest China* (338 p). Beijing: China Land Press.
48. Lehn, H., Steiner, M., & Mohr, H. (1996). *Wasser—die elementare Ressource: Leitlinien einer nachhaltigen Nutzung* (368 p). Berlin: Springer.
49. Londong, J., Hillenbrand, T., Otterpohl, R., Peters, I., & Tillman, D. (2004). Vom Sinn des Wassersparens. *KA—Abwasser, Abfall, 51*(12), 1381–1385.
50. Landu, L., & Brent, A. C. (2006). Environmental life cycle assessment of water supply in South Africa: The Rosslyn industrial area as a case study. *Water South Africa, 32*(2), 249–256.
51. Azapagic, A., & Clift, R. (1999). The application of life cycle assessment to process optimisation. *Computers & Chemical Engineering, 23*, 1509–1526.
52. Bridle, T., & Skrypski-Mantele, S. (2000). Assessment of sludge reuse options: A life cycle approach. *Water Science and Technology, 41*(8), 131–135.
53. Friedrich, E., & Buckley, C. A. (1999). *The use of life cycle assessment in the selection of water treatment processes: Final report to the Water Research Commission* (63 p). Durban.
54. Martin, K., & Sauerborn, J. (2006). *Agrarökologie* (297 p). Stuttgart: UTB.
55. Chen, X., Zhang, Q., Zhou, K., & Sun, L. (2006). Quantitative assessment and analysis on the dynamic change of ecological capital in arid areas. *Chinese Science Bulletin, 51* (Supp. I): 204–212.
56. Meadows, D., Meadows, D., & Randers, J. (1992). *Beyond the limits: Global collapse or a sustainable future* (300 p). London: Earthscan Publications.
57. Feng, Q., & Cheng, G. (1998). Current situation, problems and rational utilisation of water resources in arid Northwest China. *Journal of Arid Environments, 40*, 373–382.
58. Food and Agriculture Organization of the United Nations. (2012). Crop Water Productivity. Retrieved 25 Mar, 2012, from http://www.fao.org/landandwater/aglw/cropwater/cwp.stm.
59. AboutCivil.org. (2012). Water requirements of crops. Retrieved 25 Mar, 2012, from http://www.aboutcivil.org/water-requirements-of-crops.html.
60. Food and Agriculture Organization of the United Nations. (2012). FAO Food Price Indices. Retrieved 25 Mar, 2012, from http://www.fao.org/worldfoodsituation/wfs-home/foodpricesindex/en/.

Chapter 8
Summary and Outlook

This research project in the Urumqi Region has mainly been aimed at developing a methodological approach for the assessment of the changes in hydrological and water management systems for future megacities. Some of the inherent constraints and challenges are discussed, along with possible measures to overcome them. After evaluating and discussing the models' function and its results, a final focus is laid on its transferability and the possibilities of improvement.

Located in a narrow grassland corridor between the semi-desert Jungggar Basin and the Tianshan Mountains in Northwest China, the research area Urumqi Region is despite its semi-arid climate in a relatively favourable hydrological situation. The nearby mountains have provided water for human development, settlements and agriculture. The growing population and economy during the past 60 years, associated with increasing water consumption, has led to a demand- and population-driven water scarcity that is expected to aggravate. Concurrently, the effects of climate change and land use transformations on the hydrological system and the water availability remain uncertain. This study intends to evaluate the recent and future situation by combining a hydrological water balance model for the simulation of the water supply based on scenarios of climate and land use change with a socio-economic model for projecting the future water demand including predicted growth of population and economy.

The water balance model consists of modules for potential and actual evapotranspiration, snow accumulation, snow melt, and the partitioning of surface and groundwater flow. It was designed to simulate the long-term annual water balance calculated with average climate input variables. As it is calculated on a 30×30 m raster, the consequences of local differences of climate and land surface parameters can be evaluated. The necessary input parameters are provided as point measurements from a climatological station or from remote sensing data in a distributed raster format: the land surface characteristics are derived from Landsat images, solar radiation and possible day duration from elevation data, temperature distribution from MODIS datasets and precipitation differences from TRMM data. The model was calibrated and validated for different catchments in Urumqi Region where limited runoff data was available.

K. Fricke, *Analysis and Modelling of Water Supply and Demand Under Climate Change, Land Use Transformation and Socio-Economic Development*, Springer Theses, DOI: 10.1007/978-3-319-01610-8_8,

The averages of the measured climate variables for the long- and the short-term past and three projected climate scenarios for the year 2050 with different precipitation changes are used to assess the consequences of climate change on the hydrological system. The results of the climate scenarios show that the situation in the last decade (2001–2010) has been more favourable for the entire region than the long-term average of the last 36 years. In the future, the total water flow and surface runoff will increase in the mountain areas, but more or less decrease in the other parts of the research area. In the lower areas, the decrease is mainly due to reduced snowfall in winter and subsequently less melt water flow in spring and summer. Furthermore, groundwater recharge will decrease 4–15 % and the actual evapotranspiration will rise about 8–20 %, especially over agricultural areas, within the next 40 years. Only the climate scenario with the maximum projected precipitation will result in a total water flow equivalent to the average of the last decade.

Other results included the consequences of land use transformation in Urumqi Region that is evaluated by simulating the water balance for a land use change scenario based on the planned urban development in one of the city's districts, Midong. The projected land use changes and expansion of the sealed areas mainly affect the partitioning between surface flow and infiltration and can reduce the groundwater recharge from sealed areas up to 80 %. The annual actual evapotranspiration increases only slightly and even decreases in summer due to the changed radiation budget. In contrast, it becomes larger in winter and leads to a reduction of snow melt water similar to the climate change scenarios.

The changes of the water demand are modelled with a socio-economic model developed by Trieb [1] and calibrated with the water consumption reported by the Water Affairs Bureau Urumqi [2–5]. Six scenarios for the socio-economic development are derived from references to evaluate the possible changes of water demand until 2050. The results vary from an extremely increasing to a slightly decreasing water demand, but the most likely development will lead to an increase of 30–62.5 % of total annual water demand with a reduction of the agricultural share by 50 %.

The modelled changes of water availability and water demand show that the spatial differences of water availability in the mountains, the lower plain, and the basin will further increase, but the total water supply will not improve. The combination of socio-economic development, climate and land use change will actually worsen demand-, population- and climate-driven water scarcity. Accordingly, adaptation strategies are developed for the local geographical and hydrological conditions based on the results from both the hydrological and socio-economic model. They focus on measures to reduce water demand, especially in the agricultural sector. A shift of activity and infrastructure into areas with lower evapotranspiration and higher precipitation, even outside Urumqi Region, is proposed. In the urban areas, the use of available water and water treatment capacities should be optimised by individual and infrastructural measures. But also the natural advantage of the hydrological system, the water supply from the mountain areas and the extensive groundwater bodies, should be cautiously further exploited where suitable. The amplification of the natural cycle of surface runoff infiltration, groundwater recharge and extraction of groundwater has many benefits as it increases underground storage of water resources, water treatment through percolation in soil and underground, and reduces evaporation. However, monitoring of groundwater extraction, water quality and environmental functions is fundamental.

The limited number of measurements and observations available for the research has been enlarged with various remote sensing datasets ranging from climate to land use and information derived from elevation data. This enabled a detailed description of the water distribution, taking into account the strong spatial heterogeneities in Urumqi Region. As most of the input and calibration data are not directly measured and from secondary sources, the uncertainty of the modelling results is rather high and the model quality limited. For further improvements, additional data and modules would be necessary.

8.1 Transferability

Besides the evaluation of the simulation results, the socio-economic and the hydrological models used are evaluated and their transferability is discussed. In general, it is possible to transfer the socio-economic model to other cities, regions and countries, as it has already been transferred from the Middle East and North Africa (MENA) region to Urumqi Region which was a down-scaling from country to region level. The *ceteris paribus* assumption that water consumption mainly depends on GDP and population growth has to be fulfilled. The data availability was critical for the calibration of the socio-economic model: either several years' data about the water consumption of all sectors or specific information about the efficiency indicators have to be available to reasonably adapt the model to the local circumstances. As long as these preconditions are fulfilled, the model can be transferred to other aggregate scales, but not to an individual. The more data is available and the more detailed the modelling input, the better and probably more accurately the development and composition of water consumption can be described. Since the input parameters about population and economic growth is most probably a projection or estimation, the results of the socio-economic model can only be seen as the result of scenarios and not forecasts.

The transfer of hydrological models from one catchment to another is the main basic research focus of modelling or prediction of ungauged basins (PUB) as explained in Sect. 1.2.2. But the transferability of the model and the identified parameters to sub-catchments cannot be guaranteed even after calibrating the model at the basin outlet when they cannot be calibrated with runoff data on the sub-catchment level [6]. Nonetheless, any runoff data available for calibration and validation would improve model accuracy and reduce uncertainty when transferring the model to another catchment.

The model structure and modules focus on certain processes of the hydrological system which are representative for the intended spatial and temporal scale and local context. As discussed in Sect. 1.2.1, if these two specifications change, other processes become dominant and should be taken into account when modelling the hydrological system. Hence, that section focuses on the possibility of model transfer at a fixed scale. A transfer in time has already been conducted when using the water balance model that has been calibrated with measurements in the past for predicting the situation in 2050. Likely changes of both parameters and dominant processes are considered by Schaefli et al. [7] to be the main challenges of prediction in ungauged basins and prediction under global change. Only the possible changes of land surface parameters were included through the simulation of land use change in the Midong District.

Merz and Blöschl [8] concluded for the average model parameters of the immediate upstream and downstream basins, regionalisation by kriging to be the best way of transferring model parameters for making predictions in ungauged basins. This method is feasible when parameters are missing for some catchments within a larger runoff system. Catchments not adjacent to each other would require a different method. Fortunately, there exists a relationship between most of the model parameters and physical and climatic characteristics. When these characteristics are known for a basin, the hydrological parameters can be deducted accordingly. This approach has been used when deriving the necessary parameters for this model from land use and cover characteristics. Thirdly, a 'similar basin approach' is commonly used for predicting hydrological responses in ungauged basins and "the complete set of model parameters identified at the more in depth investigated catchment is used to predict the hydrological responses at ungauged catchments which are most similar in terms of physiogeographic and climatic characteristics" ([6]: Introduction section).

For the hydrological model used in this thesis, the last two methodological approaches are possible. Either the hydrological parameters can be deduced from local input data or they have to be substituted by transfer from similar catchments. Without adequate data or a lack of calibration and validation measurements, the prediction uncertainty is especially high and special care has to be applied to the selection of the model and parameters values as well as the model structure [6].

Since most of the **data input used for modelling in Urumqi Region was derived without extensive field research and with freely available data**, this approach can be transferred to other areas of interest. Input climate data is derived from MODIS and TRMM images which are available almost worldwide, and climate station data, whose quality has to be evaluated specifically. Elevation data is available as digital surface models from ASTER GDEM and SRTM-3. Depending on the desired resolution and availability, land use information can be classified from local datasets, Landsat images or other satellite imagery such as ASTER, Ikonos, MODIS, SPOT etc. Most of the parameters to calculate potential evapotranspiration such as albedo, stomatal resistance r_s, and canopy height h are connected to the land cover characteristics and to a lesser extent influenced by regional climate. Additional local references could be consulted to adapt the albedo of local land use characteristics. The r_s and h values used have been developed for the temperate zone or adapted for Urumqi Region; therefore, different values might be necessary for other climates from according references. For similar areas, they would be transferable. The same applies to the degree-day-factor: it has to be adjusted to each application and climate and is ideally assessed from field measurements or calibrated see [9]. Alternatively, it can be transferred from reference measurements with similar radiation, climatic situations and surface characteristics, which are widely available for the degree-day-method. It is only adequate for 'average conditions' of certain surface and snow characteristics and periods exceeding a couple of days [9].

The values of the proportion of runoff (p_{runoff}) have already been transferred from other research projects, but their accuracy could not be tested with field data in the research area. The simulation results imply that they are applicable in Urumqi Region, but do not prove this finding as too many other processes are influencing results and validation data. When transferring p_{runoff} to other catchments, special attention has to be paid to the characteristics of soil and geology and their implications on the imperviousness and infiltration capacity.

8.2 Methodological Discussion and Outlook

As already mentioned in Sect. 1.2 about the state of the research and Chap. 3 about the modelling concept, the local context and demands, spatial and temporal scale affect the choice of appropriate models and necessary parameters to represent dominant hydrological processes. Due to data and computing restrictions, not all processes of the hydrological system are accounted for. This research focussed on the most important ones and the ones that could be modelled with the available data: evapotranspiration, snow accumulation and snow melt, and the partitioning of runoff. The model proved to be sufficient for calculating an annual water balance. The main advantage of the model is that data from local sources can be used when available, but is necessary only for calibration and validation of the model. Changes in the water balance and affected areas can be easily identified. Especially the small-scale differences in very sensitive areas such as the border of the high mountain area and the low mountain areas in Urumqi Region could be modelled in a detail that would not have been possible when applying other only semi-distributed or lumped models.

When different output, scale or time steps are desired, the hydrological model described and used here has to be adapted to the according requirements. Several nearby possibilities to improve and extend the model are explained within this chapter. As Newman et al. [10] remark, most hydrological studies lump canopy interception, soil evaporation, and transpiration into one single term and module, **evapotranspiration** to estimated water budgets. This approach creates a 'black box' for biological processes that influence the hydrological cycle directly or indirectly on short (hourly, daily) or long (seasonal, interannual) time-scales. Additionally, the relationship of vegetation cover and land surface characteristics to interception, soil evaporation, and transpiration is represented in a generalised and simplified way. In this study, soil water content was assumed to be not too important similar to other studies that were facing data shortage e.g. [11]. However, relevant feedback mechanisms between soil moisture, evapotranspiration and runoff exist at the local, regional and global levels [12]. Especially considering soil water and its influence on evapotranspiration would improve the accuracy of the model especially on shorter time scales, but would require the adaptation of other model parameters.

The extrapolation of temperature and precipitation was based on remote sensing data that was used as linear extrapolation gradient of the measurements of one station. When extending the number of investigated stations, the relationship between remote sensing and station data could be averaged with a linear regression model to reduce the impact of outliers and errors in the data. The representativity of the TRMM data in the mountainous areas should be further evaluated as the model overestimated precipitation compared to other model results and data from official statistics.

In the module for **snow** accumulation and melt water, the applied degree-day-method could also be improved by shortening the time period the degree-day-factors are adjusted to from a monthly to at least weakly differentiation. Alternatively, they could be better adapted to the local conditions of degree-day-factors and threshold temperatures either by field measurements or by calibration of the spatial extent of the snow cover with other information about the snow cover area and in sub-catchments with more detailed runoff data. In the simulation results, a pronounced difference between net precipitation and total water flow could be observed. This amount was accumulated snow that was not melted by the degree-day-method and assumed to contribute to glacier accumulation and consequently glacier runoff. The accuracy of this process could not be confirmed as this amount of snow water equivalent was transferred to the glacial system which was not modelled within this research. This limitation of the model increases the need to focus on the relative and not on the absolute change on the water balance as practiced. The constraint of relative comparison could be overcome by adding a glacier module to the water balance model.

The formation of **surface flow** has been treated as a single process in the hydrological model, combining Hortonian overland flow, saturation excess overland flow, subsurface return flow or interflow. Hortonian overland flow, which is defined as infiltration excess overland flow, often occurs in arid and semi-arid regions when the infiltration capacity of the soil is unable to cope with intense rainfall events. Depending on soil type and soil moisture content, different runoff processes can occur. Also, evapotranspiration, storage or infiltration of runoff as it flows overland have not been taken into account as feedback and interaction between the raster cells was only allowed for flow accumulation. For example, the transfer of surface runoff for irrigation and its subsequent evapotranspiration is only taken into account by the increased evapotranspiration potential of the dense vegetation and not by the available water through irrigation. When modelling at a smaller temporal and spatial scale, these processes become more important for evapotranspiration and the formation of runoff and should therefore be integrated into the hydrological model. These additional processes would also improve the model's ability to simulate the monthly or even daily water balance and prediction efficiency, measured, e.g., with the Nash–Sutcliffe coefficient.

As soil and underground are treated as one zone where **groundwater flow** occurs, exchange flows between vadose and phreatic zone as well as storage and interflow and baseflow are not considered. When comparing simulated and observed monthly surface flow in Sect. 3.4, it becomes obvious that the model fails to simulate storage processes that are responsible for monthly surface and baseflow. Same as for the surface flow processes, the predictive capability of the model could be improved by adding these processes to the model structure. Especially with regard to the timing of flow peaks on a smaller time step, storage in vegetation, soil and underground as well as thresholds for the initiation of different flows should be taken into account.

The effect of water infrastructure changes such as storage in reservoirs and additional irrigation could be investigated only marginally as there was not enough data and information available to evaluate these factors in detail. Additionally, the aspects of feedback and interaction between the different models and model units deserve attention [10]. So far, the projected water consumption by agriculture, industry and population was not related to the increasing potential evapotranspiration. But water demand of irrigated crops and animal husbandry, population and urban green will grow when temperatures and evapotranspiration increase. Due to the limited data availability it seems only feasible to integrate this factor on district or city level. In other publications, land use change has been simulated driven by socio-economic scenarios. When using these scenarios in the socio-economic model and land use changes in the hydrological model the effects of socio-economic development on both water demand and supply side could be assessed simultaneously. It has to be considered, too, that the calibration and validation is only valid for observed behaviour and measurements of the past. As Wagener et al. [13] stated, in a non-stationary world and under climate and land use change, it might not be valid for the future.

References

1. Trieb, F. (2007): AQUA-CSP: Concentrating Solar Power for Seawater Desalination. DLR (279 p) (Final Report, Bundesministerium für Umwelt, Naturschutz und Reaktorsicherheit (BMU)).
2. Water Affairs Bureau Urumqi. (2003). *Water Report 2003*. Urumqi: Water Affairs Bureau Urumqi City.
3. Urumqi, Water Affairs Bureau. (2004). *Water Report 2004* (27 p). Urumqi: Water Affairs Bureau Urumqi City.
4. Urumqi, Water Affairs Bureau. (2007). *Water Report 2007* (24 p). Urumqi: Water Affairs Bureau Urumqi City.
5. Water Affairs Bureau Urumqi (2005). *Water Report 2005* (27 p). Urumqi: Water Affairs Bureau Urumqi City.

6. Hunukumbura, P. B., Tachikawa, Y., & Shiiba, M. (2011). Distributed hydrological model transferability across basins with different hydro-climatic characteristics. *Hydrological Processes,*. doi:10.1002/hyp.8294.
7. Schaefli, B., Harman, C. J., Sivapalan, M., & Schymanski, S. J. (2010). Hydrologic predictions in a changing environment: behavioral modelling. *Hydrology and Earth System Sciences Discussions, 7*, 7779–7808.
8. Merz, R., & Blöschl, G. (2004). Regionalisation of catchment model parameters. *Journal of Hydrology, 287*, 95–123.
9. Hock, R. (2003). Temperature index melt modelling in mountain areas. *Journal of Hydrology, 282*, 102–115.
10. Newman, B. D., Wilcox, B. P., Archer, S. R., Breshears, D. D., Dahm, C. N., Duffy, C. J., McDowell, N. G., Phillips, F. M., Scanlon, B. R., & Vivoni, E. R. (2006). Ecohydrology of water-limited environments: A scientific vision. *Water Resources Research, 42,* (W06302).
11. Simonneaux, V., Duchemin, B., Helson, D., Er-Raki, S., Olioso, A., & Chehbouni, A. (2008). The use of high-resolution image time series for crop classification and evapotranspiration estimate over an irrigated area in Central Morocco. *International Journal of Remote Sensing, 29*(1), 95–116.
12. Seneviratne, S. I., Corti, T., Davin, E. L., Hirschi, M., Jaeger, E. B., Lehner, I., et al. (2010). Investigating soil moisture-climate interactions in a changing climate: A review. *Earth-Science Reviews, 99*, 125–161.
13. Wagener, T., Sivapalan, M., Troch, P. A., McGlynn, B. L., Harman, C. J., Gupta, H. V., et al. (2010). The future of hydrology: An evolving science for a changing world. *Water Resources Research, 46* (W05301).

Appendix A

A.1 Calculation of Effective Heat Flux Depth

The heat flux G [W m^{-2}] is defined as the rate of heat flow I [W or J s^{-1}] per area A [m^2]:

$$G = \frac{\Delta I}{A} \tag{A.1}$$

with ΔI as the quantity of heat Q [J] flowing during the time t [s]:

$$\Delta I = \frac{\Delta Q}{\Delta t}. \tag{A.2}$$

I can also be calculated based on the heat conductivity K [W m^{-1} K^{-1}], the area A and the temperature difference ΔT [K] per depth Δz [m]:

$$\Delta I = K \cdot A \cdot \frac{\Delta T}{\Delta z}. \tag{A.3}$$

Q can be derived from the heat capacity C [J K^{-1}] and the temperature difference ΔT:

$$Q = C \cdot \Delta T, \tag{A.4}$$

where C can be calculated based on the mass specific heat capacity c_s [J kg^{-1} K^{-1}] and mass m [kg]:

$$C = c_s \cdot m, \tag{A.5}$$

or based on the volumetric heat capacity c_v [J K^{-1} m^{-3}]:

$$C = c_v \cdot V \tag{A.6}$$

$$c_v = \rho \cdot c_s \tag{A.7}$$

K. Fricke, *Analysis and Modelling of Water Supply and Demand Under Climate Change, Land Use Transformation and Socio-Economic Development*, Springer Theses, DOI: 10.1007/978-3-319-01610-8, with the volume V [m^3] and mass density ρ [kg m^{-3}]. The volume V is defined as the area A times depth z:

$$V = A \cdot z. \tag{A.8}$$

Equations (A.2, A.4, A.6, and A.8) are inserted in (A.1) leading to:

$$G = \frac{I}{A} = \frac{\Delta Q}{A \cdot \Delta t} = \frac{C \cdot \Delta T}{A \cdot \Delta t} = \frac{C \cdot \Delta T \cdot \Delta z}{\Delta t \cdot V} = \frac{c_s \cdot V \cdot \Delta T \cdot \Delta z}{\Delta t \cdot V}$$
$$G = c_s \cdot \frac{\Delta T}{\Delta t} \cdot \Delta z, \tag{A.9}$$

an equation that also used by Allen [1] to calculate soil heat flux (see Sect. 3.1.1).

The second equation for the calculation of the rate of heat flow, equation (A.3), can be used to calculate the effective depth heat can be transported to depending on the heat conductivity K and volumetric heat capacity c_s. Equation (A.3) inserted in (A.1) leads to:

$$G = K \cdot \frac{\Delta T}{\Delta z} = c_s \cdot \frac{\Delta T}{\Delta t} \cdot \Delta z, \tag{A.10}$$

which can be solved for Δz:

$$\Delta z = \sqrt{\Delta t} \cdot \sqrt{\frac{K}{c_s}}. \tag{A.11}$$

A.2 Overview of the Input Data

(Tables A.1, A.2, A.3)

Table A.1 Overview of all data sets used as input for the water balance model with description, unit, type and used dates

Name	Description and source	Unit	Type	Date
MOD11A1	MODIS land surface temperature	K	Raster, daily	10/1999–12/2000
TRMM 3B43	Precipitation, TRMM	mm hr^{-1}	Raster, monthly	10/1999–12/2000
SRTM-3	DEM, USGS	m	Raster	–
ASTER GDEM	DEM, METI & NASA	m	Raster	–
Mean air temperature	NOAA NCDC	°C	Point, daily	01/1973–12/2010 (Wulumuqi station)
Mean dew point temperature	NOAA NCDC	°C	Point, daily	01/1973–12/2010 (Wulumuqi station)

(continued)

Table A.1 (continued)

Name	Description and source	Unit	Type	Date
Minimum temperature	NOAA NCDC	°C	Point, daily	01/1973–12/2010 (Wulumuqi station)
Maximum temperature	NOAA NCDC	°C	Point, daily	01/1973–12/2010 (Wulumuqi station)
Mean wind speed	NOAA NCDC	m s^{-1}	Point, daily	01/1973–12/2010 (Wulumuqi station)
Mean station pressure	NOAA NCDC	kPa	Point, daily	01/1973–12/2010 (Wulumuqi station)
Precipitation	NOAA NCDC	mm	Point, daily	01/1973–12/2010 (Wulumuqi station)
Mean incoming extraterrestrial radiation	Calculated based on SRTM-3 and date	Wh m^{-1}	Raster, monthly	2000, 2050
Maximum possible duration of sunshine	Calculated based on SRTM-3 and date	hr	Raster, monthly	2000, 2050
Actual duration of sunshine	Urumqi Statistical Yearbooks 2001–2009	hr	Point, monthly	1999–2009
LULC	Land use and land cover classification from Landsat 7 ETM+ images	classes	Raster, Seasonally	10/1999–01/2001

Table A.2 Available and downloaded satellite images for the chosen classification dates 2000–2007 (L7 SLC-on: Landsat-7 with ETM+ with the scan line corrector on) (*Source* Landsat Program 2010)

Satellite	Path	Row	Date	Cloud cover (%)
L7 SLC-on	143	29	Oct 99	0
L7 SLC-on	143	30	Oct 99	2
L7 SLC-on	143	30	Dec 99	12
L7 SLC-on	143	29	Jan 00	8
L7 SLC-on	143	29	Mar 00	5
L7 SLC-on	143	30	Mar 00	6
L7 SLC-on	143	29	Jun 00	0
L7 SLC-on	143	30	Jun 00	0
L7 SLC-on	143	29	Sep 00	0
L7 SLC-on	143	30	Sep 00	0
L7 SLC-on	143	29	Jan 01	28
L7 SLC-on	143	30	Jan 01	25

Table A.3 MOD11A1 raster data sets used for extrapolation of temperature data (*Source* U.S. Geological Survey [2, 3]

Year	Month	MODIS-file (original, *.hdf)
2000	1	MOD11A1.A2000074.h24v04.005.2007176161215.hdf
2000	2	MOD11A1.A2000074.h24v04.005.2007176161215.hdf
2000	3	MOD11A1.A2000074.h24v04.005.2007176161215.hdf
2000	4	MOD11A1.A2000120.h24v04.005.2007183180509.hdf
2000	5	MOD11A1.A2000139.h24v04.005.2007186050937.hdf
2000	6	MOD11A1.A2000180.h24v04.005.2007195080049.hdf
2000	7	MOD11A1.A2000210.h24v04.005.2007206102647.hdf
2000	8	MOD11A1.A2000240.h24v04.005.2007212192305.hdf
2000	9	MOD11A1.A2000273.h24v04.005.2006314095731.hdf
2000	10	MOD11A1.A2000288.h24v04.005.2006318031258.hdf
2000	11	MOD11A1.A2000288.h24v04.005.2006318031258.hdf
2000	12	MOD11A1.A2000288.h24v04.005.2006318031258.hdf

A.3 Accuracy Assessment

(Tables A.4, A.5, A.6, A.7, A.8, A.9)

A.4 SWAT Input Data

The first step for modelling the water balance for the watershed of Wulapo Reservoir was the acquisition and adaption of data set for *ArcSWAT*. For the definition of hydrological response units (HRUs), soil data, type of land use and vegetation, and slope classes were required. Subsequently, climate data input was necessary for modelling the watershed.

A.4.1 Soil Data

SWAT provides only soil data for soils from the U.S. As there was only insufficient local soil data available for Urumqi region, SWAT input was created from a Soil and Terrain (SOTER) database. SOTER was compiled by *ISRIC*—World Soil Information within the framework of the Global Assessment of Land Degradation (GLADA) project as part of the FAO program Land Degradation Assessment in Drylands (LADA). Enhanced soil information was compiled in a soil and terrain database at scale 1:1 million for China amongst others [1]. The soil types given by SOTER database showed fairly good conformance with soil type information from other sources, namely a Chinese soil map. The comparison was achieved by correlating the legends of Chinese soil maps with FAO/WRB soil types. For the soil definition in SWAT, apart from the number of layers

Table A.4 Accuracy assessment of the LULC classification of Landsat ETM+ scenes 10/1999

		Ground truth, estimated													No. classified pixel
		Unclassified	Snow old	Soil	Sparse	Sealed	Snow	Agri	Water	Agri dry	Confor	Dark	Grass	Rock	
Classified in satellite image as	Unclassified	20													20
	Snow old		20												20
	Soil			20											20
	Sparse				20										20
	Sealed			6	1	11								2	20
	Snow				1		19								20
	Agri	1						19							20
	Water								20						20
	Agri dry									20					20
	Confor										17		3		20
	Dark											20			20
	Grass												20		20
	Rock	2												18	20
No. ground truth pixel		23	20	26	22	11	19	19	20	20	17	20	23	20	260

Class name	Reference totals	Classified totals	No. correct	Producers accuracy (%)	Users accuracy (%)
Unclassified	23	20	20	–	–
Snow_ ld	20	20	20	100.00	100.00
Soilnew	26	20	20	76.92	100.00
Sparse	22	20	20	90.91	100.00
Sealed	11	20	11	100.00	55.00
Snow	19	20	19	100.00	95.00
Agri	19	20	19	100.00	95.00
Water	20	20	20	100.00	100.00

(continued)

Table A.4 (continued)

Class name	Reference totals	Classified totals	No. correct	Producers accuracy (%)	Users accuracy (%)
Agri_dry	20	20	20	100.00	100.00
Confor	17	20	17	100.00	85.00
Dark	20	20	20	100.00	100.00
Grass	23	20	20	86.96	100.00
Rock	20	20	18	90.00	90.00
Totals	260	260	244		

Overall classification accuracy = 93.85 %; overall kappa statistics = 0.9333

Table A.5 Accuracy assessment of the LULC classification of Landsat ETM+ scenes 12/1999 and 01/2000

		Ground truth, estimated									No. classified pixel
		Unclassified	Soil	Sparse	Rock	Water	Confor	Snow old	Sealed	Snow	
Classified in satellite image as	Unclassified	20									20
	Soil		15					5			20
	Sparse			20							20
	Rock				17			3			20
	Water				1	19					20
	Confo			1			19				20
	Snow_old							20			20
	Sealed								20		20
	Snow									20	20
No. ground truth pixel		20	15	21	18	19	19	28	20	20	180

Class name	Reference totals	Classified totals	No. correct	Producers accuracy (%)	Users accuracy (%)
Unclassified	20	20	20	–	–
Soil	15	20	15	100.00	75.00
Sparse	21	20	20	95.24	100.00
Rock	18	20	17	94.44	85.00
Water	19	20	19	100.00	95.00
Confor	19	20	19	100.00	95.00
Snow old	28	20	20	71.43	100.00
Sealed	20	20	20	100.00	100.00
Snow	20	20	20	100.00	100.00
Totals	180	180	170		

Overall classification accuracy = 94.44 %, overall kappa statistica = 0.9375

Table A.6 Accuracy assessment of the LULC classification of Landsat ETM+ scenes 03/2000

		Ground truth, estimated											No. classified pixel
		Un-classified	Dark	Rock	Snow	Sealed	Snow old	Water	Agri dry	Sparse	Soil	Confor	
Classified in satellite image as	Unclassified	20											20
	Dark		15						1		4		20
	Rock			20									20
	Snow				20								20
	Sealed			6		13					1		20
	Snow old						19			1			20
	Water							18				2	20
	Agri dry								20				20
	Sparse									20			20
	Soil										20		20
	Confor							1	1			18	20
No. ground truth pixel		20	15	26	20	13	19	19	22	21	25	20	200

Class name	Reference totals	Classified totals	No. correct	Producers accuracy (%)	Users accuracy (%)
Unclassified	20	20	20	–	–
Dark	15	20	15	100.00	75.00
Rock	26	20	20	76.92	100.00
Snow	20	20	20	100.00	100.00
Sealed	13	20	13	100.00	65.00
Snow old	19	20	19	100.00	95.00
Water	19	20	18	94.74	90.00
Agri dry	22	20	20	90.91	100.00
Sparse	21	20	20	95.24	100.00
Soil	25	20	20	80.00	100.00
Confor	20	20	18	90.00	90.00
Totals	220	220	203		

Overall classification accuracy = 92.27 %, overall kappa statistica = 0.9150

Table A.7 Accuracy assessment of the LULC classification of Landsat ETM+ scenes 06/2000

		Ground truth, estimated														No. classified pixel
		Un-classified	Rock	Sealed	Confor	Soil	Dark	Snow	Snow old	Water	Agri	Grass	Ice	Agri dry	Sparse	
Classified in satellite image as	Unclassified	20														
	Rock		20													20
	Sealed		5	9		6										20
	Confor				20											20
	Soil					20										20
	Dark						20									20
	Snow							20								20
	Snow old								20							20
	Water									20						20
	Agri										18	2				20
	Grass											20				20
	Ice		1										19			20
	Agri dry													20		20
	Sparse				1				1						18	20
No. ground truth pixel			26	9	21	26	20	20	21	20	18	22	19	20	18	180

Class name	Reference totals	Classified totals	No. correct	Producers accuracy (%)	Users accuracy (%)
Unclassified	20	20	20	–	–
Rock	26	20	20	76.92	100.00
Sealed	9	20	9	100.00	45.00
Confor	21	20	20	95.24	100.00
Soil	26	20	20	76.92	100.00
Dark	20	20	20	100.00	100.00
Snow	20	20	20	100.00	100.00

(continued)

Table A.7 (continued)

Class name	Reference totals	Classified totals	No. correct	Producers accuracy (%)	Users accuracy (%)
Snow old	21	20	20	95.41	100.00
Water	20	20	20	100.00	100.00
Agri	18	20	18	100.00	90.00
Grass	22	20	20	90.91	100.00
Ice	19	20	19	100.00	95.00
Agri dry	20	20	20	100.00	100.00
Sparse	18	20	18	100.00	90.00
Totals	280	280	264		

Overall classification accuracy = 94.29 %, overall kappa statistica = 0.9385

Table A.8 Accuracy assessment of the LULC classification of Landsat ETM+ scenes 09/2000

		Ground truth, estimated														No. classified pixel
		Un-classified	Agri dry	Soil	Sparse	Water	Rock	Sealed	Dark	Snow	Irr	Snow old	Agri	Grass	Confor	
Classified in satellite image as	Unclassified	20														20
	Agri dry		20													20
	Soil			20												20
	Sparse				20											20
	Water					20										20
	Rock						20									20
	Sealed						9	5				6				20
	Dark								20							20
	Snow									20						20
	Irr										18	1	1			20
	Snow old											20				20
	Agri												20			20
	Grass													20		20
	Confor														20	20
No. ground truth pixel		20	20	20	20	20	29	5	20	20	18	27	21	20	20	280

Class name	Reference totals	Classified totals	No. correct	Producers accuracy (%)	Users accuracy (%)
Unclassified	20	20	20	–	–
Agri dry	20	20	20	100.00	100.00
Soil	20	20	20	100.00	100.00
Sparse	20	20	20	100.00	100.00
Water	20	20	20	100.00	100.00
Rock	29	20	20	68.97	100.00
Sealed	5	20	5	100.00	25.00

(continued)

Table A.8 (continued)

Class name	Reference totals	Classified totals	No. correct	Producers accuracy (%)	Users accuracy (%)
Dark	20	20	20	100.00	100.00
Snow	20	20	20	100.00	100.00
Irr	18	20	18	100.0	90.00
Snow old	27	20	20	74.07	100.00
Agri	21	20	20	95.24	100.00
Grass	20	20	20	100.00	100.00
Confor	20	20	20	100.00	100.00
Totals	280	280	263		

Overall classification accuracy = 93.93 %, overall kappa statistica = 0.9346

Table A.9 Accuracy assessment of the LULC classification of Landsat ETM+ scenes 01/2001

		Ground truth, estimated									No. classified pixel
		Unclassified	Confor	Sealed	Rock	Soil	Water	Snow	Snow old	Sparse	
Classified in satellite image as	Unclassified	20									20
	Confor		17			3					20
	Sealed			17	1					2	20
	Rock				19				1		20
	Soil					20					20
	Water						20				20
	Snow							20			20
	Snow old								20		20
	Sparse			1						19	20
No. ground truth pixel		20	17	18	20	23	20	20	21	21	180

Class name	Reference totals	Classified totals	No. correct	Producers accuracy (%)	Users accuracy (%)
Unclassified	20	20	20	–	–
Confor	17	20	17	100.00	85.00
Sealed	18	20	17	94.44	85.00
Rock	20	20	19	95.00	95.00
Soil	23	20	20	86.96	100.00
Water	20	20	20	100.00	100.00
Snow	20	20	20	100.00	100.00
Snow old	21	20	20	95.25	100.00
Sparse	21	20	19	90.48	100.00
Totals	180	180	172		

Overall classification accuracy = 95.56 %, overall kappa statistica = 0.9500

Table A.10 Description, field name and units of soil parameters required by SWAT

Description	Field Name	Units	Source
Hydrologic soil group	HYDGRP		See Sect. A.4.1, Table A.12
Max. rooting depth	SOL_ZMX	mm	[2], field survey
Fraction of porosity from which anions are excluded	ANION_EXCL	Fraction of porosity	Standard value 0.5 [3]
Potential or maximum crack volume	SOL_CRK	m^3/m^3	Standard value 0.5 [3]
Texture	TEXTURE		optional
Soil depth*	SOL_Z	mm	Standard value [4], modified according to [2] and field survey
Bulk density*	SOL_BD	g/cm^3	BULK from [4]
Available water content*	SOL_AWC	mm/mm	AWC3 from [4]
Organic carbon content*	SOL_CBN	% wt.	ORGC from [4]
Saturated hydraulic conductivity*	SOL_K	mm/hr	[5], www.pedosphere.com/resources/texture/worktable_us.cfm, 15.10.2011
Clay content*	CLAY	% wt.	CLAY from [4]
Silt*	SILT	% wt.	SILT from [4]
Sand*	SAND	% wt.	SAND from [4]
Rock/gravel content*	ROCK	% wt.	GRAVEL from [4]
Albedo*	SOL_ALB	fraction	[6], http://agsys.cra-cin.it/tools/solarradiation/help/Albedo.html, accessed 17.10.2011
USLE soil erodibility factor*	USLE_K		Williams 1995 in [3]
Electric conductivity*	SOL_EC	dS/m	ECE from [4]

* Required for each layer (*source* Own design; [3])

Table A.11 Dominant soil type (DOMSOIL) with FAO soil unit codes (7), name and determined hydrologic soil group (HSG) located in the research area

DOMSOIL	Name	Hydrologic soil group (HSG)
ATa	Aric Anthrosol	C
FLc	Calcaric Fluvisol	C
CHk	Calcic Chernozem	C
GYk	Calcic Gypsisol	C
KSk	Calcic Kastanozem	C
LPe	Eutric Leptosol	C
LPi	Gelic Leptosol	A
GRh	Haplic Greyzem	C
KSh	Haplic Kastanozem	C
CLl	Luvic Calcisol	C
GYl	Luvic Gypsisol	B
KSl	Luvic Kastanozem	C
LPm	Mollic Leptosol	B
SCm	Mollic Solonchak	C

Source [4, 7, 1, 8]

Table A.12 Hydrologic soil groups and determination factors

Criteria	A	B	C	D
Water transmission	Freely	Unhindered	Somewhat restricted	(very) restricted
Clay (%)	<10	10–20	20–40	>40
Sand (%)	>90	50–90	<50	<50
Saturated hydraulic conductivity	>10 μm/s	4–10 μm/s	0.4–4 μm/s	<0.4 m/s
Depth to a water impermeable layer (cm)	>50	>50	>50	<50
Depth to water table (cm)	>60	>60	>60	<60

Source NRCS [8]

NLAYERS, following parameters were required as data input in the database sheet "usersoil" (see Tables A.10, A.11).

A.4.2 HYDGRP Hydrologic Soil Groups

The hydrologic soil group (HSG) A, B, C or D describes groups of soils that have similar runoff potential under similar storm and cover conditions [9]. The HSG is important for the assignment of SCS curve numbers. The group a soil belongs to depends on infiltration, saturated hydraulic conductivity, and groundwater depth. The hydrologic soil group, along with land use, management practices, and hydrologic conditions assigns a soil's associated runoff curve number, which are used to estimate direct runoff from rainfall. The classes are based on the potential intake and transmission of water under when thoroughly wet, unfrozen soil, bare soil surface and maximum swelling of expansive clays, the slope is not considered [8]. The soils were assigned to the groups using the National Engineering Handbook Hydrology, published by the NRCS [8]. If saturated hydraulic conductivity data is not available, other soil properties such as texture, bulk density, and strength of soil structure, clay mineralogy, and organic matter are used to assign the soil to a HSG. As the saturated hydraulic conductivity was already calculated with the proportion of sand and clay, it was decided to use the same approach (based on soil texture) to determine the HSGs. However, the calculated saturated hydraulic conductivity was taken as additional information when the texture and the limits for sand and clay produced unclear results. All criteria in the next four paragraphs are based on the National Engineering Handbook Hydrology and the SWAT Theoretical Documentation [8, 9]. The assigned HSGs coincided not all with the HSGs determined by other references [10], which may be caused by different soil textures of the investigated soils.

- SOL_ZMX: Maximum rooting depth, total depth of soil profile

As a standard root depth, 100 cm were used "because it corresponds with the effective rooting depth of most annual crops" (cf. Doorenbos and Kassam 1978 in [2]). For Lithosols (FAO classification: Leptosols[1]) and Rankers and Rendzinas (FAO classification: Regosols) a shallower depth was applied, 10 and 30 cm respectively [2].

The standard depth of some soil types was corrected with field data that was collected during a field trip in spring and early summer 2011 by Klenk et al. (Institute for Environmental Physics, Heidelberg University). The adapted soils include Cambic Arenosols (130 mm), Haplic Calcisols (600 mm), Haplic Chernozems (500 mm) and Calcic and Haplic Kastanozems (800 mm).

- SOL_Z

SOL_Z is the soil layer depth. Following Batjes [4], the standard value for the top soil was 0–30 cm and for the bottom soil 30–100 cm. The soil layer depth for Leptosols and Regosols were taken from the soil depths (see SOL_ZMX above). For the soil types corrected with field data, the soil layer depths were modified as well.

- SOL_AWC

SOL_AWC is the available water content/capacity of the soil layer in mm H_2O/mm soil. The available water content is calculated by subtracting the amount of water at permanent wilting point from the water content at field capacity (cf. [2, 3]).

The information about available water content was taken from the variable water retention AWC3_TM and AWC3_BM [% v/v] in SUMTAB90.dbf [4]. AWC_3 represents the available water from −33 to −1,500 kPa or pF 2.5–4.2 [4]. "The

[1] Encyclopedia of Soil Science online, 35; World reference base for soil resources, Ausgabe 94, 6.

commonly used pF intervals for defining AWC include pF 1.7–4.2 in the United Kingdom, pF 2.0–4.2 in the Netherlands, and pF 2.5–4.2 in the USA“ [2: 33]. As *ArcSWAT* is developed in the USA, the according limits as described by Batjes [2] were used (cf. also [3]).

Pedotransfer functions (PTFs) are used to predict the AWC from measured silt, clay and organic matter content [2]. Batjes [2: 49] considers the resolution of the AWC data set as appropriate “for one-layer water balance models, with monthly time steps [...]”. Therefore the data set was regarded as useful for the generation of soil input data for the *ArcSWAT* simulation. Batjes [2] also limits the scale of the simulations to global and continental studies due to the present spatial resolution $0.5 \times 0.5°$ of the WISE data set. Fortunately, for the *ArcSWAT* simulation of the Urumqi River catchment area, a soil data set with a higher resolution was available. However, Batjes also states that it is “difficult to quantify the precision of the TAWC data presented in this paper [...]” and “the comparability of soil analytical data remains a critical issue in data bases compiled from different sources” [2: 50]. The issue of the spatialisation of soil hydraulic characteristics, but also of ‘representative’ soils and their failure to represent spatial soil heterogeneities and variability of hydraulic properties is also stressed by [11]. Their suggestion is the continuous effort to update the spatial information about soil variability to improve the interaction between pedology and hydrology [11].

- SOL_K

The soil database did not provide the saturated hydraulic conductivity K_h (SOL_K in SWAT) as the number of measured K_h data is too limited for developing pedotransfer functions [2]. Hence, K_h is estimated based on soil texture according to the method of Saxton et al. [5] using a web-based calculation program (http://www.pedosphere.com/resources/texture/worktable_us.cfm, accessed 15.10.2011). However, for clay contents above 60 % and below ca. 5 % as well as silt contents below 5%, the hydraulic conductivity could not be calculated. Inside these boundaries, the calculated values coincided with the measured values from other publications (cf. [12]).

- SOL_ALB

SOL_ALB is the moist soil albedo (fraction of the amount of solar radiation reflected by the soil body) of the top layer [cf. 3]. Berge [6] has published the albedo for various soil types in wet and dry state. The average albedo value for one soil type was used for the SWAT soil input data (see Table A.13). The dependence of albedo on solar elevation [cf. 6] could not be considered for the input data as there was only one value possible.

- USLE_K

The USLE equation soil erodibility factor *K* was calculated using an equation by Williams (1995) published in [3]. The factors of this equation use the percent sand content m_s, the percent silt content m_{silt}, the percent clay content m_c, and the percent organic carbon content $orgC$ of the layer. All parameters could be taken

Table A.13 Albedo for different soil types and moisture contents

Soil type	Wet	Dry
Dune sand	0.24	0.37
Sand	0.18–0.22 (0.20)	0.38–0.42
Grey sand	0.09	0.18
Clay	0.08	0.14
Silty clay	0.15–0.17 (0.16)	0.23–0.32
Loam	0.12–0.16 (0.14)	0.21–0.30
Clay loam	0.10–0.14 (0.12)	0.18–0.23
Silt loam	0.13	0.31
Sandy loam	0.08–0.19 (0.14)	0.17–0.33

Source Summarised according to [6]

from the already derived soil parameters SOL_SAND, SOL_SILT, SOL_CLAY and SOL_CBN.

$$K_{USLE} = f_{csan} * f_{cl-si} * f_{orgc} * f_{hisand}, \tag{A.12}$$

with f_{csand} as the factor that gives low soil erodibility factors for soils with high coarse-sand contents and high values for soil with little sand:

$$f_{csand} = \left(0.2 + 0.3 * \exp\left[-0.256 * m_s * \left(1 - \frac{m_{silt}}{100}\right)\right]\right), \tag{A.13}$$

$f_{cl\text{-}si}$ as the factor that gives low soil erodibility factors to soils with high clay to silt ratios:

$$f_{cl-si} = \left(\frac{m_{silt}}{m_c + m_{silt}}\right)^{0.3}, \tag{A.14}$$

f_{orgc} as the factor that reduces soil erodibility for soils with high organic content:

$$f_{orgc} = \left(1 - \frac{0.25 * orgC}{orgC + \exp[3.72 - 2.95 * orgC]}\right), \tag{A.15}$$

and f_{hisand} as the factor that reduces soil erodibility for soils with extremely high sand contents:

$$f_{hisand} = \left(1 - \frac{0.7 * \left(1 - \frac{m_s}{100}\right)}{\left(1 - \frac{m_s}{100}\right) + \exp\left[-5.51 + 22.9 * \left(1 - \frac{m_s}{100}\right)\right]}\right). \tag{A.16}$$

- Others

SOL_BD is the moist bulk density of the soil layer. The values for moist bulk density were taken from the parameters bulk density BULK_TM and BULK_BM [g cm^{-3}] of SUMTAB90.dbf, respectively [4].

SOL_CBN is the organic carbon content of first soil layer in % soil weight [3]. The information was derived from the parameters organic matter content ORGC_TM and ORGC_BM, also % by weight, from SUMTAB90.dbf [cf. 4].

SOL_CLAY is the clay content (soil particles < 0.002 mm in equivalent diameter) of the soil layer in % soil weight and taken from the parameter clay CLAY_TM and CLAY_BM (%) in SUMTAB90.dbf [cf. 3, 4].

SOL_SILT is the silt content (soil particles between 0.05 and 0.002 mm in equivalent diameter) of the soil layer in % soil weight and taken from the parameter silt SILT_TM and SILT_BM (%) in SUMTAB90.dbf [cf. 3, 4].

SOL_SAND is the sand content (soil particles between 2.0 and 0.05 mm in equivalent diameter) of the soil layer in % soil weight and taken from the parameter sand SAND_TM and SAND_BM (%) in SUMTAB90.dbf [cf. 3, 4].

SOL_ROCK is the rock fragment content (soil particles >2 mm in equivalent diameter) in % total weight of the soil layer. The information is taken from the parameter gravel content [% v/v] GRAVEL_TM and GRAVEL_BM of SUMTAB90.dbf [cf. 3, 4]. Gravel content refers to the percentage of fragments with a diameter less than 2 mm (cf. 2).

The electric conductivity SOL_EC [dS m^{-1}] [3] was taken from the *parameter electrical conductivity* [dS m^{-1}] (ECE_TM and ECE_BM) from SUMTAB90.dbf [cf. 4].

A.4.3 Slope, Watershed and Stream Data

Di Luzio et al. [13] state that the choice of DEM was “critical for a realistic definition of watershed and sub watershed boundaries and topographic input, and consequently simulated output”. According to Gassmann et al. [14], several studies stated that the DEM resolution was significant for SWAT stream flow estimates, however did the most accurate results not coincide with the highest DEM resolutions, contrary to expectations. Due to the previous experiences with the quality of DEMs in Urumqi Region, the watershed and stream data was calculated for ASTER and SRTM DEM beforehand and compared with rivers digitised from maps. The best congruence showed the streams and watershed calculated from ASTER GDEM. Using this calculation for modelling with *ArcSWAT*, Urumqi River at Wulapo Reservoir has a catchment area of 2,536.95 km^2. As the ASTER GDEM showed significant irregularities when calculating the derivation of the elevation information (see Sect. 3.3.1.2), the SRTM-3 DEM was used as input for the calculation of slope values. For the HRU analysis, five slope classes had to be specified. The slope of an area is influencing the runoff potential.

As the research area includes both very flat as well as steep mountainous areas, the wanted slope classification could not take the usual characteristics and requirements of slope classifications for one of these regions, differentiation in the low or high altitudes, into account. Instead, a comprehensive classification as used by the U.S. Geological Survey was chosen for the HRU analysis (Table A.14).

Table A.14 USGS slope class

USGS slope class (%)
<= 2
>2–5
>5–10
>10–20
>20–50
>50

Source http://nc.water.usgs.gov/reports/ofr01496/index.html, accessed 24 March 2011

A.4.4 Climate Data

The input climate data was taken from the data obtained from the National Oceanic and Atmospheric Administration's (NOAA) National Climatic Data Centre (NCDC) for Wulumuqi Station. Monthly values for the climate variables average or mean daily dewpoint, maximum and minimum temperature, monthly precipitation, solar radiation for month and average daily wind speed were required by *ArcSWAT*. With regard to the climate parameters, SWAT uses a lumped concept with each (sub-) watershed taking the climate data from the station nearest to its centroid [15]. Lapse rates for temperature and precipitation are only used to account for the change in temperature with elevation when sub-basins with different elevation bands are assigned, otherwise the climate data is applied without adjustment [3]. Weather station data and climate variables were input into the SWAT database "userwgn" and "weatherstations".

A.4.5 Land Use Data

The land use data was taken from the land use/cover classification already done for the hydrological model. The LUC files had to be reclassified according to the SWAT Land Use Code (see Table A.15) and then used in the land use definition step. The SWAT Land Use class BRR6 was created in the SWAT database "crop" for barren land according to suggestions from *ArcSWAT* forum threads.[2]

Rock, ice, snow and old snow were created based on the values for different urban land use codes in the SWAT database "urban". *ArcSWAT* uses one data set of land use for one simulation. To be able to calculate the year 2000 in one run, one land use classification had to be selected. The classifications for March and June 2000 were considered as appropriate as they represent best the average distribution of land use and cover classes in the research area. Both were tested in simulation runs and finally, the classification from June 2000 was chosen as the

[2] http://groups.google.com/group/arcswat/browse_frm/thread/6dffb8b4287255f9/7655e30d730d72ce?lnk=gst&q=barren+land#7655e30d730d72ce , accessed 19.10.2011.

Table A.15 Land use and cover class, SWAT land use code and description

Class	SWAT land use code	Description
Agriculture	AGRC	Agricultural land-close grown
Agriculture dry	AGRL	Agricultural land-generic
Sparse vegetation	RNGE	Range-grasses
Soil	BRR6	Barren land
Coniferous forest	FRSE	Forest-evergreen
Grass	PAST	Pasture
Rock	ROCK	Rock
Sealed area	URML	Residential med/low density
Ice	ICEF	Ice
Snow	SNOW	Snow
Snow old	SNOO	Snow old
Irrigated areas	WETN	Wetlands-non forested
Water	WATR	Water
Dark areas	UIDU	Industrial

Source own design and [3]

results were closest to the observed runoff data. Soil, land use and slope (see Fig. A.1) were used to define overall 205 HRUs in the modelled watershed.

A.4.6 Calibration and Results

The *ArcSWAT* simulation was run from January until December 2000. To improve the simulated results, the most important parameters were identified with a sensitivity analysis and a manual calibration was conducted. For the latter, the observed total annual runoff, surface and base flows were compared with the total water yield (WYD), surface flow (SURQ) and lateral and groundwater flow (LATQ and GWQ) contribution to reach simulated by *ArcSWAT*. Two aquifers are simulated per subbasin by SWAT: one unconfined shallow aquifer contributing to the main channel or reach of the subbasin (interflow) and one deep confined aquifer contributing to streamflow somewhere outside the watershed [9]. Parameters for overland flow, channel routing, groundwater, soil storage/routing and interception were available for calibration [16].

The program *ArcSWAT* provides two types of sensitivity analysis, the output value and the objective function. The output value analysis helps to identify parameters that have an impact on some measure of simulated output, such as averaged stream flow, without using observed data. The objective function identifies the parameters to which the given project is most sensitive with regard to the measured time series and gives overall "goodness of fit" estimation between the modelled and the measured time series [17].

There are also several parameters for overland flow, channel routing, groundwater, soil storage/routing and interception that can be calibrated to

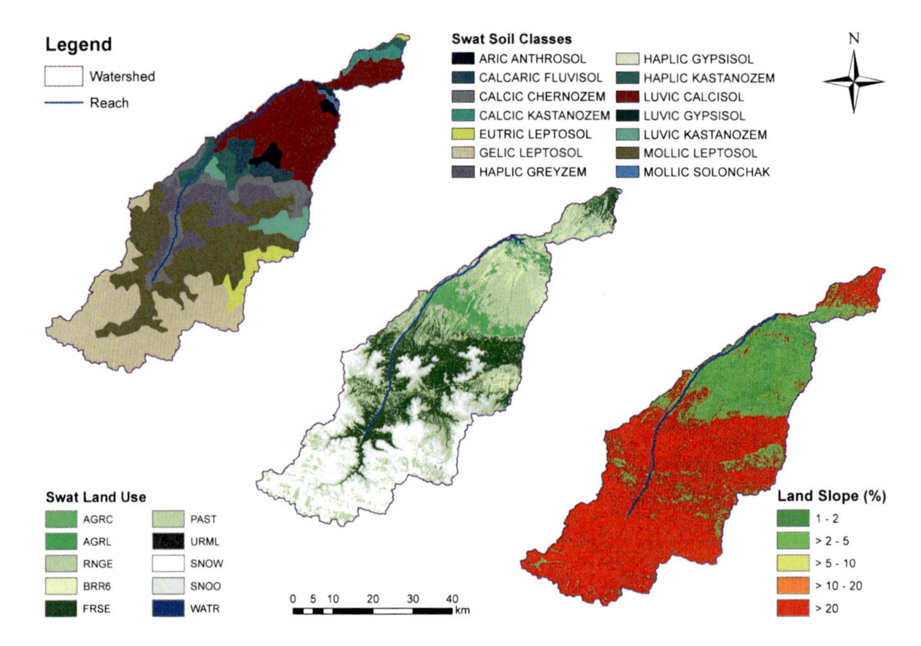

Fig. A.1 Soil, land use and slope for the watershed of Wulapo Reservoir (*source ArcSWAT*, own calculations)

optimize simulation results [16]. A mean value of 0.2 as threshold seems reasonable from Fig. A.2 for choosing the most important parameters. The results of the *output value* sensitivity analysis were of less interest as the goal was to improve the simulation results compared to measured runoff data. After evaluating the sensitivity analysis *objective function* with observed data, the most sensitive parameters for average stream flow were Cn2[3], Timp[4], Gwqmn[5], Alpha_BF[6], Sol_K, Canmx[7], Sol_Z[8], Slope, Blai[9], Sol_AWC[10], and Esco.[11]

---

[3] "Initial SCS runoff curve number for moisture condition II." i.e. the higher Cn2 the more runoff occurs [3: 240].

[4] "Snow pack temperature lag factor" [Neitsch et al. 3: 88].

[5] "Threshold depth of water in the shallow aquifer required for return flow to occur [mm H_2O]. Groundwater flow to the reach is allowed only if the depth of water in the shallow aquifer is equal to or greater than GWQN" [3: 319].

[6] "Baseflow alpha factor [days] (…) direct index of groundwater flow response to changes in recharge" [3: 319].

[7] "Maximum canopy storage [mm H_2O]" [3: 227f].

[8] "Maximum root depth of soil profile [mm]" [3: 298].

[9] "Maximum potential leaf area index (…) use to quantify leaf area development of a plant species during growing season" [3: 187].

[10] "Available water capacity oft he soil layer [mm H_2O/mm soil]" [3: 299].

[11] "Soil evaporation compensation factor (…) to allow (…) to modify the depth distribution used to meet the soil evaporative demand" [3: 231].

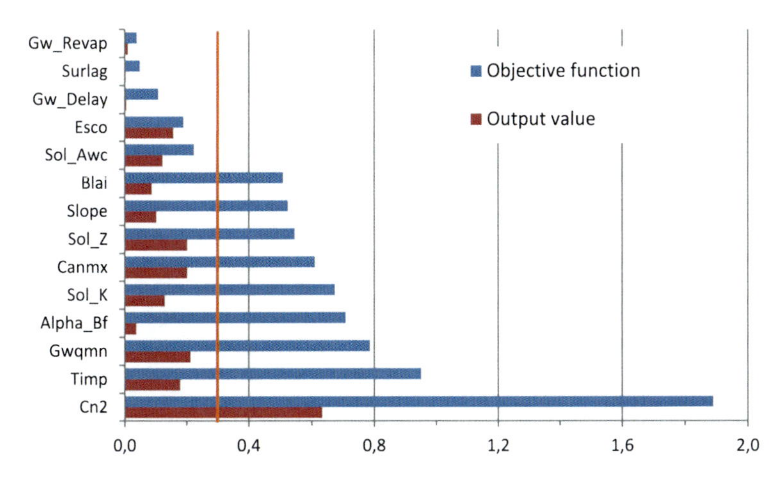

Fig. A.2 Results of the sensitivity analysis output value (without observed data) and objective function (with observed data) (*source* own design, *ArcSWAT*)

Additional parameters were tested for the calibration of surface runoff, base flow and temporal flow calibration according to the suggestions of the SWAT User Manual [9]: GW_Revap[12], Revapmn[13], Ch_K[14], Smfmx[15], Smfmn[16], and Tlaps.[17] The base flow of the measured runoff data was calculated using the graphical separation method assuming constant baseflow discharge, where the base flow is determined by the lowest runoff value during a period without recharge (see Fig. A.3). Due to the lack of detailed (daily) runoff data, filtering methods could not be applied and the almost constant runoff value from November to May suggests a high probability of a nearly constant base flow.

For the evaluation and calibration of the SWAT simulation, *NSE* for monthly values and total water yield, surface and base flow for annual values were used as control measures: The annual totals for measured surface and base flow were compared to simulated surface runoff and lateral flow and contributions from the shallow aquifer to stream runoff. Additionally, the NSE for monthly total water

[12] "Groundwater "revap" coefficient, upward diffusion of water from the shallow aquifer" [3: 320].

[13] Threshold depth of water in the shallow aquifer for "revap" or percolation to the deep aquifer to occur [mm H_2O] [3: 320].

[14] "Effective hydraulic conductivity in main channel alluvium [mm hr^{-1}]" [3: 327].

[15] Melt factor for snow on June 21st [mm H_2O/°C day] [3: 87].

[16] Melt factor for snow on December 21st [mm H_2O/°C day] [3: 87].

[17] "Temperature lapse rate (for) an increase in temperature with an increase in elevation [°C km^{-1}]" [3: 122].

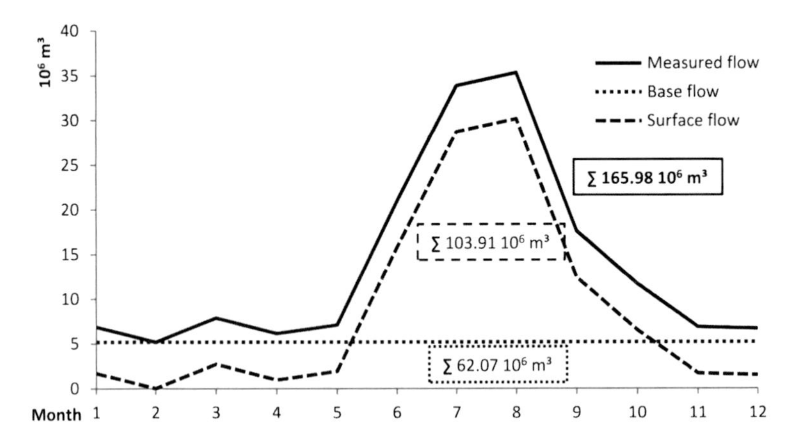

Fig. A.3 Observed runoff at Wulapo Reservoir for the year 2000, divided in surface and base flow, with annual sums (*source* Xinjiang Agricultural University, own calculations)

yield vs. observed monthly runoff were considered when calibration the SWAT simulation. Before the calibration, the total water yield of the default simulation was too high and the flow peaks were not reasonable. *ArcSWAT* offers a manual calibration helper to adjust certain parameters and to achieve better congruence between observed and simulated results. All parameters considered as potentially influencing the components of total water yield were tested and then chosen based on their effect on the results. The chosen SWAT input parameters were then rewritten (see Table A.16) and the simulation was run again. Other tested parameters did not show the desired effects on the simulation output or were already at their extreme values. These steps were repeated until the result for total water yield, surface and base flow showed the best achievable congruence with the observed runoff data.

The calibration of the annual values for surface and subsurface flow were successful. However, it was not possible to adjust the temporal flow completely to the observed runoff dynamics [14] cite several studies where SWAT underpredicted winter stream flow and was not able to predict snow melt dominated runoff. Other factors are inadequate representation of rainfall inputs, lack of model calibration and inaccuracies in measured streamflow [14]. Still, when evaluation the results from the *ArcSWAT* simulation against the observed runoff data, the *NSE* could be improved to 0.09 for total water yield (see also Fig. A.4).

Table A.16 Parameters used for manual calibration of *ArcSWAT* simulation (mgt—land and water management practices; sol—soil data; gw—groundwater data; bsn—basin file, general watershed attributes)

Parameter	Explanation	Changed value	Effects
Calibration of water balance and total flow			
Cn2 (mgt)	Soil curve number	*0.8	Runoff − , infiltration +
SOL_K (sol)	Saturated hydraulic conductivity	*0.7	Lateral runoff −, groundwater recharge +
SOL_Z (sol)	Soil depth	*0.75	Surface runoff +, groundwater recharge +
SOL_AWC (sol)	Available water content	−0.04	Evapotranspiration −, groundwater recharge +
REVAPMN (gw)	Threshold depth of water in the shallow aquifer for "revap" to occur	= 2	Groundwater recharge +
GWQMN (gw)	Threshold depth of water in the shallow aquifer required for base flow to occur	= 5	Total water yield −
RCHRG_DP (gw)	Deep aquifer percolation fraction	= 0.7	Deep aquifer recharge +
Temporal flow calibration			
SMFMX (bsn)	Maximum melt rate for snow	= 2	Melt water peaks in spring −
SMFMN (bsn)	Minimum melt rate for snow	= 4	Melt water peaks in spring and fall −
TIMP (bsn)	Snow pack temperature lag factor 0.01 − 1	= 0.165	Melt water peaks −

Source [3], own calculations

A.5 Data for climate scenarios

(Table A.17, A.18, A.19, A.20, A.21, A.22, A.23).

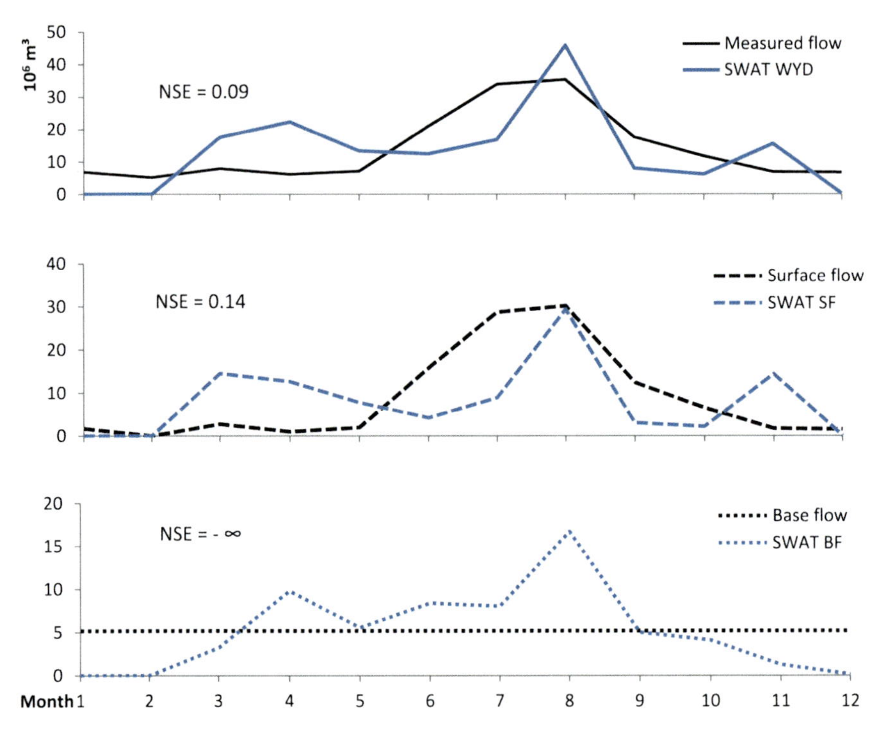

Fig. A.4 Measured and simulated total, surface and base flow. The total measured and surface flow are simulated quite well, but *ArcSWAT* failed to correctly model the assumed baseflow (*source ArcSWAT*, own calculations)

Table A.17 Linear regression for monthly grouped values of mean air temperature, smoothed (unit: °C)

	y-intercept	Slope	R^2	Mean 1975–2010	1975	2010	2050
Year	−81.96	0.04	0.83	7.3	6.5	8.1	9.9
Jan	−43.55	0.02	0.08	−12.3	−12.6	−12.0	−11.4
Feb	−107.23	0.05	0.23	−9.6	−10.5	−8.7	−6.8
Mar	−160.34	0.08	0.53	−0.5	−1.9	0.9	4.1
Apr	−98.39	0.05	0.64	10.5	9.5	11.4	13.6
May	−27.07	0.02	0.18	17.1	16.7	17.5	18.4
Jun	−40.14	0.03	0.24	22.1	21.6	22.7	23.9
Jul	40.85	−0.01	0.07	24.1	24.3	24.0	23.6
Aug	31.86	0.00	0.01	22.8	22.9	22.7	22.6
Sep	−28.11	0.02	0.19	17.1	16.7	17.5	18.4
Oct	−122.80	0.07	0.60	8.3	7.2	9.5	12.1
Nov	−260.14	0.13	0.89	−2.0	−4.2	0.3	5.5
Dec	−124.30	0.06	0.19	−9.8	−10.8	−8.8	−6.5

Source NOAA NCDC Wulumuqi station 1975–2010 [19], own calculations

Table A.18 Linear regression for monthly grouped values of mean dew point temperature, smoothed (unit: °C)

	y-intercept	Slope	R^2	Mean 1975–2010	1975	2010	2050
Year	−62.57	0.03	0.53	−2.54	−3.1	−2.0	−0.8
Jan	−15.5	−45.38	0.02	0.10	−15.8	−15.2	−14.6
Feb	−13.1	−107.42	0.05	0.22	−13.9	−12.2	−10.3
Mar	−6.3	−49.74	0.02	0.08	−6.7	−6.0	−5.1
Apr	−2.4	−23.51	0.01	0.24	−2.5	−2.2	−1.7
May	1.9	−54.15	0.03	0.12	1.4	2.4	3.5
Jun	6.9	75.23	−0.03	0.42	7.5	6.3	4.9
Jul	9.0	−53.71	0.03	0.16	8.4	9.5	10.8
Aug	7.1	−97.20	0.05	0.36	6.2	8.0	10.1
Sep	2.9	−0.19	0.00	0.00	2.8	2.9	2.9
Oct	−1.1	−31.70	0.02	0.11	−1.4	−0.8	−0.2
Nov	−6.8	−213.92	0.10	0.84	−8.6	−4.9	−0.8
Dec	−13.1	−112.22	0.05	0.17	−14.0	−12.2	−10.3

Source NOAA NCDC Wulumuqi station 1975–2010 [19], own calculations

Table A.19 Linear regression for monthly grouped values of mean daily maximum temperature, smoothed (unit: °C)

	y-intercept	Slope	R^2	mean	1975	2010	2050
Year	−35.36	0.02	0.54	12.4	11.9	12.8	13.7
Jan	60.57	−0.03	0.23	−7.0	−6.4	−7.6	−8.9
Feb	−31.19	0.01	0.02	−4.9	−5.1	−4.6	−4.1
Mar	−167.17	0.09	0.49	3.6	2.1	5.1	8.5
Apr	−45.56	0.03	0.25	15.7	15.2	16.3	17.5
May	28.41	0.00	0.00	22.3	22.4	22.3	22.1
Jun	−9.11	0.02	0.09	27.3	27.0	27.6	28.4
Jul	98.80	−0.03	0.60	29.4	30.0	28.8	27.4
Aug	89.51	−0.03	0.36	28.5	29.0	27.9	26.7
Sep	26.28	0.00	0.00	22.7	22.7	22.6	22.6
Oct	−90.12	0.05	0.44	13.4	12.5	14.3	16.4
Nov	−227.57	0.12	0.84	2.3	0.3	4.4	9.0
Dec	−109.87	0.05	0.15	−5.0	−6.0	−4.1	−2.0

Source NOAA NCDC Wulumuqi station 1975–2010 [19], own calculations

Table A.20 Linear regression for monthly grouped values of mean daily minimum temperature, smoothed (unit: °C)

	y-intercept	Slope	R^2	mean	1975	2010	2050
Year	−116.43	0.06	0.79	2.4	1.4	3.5	5.9
Jan	−28.21	0.01	0.01	−15.6	−15.7	−15.5	−15.2
Feb	−129.09	0.06	0.31	−13.3	−14.3	−12.3	−10.0
Mar	−195.78	0.10	0.64	−4.7	−6.4	−3.0	0.8
Apr	−126.26	0.07	0.68	4.7	3.6	5.9	8.5
May	−93.29	0.05	0.51	11.0	10.1	11.9	14.0

(continued)

Table A.20 (continued)

	y-intercept	Slope	R^2	mean	1975	2010	2050
Jun	−89.12	0.05	0.38	16.1	15.2	17.0	19.1
Jul	−47.13	0.03	0.32	18.2	17.6	18.8	20.1
Aug	−57.56	0.04	0.29	16.7	16.1	17.4	18.9
Sep	−98.71	0.06	0.40	11.3	10.3	12.2	14.4
Oct	−186.19	0.10	0.66	3.4	1.7	5.0	8.8
Nov	−297.29	0.15	0.92	−5.6	−8.1	−3.0	2.9
Dec	−51.71	0.02	0.05	−12.9	−13.3	−12.6	−11.8

Source NOAA NCDC Wulumuqi station 1975–2010 [19], own calculations

Table A.21 Extrapolation of the monthly grouped values of precipitation with maximum precipitation response

Month	y_1 Mean 1980–1999 (mm)	Diff (%)	y_2 Mean 2080–2099 (mm)	y-intercept	Slope	2050
Year	365.15	0.28	467.33	−1668.73	1.02	426.95
1	9.71	0.36	13.20	−59.83	0.03	11.82
2	12.78	0.36	17.38	−78.73	0.05	15.56
3	23.69	0.34	31.75	−136.57	0.08	28.57
4	41.15	0.34	55.14	−237.20	0.14	49.61
5	54.55	0.34	73.09	−314.42	0.19	65.77
6	34.56	0.28	44.24	−157.98	0.10	40.42
7	37.77	0.28	48.35	−172.64	0.11	44.17
8	42.46	0.28	54.34	−194.05	0.12	49.65
9	40.73	0.21	49.28	−129.43	0.09	45.90
10	28.39	0.21	34.35	−90.21	0.06	31.99
11	20.99	0.21	25.40	−66.72	0.04	23.66
12	18.38	0.36	24.99	−113.24	0.07	22.38
Sum	365.15		471.5			429.50

Source NOAA NCDC Wulumuqi station 1975–2010 [19, 20], own calculations

Table A.22 Extrapolation of the monthly grouped values of precipitation with 75 % precipitation response

Month	y_1 Mean 1980–1999 (mm)	Diff (%)	y_2 Mean 2080–2099 (mm)	y-intercept	Slope	2050
Year	365.15	0.13	412.62	−579.26	0.47	393.87
1	9.71	0.26	12.23	−40.51	0.03	11.24
2	12.78	0.26	16.10	−53.31	0.03	14.79
3	23.69	0.14	27.01	−42.30	0.03	25.70
4	41.15	0.14	46.91	−73.46	0.06	44.63
5	54.55	0.14	62.18	−97.38	0.08	59.17
6	34.56	0.10	38.02	−34.20	0.03	36.66
7	37.77	0.10	41.55	−37.37	0.04	40.06

(continued)

Table A.22 (continued)

Month	y_1Mean 1980–1999 (mm)	Diff (%)	y_2Mean 2080–2099 (mm)	y-intercept	Slope	2050
8	42.46	0.10	46.70	−42.01	0.04	45.03
9	40.73	0.14	46.43	−72.71	0.06	44.18
10	28.39	0.14	32.36	−50.68	0.04	30.79
11	20.99	0.14	23.93	−37.48	0.03	22.77
12	18.38	0.26	23.15	−76.68	0.05	21.27
Sum	365.15		416.57			396.29

Source NOAA NCDC Wulumuqi station 1975–2010, [20], own calculations

Table A.23 Extrapolation of the monthly grouped values of precipitation with 50 % precipitation response

Month	y_1 Mean 1980–1999 (mm)	Diff (%)	y_2 Mean 2080–2099 (mm)	y-intercept	Slope	2050
Year	365.15	0.10	401.66	−361.31	0.37	387.24
1	9.71	0.19	11.55	−26.99	0.02	10.83
2	12.78	0.19	15.20	−35.52	0.02	14.24
3	23.69	0.10	26.06	−23.44	0.02	25.13
4	41.15	0.10	45.26	−40.72	0.04	43.64
5	54.55	0.10	60.00	−53.97	0.05	57.85
6	34.56	0.04	35.95	7.06	0.01	35.40
7	37.77	0.04	39.28	7.71	0.02	38.69
8	42.46	0.04	44.15	8.67	0.02	43.48
9	40.73	0.08	43.99	−24.10	0.03	42.70
10	28.39	0.08	30.66	−16.79	0.02	29.76
11	20.99	0.08	22.67	−12.42	0.02	22.01
12	18.38	0.19	21.87	−51.09	0.03	20.49
Sum	365.15		396.64			384.22

Source NOAA NCDC Wulumuqi station 1975–2010 [19, 20], own calculations

A.6 Calibration Datasets and Modelling Results

(Tables A.24, A.25, A.26, A.27, A.28, A.29).

Table A.24 Available datasets for calibration and validation

Data set	Area	Source	Date
Surface runoff/ inflow	Wulapo Reservoir	Xinjiang Agricultural University	2000, monthly
Precipitation	Urumqi River	Water Affairs Bureau Urumqi	2003–2007, average
Surface runoff	Toutun River		of several years
	Urumqi Region		

Table A.25 Water balance components in the catchments of Urumqi Region simulated for the long- and short-term climate scenario (units: 10^8 m^3): precipitation (P), actual evapotranspiration (ET_a), net precipitation (P_n), melt water (Q_m), total water flow (Q_n), surface water (Q_{sf}), and groundwater flow (Q_{gw})

	Catchment	Urumqi Mountain	Wulapo Reservoir	Urumqi City	Urumqi River	Toutun Mountain	Toutun River	Midong	Urumqi Region
SC_{LT}	P	5.17	10.90	1.91	12.81	3.47	10.40	6.49	29.71
(1975–2010)	ET_a	2.93	6.98	1.42	8.40	2.11	7.13	4.62	2.02
	P_n	2.24	3.93	0.48	4.41	1.36	3.27	1.88	9.56
	Q_m	0.81	1.75	0.37	2.12	0.50	1.66	1.29	5.08
	Q_n	1.62	2.92	0.40	3.32	1.00	2.53	1.47	7.32
	Q_{sf}	0.83	1.51	0.21	1.72	0.53	1.35	0.70	3.78
	Q_{gw}	0.79	1.42	0.19	1.61	0.47	1.17	0.77	3.54
SC_{ST}	P	5.36	11.31	1.98	13.29	3.64	10.92	6.72	30.93
(2001–2010)	ET_a	2.92	7.03	1.45	8.48	2.12	7.31	4.70	20.49
	P_n	2.44	4.28	0.53	4.81	1.52	3.60	2.03	10.44
	Q_m	1.02	2.30	0.44	2.74	0.66	2.20	1.69	6.62
	Q_n	2.01	3.85	0.54	4.39	1.30	3.47	2.03	9.89
	Q_{sf}	1.02	1.96	0.28	2.24	0.68	1.84	0.96	5.04
	Q_{gw}	0.99	1.89	0.26	2.15	0.62	1.63	1.07	4.85

Source own calculations

Table A.26 Water balance components in the catchments of Urumqi Region in the past (units: mm and % of total water flow): precipitation (P), actual evapotranspiration (ET_a), net precipitation (P_n), melt water (Q_m), total water flow (Q_n), surface water (Q_{sf}), and groundwater flow (Q_{gw}) simulated for SC_{LT} and SC_{ST}

	Catchment	Urumqi Mountain		Wulapo Reservoir		Urumqi City		Urumqi River		Toutun Mountain		Toutun River		Midong		Urumqi Region	
		mm	%	mm	%	mm	%	mm	%	mm	%	mm	%	mm	%	mm	%
SC_{LT} (1975–2010)	P	482		426		337		410		492		415		339		393	
	ET_a	273		272		251		269		300		285		241		267	
	P_n	209		153		86		141		192		130		98		127	
	Q_m	76	49.9	68	60.0	65	91.7	68	63.8	71	50.4	66	65.9	67	88.0	67	69.4
	Q_n	151		114		71		106		141		101		77		97	
	Q_{sf}	78	51.4	59	51.5	37	52.8	55	51.7	75	52.8	54	53.6	37	47.9	50	51.6
	Q_{gw}	74	48.6	55	48.5	33	47.2	51	48.3	67	47.2	47	46.4	40	52.1	47	48.4
SC_{ST} (2001–2010)	P	500		442		350		425		517		435		350		410	
	ET_a	272		275		257		271		301		292		245		271	
	P_n	227		167		93		154		216		144		106		138	
	Q_m	95	50.5	90	59.8	77	81.6	88	62.4	93	50.7	88	63.4	88	83.0	88	67.0
	Q_n	187		150		95		140		184		138		106		131	
	Q_{sf}	95	50.6	76	50.8	49	52.1	72	51.0	96	52.3	73	53.0	50	47.5	67	50.9
	Q_{gw}	93	49.4	74	49.2	46	47.9	69	49.0	88	47.7	65	47.1	56	52.6	64	49.1

Source own calculations

Table A.27 Water balance components in the catchments of Urumqi Region simulated for the future climate scenarios (units: 10^8 m^3): precipitation (P), actual evapotranspiration (ET_a), net precipitation (P_n), melt water (Q_m), total water flow (Q_n), surface water (Q_{sf}), and groundwater flow (Q_{gw})

	Catchment	Urumqi Mountain	Wulapo Reservoir	Urumqi City	Urumqi River	Toutun Mountain	Toutun River	Midong	Urumqi Region
2050 50%	P	5.60	11.78	2.05	13.83	3.74	11.17	6.97	31.97
	ET_a	3.12	7.59	1.60	9.20	2.24	7.81	5.15	22.15
	P_n	2.49	4.19	0.44	4.63	1.50	3.37	1.82	9.82
	Q_m	0.78	1.58	0.27	1.84	0.52	1.48	1.04	4.36
	Q_n	2.02	3.55	0.43	3.97	1.25	0.30	1.72	8.66
	Q_{sf}	1.08	1.86	0.23	2.09	0.67	1.61	0.83	4.53
	Q_{gw}	0.95	1.68	0.20	1.88	0.57	1.37	0.89	4.14
2050 75%	P	5.78	12.15	2.11	14.26	3.86	11.52	7.18	32.97
	ET_a	3.19	7.80	1.65	9.45	2.29	8.02	5.30	2.28
	P_n	2.59	4.35	0.46	4.81	1.57	3.50	1.89	10.20
	Q_m	0.80	1.61	0.28	1.89	0.53	1.52	1.08	4.48
	Q_n	2.10	3.68	0.45	4.12	1.30	3.08	1.78	8.99
	Q_{sf}	1.12	1.94	0.24	2.17	0.70	1.67	0.86	4.70
	Q_{gw}	0.98	1.74	0.21	1.95	0.59	1.41	0.92	4.28
2050 max	P	6.28	13.18	2.29	15.47	4.20	12.52	7.78	35.77
	ET_a	3.37	8.34	1.78	10.12	2.42	8.63	5.74	24.49
	P_n	2.91	4.84	0.51	5.35	1.79	3.89	2.04	11.29
	Q_m	0.83	1.70	0.29	1.99	0.56	1.60	1.15	4.74
	Q_n	2.32	4.05	0.49	4.54	1.45	3.41	1.94	9.89
	Q_{sf}	1.24	2.14	0.27	2.41	0.79	1.86	0.94	5.21
	Q_{gw}	1.08	1.91	0.22	2.16	0.66	1.55	0.99	4.67

Source own calculations

Table A.28 Water balance components in the catchments of Urumqi Region simulated for the future climate scenarios (units: mm and % of total water flow): precipitation (P), actual evapotranspiration (ET_a), net precipitation (P_n), melt water (Q_m), total water flow (Q_n), surface water (Q_{sf}), and groundwater flow (Q_{gw})

	Catchment	Urumqi Mountain		Wulapo Reservoir		Urumqi City		Urumqi River		Toutun Mountain		Toutun River		Midong		Urumqi Region	
		mm	%	mm	%	mm	%	mm	%	mm	%	mm	%	mm	%	mm	%
2050 50%	P	523		460		362		442		531		446		363		423	
	ET_a	291		297		283		294		318		311		268		293	
	P_n	232		164		78		148		213		134		95		130	
	Q_m	73	38.7	62	44.4	47	62.3	59	46.4	74	41.6	59	50	54	60.3	58	50
	Q_n	189		138		76		127		177		118		90		115	
	Q_{sf}	101	53.2	73	52.5	40	53.1	67	52.6	95	53.9	64	54	43	48.3	60	52
	Q_{gw}	88	46.8	66	47.5	35	46.9	60	47.4	81	46.1	54	46	46	51.7	55	48
2050 75%	P	539		474		373		456		547		460		374		437	
	ET_a	298		304		292		302		325		320		276		301	
	P_n	242		170		82		154		222		140		98		135	
	Q_m	74	37.9	63	43.9	49	61.9	60	45.9	75	40.8	61	49	56	60.3	59	50
	Q_n	196		144		79		132		184		123		93		119	
	Q_{sf}	105	53.4	76	52.6	42	53.3	69	52.7	99	54.1	67	54	45	48.3	62	52
	Q_{gw}	91	46.6	68	47.4	37	87.6	62	89.7	84	84.9	56	85	48	106.9	57	91
2050 max	P	586		515		404		495		596		499		406		474	
	ET_a	314		326		315		324		343		344		299		324	
	P_n	272		189		89		171		253		155		107		149	
	Q_m	78	36	66	42	51	59.4	64	43.9	79	38.5	64	47	60	59.4	63	48
	Q_n	217		158		87		145		206		136		101		131	
	Q_{sf}	116	53.6	84	52.9	47	54.3	77	53	112	54.5	74	55	49	48.8	69	53
	Q_{gw}	101	46.4	75	47.1	40	45.7	68	47	94	45.5	62	45	52	51.2	62	47

Source own calculations

Table A.29 Projected water demand of Urumqi City under different trends for economic and demographic development (unit: 10^6 m^3)

Scenarios	Low		Medium		High	
	GDP_{pwc}		$GDP_{Wilson\&Stupnytska}$		GDP_{Keidel}	
Year	Pop_{WPP}	Pop_{regr}	Pop_{WPP}	Pop_{regr}	Pop_{WPP}	Pop_{regr}
2010	1,020	1,026	1,038	1,044	1,044	1,050
2020	963	1,024	1,142	1,205	1,265	1,330
2030	905	1,018	1,164	1,283	1,581	1,708
2040	852	1,010	1,161	1,330	1,977	2,167
2050	807	1,003	1,110	1,320	2,400	2,648

Source own calculations

References

1. International Soil Reference and Information Centre (ISRIC) (2011). Soil and terrain database for China. Version 1.0 - scale 1:1 million. ISRIC Wageningen. Retrieved from 5 Oct, 2011. http://www.isric.org/data/soil-and-terrain-database-china.
2. Batjes, N. (1996). Development of a world data set of soil water retention properties using pedotransfer rules. *Geoderma, 71*, 31–52.
3. Neitsch, S. L., Arnold, J. G., Kiniry, J. R., Srinivasan, R., & Williams, J. R. (2009). *Soil and water assessment tool: Input/output file documentation Version 2009* (604 p). Temple: Grassland, Soil and Water Research Laboratory.
4. Batjes, N. (2002). *Soil parameter estimates for the soil types of the world for use in global and regional modelling* (Version 2.1., 46 p). Wageningen: ISRIC Report, 2002/02c, International Food Policy Research Institute and International Soil Reference and Information Centre. Retrieved 25 Mar, 2011 , from http://www.isric.eu/isric/webdocs/docs/ISRIC_Report_2002_02c.pdf.
5. Saxton, K. E., Rawls, W. J., Romberger, J. S. & Papendick, R. I. (1986). Estimating Generalised Soil-water Characteristics from Texture. *Soil Science Society of America Journal 50*, 1031–1036.
6. Ten Berge, H. F. M. (1986). *Heat and water treansfer at the bare soil surface*. Aspects affecting thermal imagery. Dissertation. p. 212, Wageningen Agricultural University, Wageningen.
7. Food agricultural organization (FAO) (1995). Multilingual soil database. FAO, Rome (World Soil Resources Reports, 81), 95 p.
8. Natural resources conservation service (NRCS) (2004). *National Engineering Handbook. Hydrology* (630 p). U.S. Department of Agriculture, http://www.mi.nrcs.usda.gov/technical/engineering/neh.html, 24 Mar, 2011.
9. Neitsch, S. L., Arnold, J. G., Kiniry, J. R. & Williams, J. R. (2005). *Soil and water assessment toolTheoretical Documentation: Version 2005* (476 p). Temple, Texas: Grassland, Soil and Water Research Laboratory, Blackland Research Centre.
10. Merz, R., Blöschl, G., & Parajka, J. (2006). Spatio-temporal variability of event runoff coefficients. *Journal of Hydrology,331*(3–4), 591–604.
11. Terribile, F., Coppola, A., Langella, G., Martina, M., & Basile, A. (2011). Potential and limitations of using soil mapping information to understand landscape hydrology. *Hydrology and Earth System Sciences Discussions,8*, 4927–4977.
12. Oosterbaan, R. J. & Nijland, H. J. (1994). Determining the Saturated Hydraulic Conductivity. In: H. P. Ritzema (Ed.) *Drainage principles and applications*. ILRI, Wageningen: 435–476.

13. Di Luzio, M., Arnold, J. G., & Srinivasan, R. (2005). Effect of GIS data quality on small watershed stream flow and sediment simulations. *Hydrological Processes,19*(3), 629–650.
14. Gassmann, P. W., Reyes, M. R., Green, C. H., & Arnold, J. G. (2007). The soil and water assessment tool: Historical development, applications, and future research directions. *Transactions of the American Society of Agricultural and Biological Engineers,50*(4), 1211–1250.
15. Aguilar, C., Herrero, J., & Polo, M. J. (2010). Topographic effects on solar radiation distribution in mountainous watersheds and their influence on reference evapotranspiration estimates at watershed scale. *Hydrology and Earth System Sciences,14*(12), 2479–2494.
16. Milewski, A., Sultan, M., Yan, E., Becker, R. A., Soliman, F. & Gelil, K. A. (2009). A remote sensing solution for estimating runoff and rechare in arid environments. *Journal of Hydrology*, *373*(1–2), 1–14.
17. Van Griensven, A. (2005). Sensitivity, auto-calibration, uncertainty and model evaluation in SWAT2005.
http://www.biomath.ugent.be/~ann/swat_manuals/SWAT2005_manual_sens_cal_unc.pdf, accessed 11 Feb 2012.
18. Neitsch, S. L., Arnold, J. G., Kiniry, J. R., Srinivasan, R., & Williams, J. R. (2002). *Soil and water assessment tool: Input/Output File Documentation Version 2009* (472 p). Temple,Texas: Texas Water Resources Institute, College Station.
19. National Oceanic and Atmospheric Administration of the U.S. Department of Commerce National Climatic Data Centre (NOAA NCDC) (2011). *Global Summary of the Day Wulumuqi Station*, Station Number 514630, 22.08.1956–31.12.2010. Retrieved 16 May, 2011, from http://www7.ncdc.noaa.gov/CDO/cdo.
20. Christensen, J. H., Hewitson, B., Busuioc, A., Chen, A., Gao, X., Held, I., Jones, R., Kolli, R. K., Kwon, W.-T., Laprise, R., & Magaña Rueda, V. (2007). Regional climate projections. In S. Solomon, D. Qin, M. Manning, Z. Chen, M. Marquis, K. B. Averyt, & H. L. Miller (Eds.), *Climate change 2007: The physical science basis. contribution of working group I to the fourth assessment report of the Intergovernmental Panel on Climate Change* (pp. 847–940). Cambridge, UK: Cambridge University Press.

Printed by Publishers' Graphics LLC
DBT140604.23.34.111